SEDIMENTARY PETROLOGY

An Introduction to the Origin of Sedimentary Rocks

Maurice E. Tucker
BSc, PhD, FGS, CGeol
Department of Geological Sciences
University of Durham

THIRD EDITION

**Blackwell
Science**

To Vivienne (again)

© 1981, 1991, 2001 by
Blackwell Science Ltd
Editorial Offices:
Osney Mead, Oxford OX2 0EL
25 John Street, London WC1N 2BS
23 Ainslie Place, Edinburgh EH3 6AJ
350 Main Street, Malden
    MA 02148-5018, USA
54 University Street, Carlton
    Victoria 3053, Australia
10, rue Casimir Delavigne
    75006 Paris, France

Other Editorial Offices:
Blackwell Wissenschafts-Verlag
    GmbH
Kurfürstendamm 57
10707 Berlin, Germany

Blackwell Science KK
MG Kodenmacho Building
7–10 Kodenmacho Nihombashi
Chuo-ku, Tokyo 104, Japan

Iowa State University Press
A Blackwell Science Company
2121 S. State Avenue
Ames, Iowa 50014-8300, USA

First published 1981
Reprinted 1982, 1984, 1985, 1986,
    1987
Second edition 1991
Reprinted 1992, 1994 (twice), 1995,
    1996, 1998, 1999
Third edition 2001

Set by Best-set Typesetter Ltd., Hong
    Kong
Printed and bound in Great Britain
at the Alden Press, Oxford and
    Northampton

A catalogue record for this title is
available from the British Library

ISBN 0-632-05735-1

Library of Congress
Cataloging-in-Publication Data

Tucker, Maurice E.
    Sedimentary petrology : an
introduction to the origin of
sedimentary rocks / Maurice E.
Tucker. — 3rd ed.
    p. cm.
    Includes bibliographical
references and index.
    ISBN 0-632-05735-1 (pbk.)
    1. Rocks, Sedimentary.
I. Title.

QE471.T827 2001
552'.5—dc21   2001025550

DISTRIBUTORS

Marston Book Services Ltd
PO Box 269
Abingdon, Oxon OX14 4YN
(Orders: Tel: 01235 465500
          Fax: 01235 465555)

USA
Blackwell Science, Inc.
Commerce Place
350 Main Street
Malden, MA 02148-5018
(Orders: Tel: 800 759 6102
          781 388 8250
          Fax: 781 388 8255)

Canada
Login Brothers Book Company
324 Saulteaux Crescent
Winnipeg, Manitoba R3J 3T2
(Orders: Tel: 204 837 2987)

Australia
Blackwell Science Pty Ltd
54 University Street
Carlton, Victoria 3053
(Orders: Tel: 3 9347 0300
          Fax: 3 9347 5001)

For further information on
Blackwell Science, visit our website:
www.blackwell-science.com

Contents

Preface to the third edition

The time has come for another revision of this book. In the 10 years since the second edition, there has been an enormous amount of material published on sediments and I have amended the text where necessary and added recent references to take this into account. Many very useful scientific papers from the 1970s and 1980s cited in the last edition have been removed and replaced by 1990s and 2000 references; students should be able to find their way into the modern literature through these. Much emphasis in the last 10 years in soft-rock research has been in the area of sequence stratigraphy, recognizing key surfaces, sedimentary cycles and their stacking patterns, and relating depositional environments, facies and diagenesis to relative changes in sea-level. However, in all this work, a proper understanding of the composition, textures, structures and origins of sedimentary rocks is essential; this book aims to provide students with that basic knowledge. In the next 10 years the sequence stratigraphic approach may go out of fashion or be replaced by some other paradigm; it will still be essential to know how to describe and interpret sediment.

To help with the petrographic side of sedimentary studies, this edition includes 74 colour photomicrographs of sedimentary rocks in thin-section. In addition several tables are included to help with the description and interpretation of sandstones and limestones.

This author derives enormous pleasure from studying sedimentary rocks; seeing them in the field, looking at them down the microscope, and puzzling over their origin. I hope the reader can also appreciate the excitement of sediments and the stories they can tell, and will enjoy learning about these fascinating rocks.

I am grateful to Alison Jones, Rob Raiswell and Stuart Jones for reading parts of the text, and also to lecturers who use this book in their classes for reviews and suggestions for this edition. As always, thanks to Vivienne for patience, understanding, and enduring support.

Maurice Tucker
Durham
February 2001

Preface to the second edition

In the 10 years that have passed since the first edition of this book was published, there have been advances in our understanding of many aspects of sedimentary petrology, so that a new edition is now required to bring the book up to date. In fact, at the level at which this book is written, much of the original material is still correct, but there are new ways of looking at rocks, new terms and interpretations have come in and more recent references to the literature are required. As to be expected, there are areas of the subject where there is still much controversy and few new ideas have come forward in the last 10 years. The book has expanded in text and figures by about 30% throughout, and many of the original photomicrographs are a little bigger in the new double-column format. The depositional environments and facies sections of several chapters have been enlarged a little too, to make the book more complete. Descriptions of specific examples are not given, however, but simply noted. Many of the references in the first edition have been replaced by 1980s papers.

Sedimentary petrology is still a most important branch of the earth sciences. A knowledge of a sediment's depositional and diagenetic history is essential for an understanding of facies geometry and porosity evolution, critical factors in hydrocarbon exploration. The last few years have seen the concepts of sequence stratigraphy, derived from seismic stratigraphy, applied more and more to surface, as well as subsurface, sedimentary rocks, to help elucidate the larger-scale controls (tectonics versus eustatic sea-level changes) and to provide a predictive tool in frontier basins. However, an understanding and consideration of the deposition of the sedimentary rocks themselves are essential to the sequence stratigraphy approach, if the latter is not to give misleading information and erroneous interpretations and correlations.

Acknowledgements for the second edition

I am indebted to John Aggett, Peter Balson and Mic Jones for comments on the original chapters. The reviews of the book by sed. pet. course professors in the early eighties and in the last year proved very useful indeed in the revision. I am most grateful to Karen Atkinson for drafting the new figures; to Gerry Dresser for the new photomicrographs; to George Randall and Ron Lambert for thin sections and Emmie Williamson at Blackwells for handling the manuscript and proofs so efficiently. As ever, I must thank Vivienne for her inexhaustible support and for typing the text and references. A and Z (Fig. 2.39) are to be thanked for being a great help rather than a hindrance this time, and for helping with the references too.

Maurice Tucker
Durham
March 1991

Preface to the first edition

The study of sedimentary rocks—sedimentary petrology—goes back to the last century and beyond. It is only in the past few decades, however, that we have begun to understand and appreciate the processes by which these rocks are formed. Many of the latest steps forward have come from research on modern sediments and material from shallow and deep bore-holes. The advent of sophisticated instruments, such as the electron microscope, has also been important. Some of the impetus for investigating sedimentary rocks has come from their economic importance: the fossil fuels coal and petroleum, and many essential minerals and raw materials are contained in these rocks. In spite of recent advances, there is still much that can be done just with a hammer in the field and simple microscope in the laboratory and a pair of sharp eyes.

This book attempts to present a concise, up-to-date account of sedimentary petrology. In recent years, many texts have been published dealing more with the depositional environments and facies of sediments, with less attention being given to features of the rocks themselves. This book approaches the subject from the other direction, examining each rock group in turn, with discussions of composition, petrography, sedimentary structures, diagenesis, and depositional environments and facies.

This book has been written with undergraduate students in mind. Because of this, references to the literature have been critically chosen. By and large, students do not want (or need) to consult original papers on a topic written many years ago. Students require up-to-date information, the latest ideas, and reviews. Good review papers of course do cite the early literature so that the keen student can soon delve back and locate important papers. All the references cited in this book should be readily available in university and institute libraries.

Acknowledgements

Many friends and colleagues have willingly read early drafts of chapters and made very valuable and useful comments. I am particularly grateful to Hugh Battey, Colin Braithwaite, Paul Bridges, Trevor Elliott, John Hemingway, Mic Jones, Duncan Murchison, Andrew Parker, Tim Pharoah, Alastair Robertson, Colin Scrutton and Bruce Sellwood. I should also thank the many people (acknowledged in the text) who have supplied specimens, photographs or thin-sections for text figures. I am indebted to Mrs K. Sales of Newcastle University's Photographic Department for all her efforts in printing the photographs. My deepest gratitude must go to my wife, Vivienne, for doing much of the donkey-work (typing, etc., etc.), for giving up so many evenings and weekends without too much complaint, and for keeping the little horrors of Fig. 2.40 quiet during the day.

Maurice Tucker
Durham
January 1981

1   Introduction: basic concepts and methodology

1.1 Introduction

Some 70% of the rocks at the Earth's surface are sedimentary in origin, and these include the familiar sandstones, limestones and shales, and the less common but equally well-known salt deposits, ironstones, coal and chert.

Sedimentary rocks of the geological record were deposited in the whole range of natural environments that exist today. The study of these modern environments and their sediments and processes contributes much to the understanding of their ancient equivalents. There are some sedimentary rock types, however, for which there are no known modern analogues, or their inferred depositional environments are only poorly represented at the present time.

Once deposited, sediments are subjected to the processes of diagenesis, that is, physical, chemical and biological processes which bring about compaction, cementation, recrystallization and other modifications to the original sediment, and form rocks.

There are many reasons for studying sedimentary rocks, not least because of the wealth of economic minerals and materials contained within them. The fossil fuels oil and gas are derived from the maturation of organic matter in sediments and these then migrate to a suitable reservoir, mostly a porous sedimentary rock. The other fossil fuel, coal, is also contained within sedimentary sequences of course. Sedimentological and petrological techniques are increasingly used in the search for new reserves of these fuels and other natural resources. Sedimentary rocks supply much of the world's iron, potash, salt, building materials and many, many other essential raw materials.

Environments and processes of deposition and palaeogeography and palaeoclimatology can all be deduced from studies of sedimentary rocks. Such studies contribute much towards a knowledge and understanding of the Earth's geological history. Sedimentary rocks contain the record of life on Earth, in the form of fossils, and these are the principal means of stratigraphic correlation in the Phanerozoic.

1.2 Basic concepts

1.2.1 Classification of sedimentary rocks

Sedimentary rocks are formed through physical, chemical and biological processes. On the basis of the dominant process(es) operating, the common sediment lithologies can be grouped into four broad categories (Table 1.1). The siliciclastic sediments (also referred to as terrigenous or epiclastic deposits) are those consisting of fragments (clasts) of pre-existing rocks, which have been transported and deposited by physical processes. The conglomerates and breccias, sandstones and mudrocks, discussed in Chapters 2 and 3 belong to this group. Sediments largely of biogenic, biochemical and organic origin are the limestones, which may be altered to dolomite (Chapter 4), phosphate deposits (Chapter 7), coal and oil shale (Chapter 8) and cherts (Chapter 9). Sedimentary rocks largely of chemical origin are the evaporites (Chapter 5) and ironstones (Chapter 6). Volcaniclastic deposits (Chapter 10) constitute a fourth category and consist of lava and rock fragments derived from penecontemporaneous volcanic activity. Each of these various sedimentary rock types can be divided further, usually on the basis of composition. In addition, many rock types grade laterally or vertically into others through intermediate lithologies. A scheme to help with the identification and description of sedimentary rock types is presented in Table 1.2.

1.2.2 Sedimentary environments and facies

Sedimentary environments vary from those where erosion and transportation dominate to those where deposition prevails. Most weathering and erosion, liberating sediment grains and ions in solution, takes place in continental areas, and climate, local geology

Table 1.1 Principal groups of sedimentary rock

Siliciclastic sediments	Biogenic, biochemical and organic sediments	Chemical sediments	Volcaniclastic sediments
conglomerates and breccias, sandstones, mudrocks	limestones (and dolomites), cherts, phosphates, coal and oil shale	evaporites, ironstones	e.g. ignimbrites, tuffs and hyaloclastites

Table 1.2 Scheme for the identification and description of sedimentary rocks in hand specimen

Examine the rock for colour, texture, composition, sedimentary structures and fossils and then identify the sedimentary rock type. If there is enough evidence, give an interpretation of the depositional environment and diagenesis of the sediment

Colour
It should be easy to describe the colour. The colour is usually a reflection of the organic content (grey to black with increasing organic matter) and oxidation state of iron: ferrous iron, occurring in clay minerals (e.g. chlorite) and iron minerals (such as berthierine–chamosite) gives a green colour; ferric iron, occurring in iron minerals, gives red (in hematite) and yellow–brown colours (in goethite–limonite). Some sedimentary minerals may have a particular colour, such as the white of pure anhydrite and gypsum

Texture
Determine the grain size of the rock with a hand-lens; look at the grain shape: rounded or angular? Look at the grain sorting, is it well or poorly sorted? Look for the nature of the contacts between the grains (if visible), and for any preferred orientation of the grains (fabric)

Composition
Identify the composition of the sediment using a hand-lens
Is it sandstone?—made of quartz, feldspar, rock fragments. If so, is it a quartz arenite, litharenite, arkose or greywacke (the four common types)?
Is it a limestone (fizzes with acid)?—made of bioclasts (fossils), ooids, peloids. If so, is it a grainstone, packstone, wackestone, mudstone or boundstone?
Is it a dolomite (dolomitized limestone, fizzes little)?—crystalline, poorly preserved fossils and structures, pale brown–buff colour
Is it a mudrock? If so, is it fissile (a shale) or not (mudstone)? Any nodules present? Composition?
Is it a conglomerate? Determine whether monomictic or polymictic (from clast composition), orthoconglomerate or paraconglomerate (from texture)
Less common sedimentary rock types are evaporites (may be salty or soft), cherts (hard and splintery) and ironstones (red or green, heavy, oolitic)

Sedimentary structures
Look for structures such as bedding, lamination, cross-bedding, cross-lamination, parting lineation, sole structures, burrows, nodules, stylolites, etc.

Fossils
If present (a hand-lens may be needed to see them) try and identify them to phylum level (further if you can). Also look for the preservation of the fossils (shells articulated, broken, bored, dissolved, etc.)

Interpretation
From all the evidence gathered, suggest a rock type and possibly depositional environment. There may be several alternatives. Comment on the rock's diagenesis: cementation, compaction, replacement, etc., and near-surface versus burial diagenetic effects

and topography control the type and amount of material released. The main continental depositional environments are fluvial and glacial systems, lakes and the aeolian sand seas of deserts. Most shoreline environments, deltas, lagoons, tidal flats, sabkhas, beaches and barriers, and open marine environments, shallow shelves and epeiric seas, and bathyal–abyssal sites of pelagic, hemipelagic and turbidite sedimentation, are areas of net deposition, involving the whole range of sediment lithologies. Many of these sediments possess distinctive characteristics, which can be used to recognize their equivalents in the geological record.

Facies

With sedimentary rocks, once they have been described and identified (the theme of this book), and their stratigraphic relationships elucidated, then the concept of facies is applied. A facies is a body or packet of sedimentary rock with features that distinguish it from other facies. A facies is the product of deposition, and it may be characteristic of a particular depositional environment, or a particular depositional process. Features used to separate facies are sediment composition (lithology), grain size, texture, sedimentary structures, fossil content and colour. Lithofacies are defined on the basis of sedimentary characteristics, whereas biofacies rely on palaeontological differences. With detailed work, subfacies can be recognized, and microfacies if microscope studies are used to distinguish between rocks that in the field appear similar (often the case with limestones). Facies can be described in terms of (a) the sediment itself (e.g. cross-bedded sandstone facies), (b) the depositional process (e.g. stream-flood facies) and (c) the depositional environment (e.g. tidal-flat facies). Only (a) is objective and, hopefully, unequivocal; (b) and (c) are both interpretative. Different facies commonly occur together and so form *facies associations* or *facies assemblages*. Repetitions of facies sequences are common and give rise to small-scale cycles a few metres thick. Some cycles develop naturally within the sedimentary environment without any changes in external factors.

There are many factors that control and affect the sediments deposited and determine the sedimentary rock type and facies produced. On a gross scale, over-riding controls are (a) the depositional processes, (b) the depositional environment, (c) the tectonic context and (d) the climate.

Depositional processes and environments

Sediments can be deposited by a wide range of processes including the wind, flowing water as in streams, tidal currents and storm currents, waves, sediment + water flows such as turbidity currents and debris flows, the *in situ* growth of animal skeletons as in reefs and the direct precipitation of minerals as in evaporites. The depositional processes leave their record in the sediment in the form of sedimentary structures and textures. Some depositional processes are typical of a particular environment, whereas others operate in several or many environments. Environments are defined on physical, chemical and biological parameters and they can be sites of erosion, non-deposition or sedimentation. Water depth, degree of agitation and salinity are important physical attributes of subaqueous environments and these affect and control the organisms living on or in the sediment or forming the sediment. Chemical factors such as Eh (redox potential) and pH (acidity–alkalinity) of surface waters and pore waters affect organisms and control mineral precipitation.

Tectonic context

This is of paramount importance because it determines the depositional setting, whether it is, for example, a stable craton, back-arc basin or rift. There have been many studies in recent years of modern and ancient sedimentary basins and the main categories are given in Table 1.3. Each basin has a particular pattern of sedimentary fill, some with distinctive facies or even characteristic lithologies. The deposits of many ancient passive margins, back-arc/fore-arc basins and ocean floors, commonly much deformed, occur in mountain belts, produced by plate collisions. Rates of subsidence and uplift, level of seismic activity and occurrence of volcanoes are also dependent on the tectonic context and are reflected in the sediments deposited.

Climate

This is a major factor in subaerial weathering and erosion and strongly affects the composition of terrigenous clastic sediments. Climate is instrumental in the formation of some lithologies, evaporites and limestones, for example, and there is a strong palaeolatitudinal control on some rock types (Fig. 1.1). Two other factors controlled by climate and tectonic context are sediment supply and organic productivity. Sediment supply is important in so far as low rates favour limestone, evaporite, phosphate and ironstone formation. High levels of organic productivity are important in the formation of limestones, phosphates, cherts, coal and oil shale.

Role of relative sea-level

Another major factor affecting sedimentary facies is

Table 1.3 Plate-tectonic classification of sedimentary basins and their typical rock types

Spreading-related or passive settings
1 Intracratonic rifts (e.g. East Africa). Mostly filled by alluvial fan, fluvial and lacustrine facies
2 Failed rifts or aulacogens (e.g. Benue Trough). Thick successions from deep-sea fan to fluvial
3 Intercontinental rifts:
   (a) early (e.g. Red Sea)—evaporites, carbonates, siliciclastics; fluvial to deep marine
   (b) late (e.g. Atlantic)—fluvial–deltaic, clastic shelf, carbonate platform on passive margin, passing to turbidites, hemipelagites and pelagites on ocean floor
4 Intracratonic basins (e.g. Chad, Zechstein, Delaware, Michigan). Terrestrial to marine clastic, carbonate and evaporite facies

Active settings
1 Continental collision-related:
   (a) remnant ocean basins (e.g. Bay of Bengal, Mediterranean)—sediments variable, turbidites, anoxic muds, evaporites
   (b) foreland basins (e.g. sub-Himalayas, Alpine molasse basins, Western Canada)—terrestrial to shallow to deep marine clastics and carbonates
2 Strike-slip/pull-apart basins (e.g. California). Thick successions, deep-sea fan to fluvial
3 Subduction-related settings:
   (a) continental margin magmatic arcs (e.g. Andes)
     (i) fore-arc basins—thin to thick successions, fluvial to deep-sea fan and volcaniclastics
     (ii) back-arc/retro-arc basins—mostly terrestrial facies and volcaniclastics
   (b) intra-oceanic arcs (e.g. Japan, Aleutians)
     (i) fore-arc basins—turbidites, hemipelagites, pelagites, volcaniclastics
     (ii) back-arc basins—marine and volcanic facies; terrigenous influences

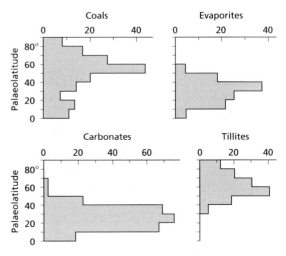

Fig. 1.1 Palaeolatitude zonation of climate-sensitive deposits. Frequency (number of occurrences) against palaeolatitude.

Table 1.4 The orders of sea-level change and possible mechanisms. There is still much discussion over the mechanism(s) behind global eustasy as a cause of sea-level changes on the scale of 1–10 Myr (second- to third-order), and the significance of changes of the in-plane stress regime within plates in terms of relative sea-level change has still to be evaluated

First-order, 10^8 years	tectono-eustasy
Second-order, 10^7 years	rifting and thermal subsidence ⎫ eustasy
Third-order, 10^6 years	in-plane stress
Fourth-order, 10^5 years	glacio-eustasy, tectonics, sedimentary
Fifth-order, 10^4 years	processes

the global position of sea-level and relative sea-level changes. The position of sea-level is determined principally by climatic and tectonic factors, and processes causing change in relative sea-level may be regional (e.g. changes in rates of uplift/subsidence and sediment supply), or global, referred to as eustatic (e.g. variations in climate causing changes in size of the polar ice-caps, opening/closing of oceans and fluctuations in rates of sea-floor spreading causing ocean-basin volume changes). Various orders of sea-level change can be distinguished (Table 1.4), although there has been much discussion over the mechanisms responsible. The first-order global sea-level curve, shown in Figs 4.4 & 8.7, is the result of the opening and closing of the Iapetus Ocean in Palaeozoic times and the opening of the Atlantic and Tethys and closing of Tethys in Mesozoic and Cenozoic times. The occurrence of limestones (more abundant at times of relative sea-level high-stand) and their primary mineralogy (see Fig. 4.4), the development of hydrocarbon source rocks (Fig. 8.7) and, to a certain extent, the abundance of dolomites, phosphorites and ironstones all broadly correlate with this first-order sea-level curve.

Sequence stratigraphy

Second- to third-order relative sea-level changes are responsible for the formation of *depositional se-*

quences, packages of genetically related strata, generally several hundreds of metres thick, bound by unconformities and their correlative conformities (i.e. sequence boundaries). Carbonate platforms, for example, are generated during such periods of time (1–10 Myr). The stratigraphic record is made up of depositional sequences, and within them there is commonly a regular and predictable arrangement of sedimentary facies (the product of lowstand, transgressive and highstand *systems tracts*) deposited at specific intervals on the curve of relative sea-level change (see Fig. 2.86). Systems tracts are separated by *key surfaces*: sequence boundaries, transgressive surfaces and maximum flooding surfaces. Fourth- to fifth-order relative sea-level changes are responsible for the repetition of small-scale cycles (*parasequences*), 1–10 m thick, which are a feature of many sedimentary formations. Systematic vertical changes in the nature of the parasequences reflect long-term onlap (transgression/relative sea-level rise) or offlap (regression/relative sea-level fall), i.e. longer-term relative sea-level changes. Orbital forcing in the Milankovitch band is a popular but contentious explanation for parasequences, especially carbonate ones, but other explanations include sedimentary processes (autocyclicity) and periodic tectonic subsidence.

Facies models

Many attributes of a facies are reflections of the depositional processes and environment. There is a finite number of environments so that similar facies and facies associations are produced wherever and whenever a particular environment existed in the geological past. Differences do arise of course, from variations in provenance (the source of the sediment), the nature of the fossil record at the time and climatic and tectonic considerations. From studies of modern and ancient sedimentary environments, processes and facies, generalized *facies models* have been proposed to show the lateral and vertical relationships between facies. These models facilitate interpretations of sedimentary formations and permit predictions of facies distributions and geometries. However, facies models are just snapshots of an environment; depositional systems are dynamic and a facies model may only relate to a particular state of relative sea-level change. The importance of the vertical succession of facies was first appreciated by Johannes Walther at the end of the nineteenth century in his 'Law of the Correlation of Facies': different facies in a vertical succession reflect environments that originally were adjacent to each other, providing there were no major breaks in sedimentation. Vertical changes in facies result from the effects of internal and external processes. Familiar examples of the former are the progradation (building out) of deltas and tidal flats into deeper water, and the combing of a river across its floodplain. External processes are again chiefly tectonic movements, acting on a regional or global scale, and climatic changes. Both of these affect the relative position of sea-level, a major factor in facies development, and the supply of sediment, as noted above.

1.2.3 Diagenesis

Considerations of sedimentary rocks do not stop with environmental interpretations. There is a whole story to be told of events after deposition, that is during *diagenesis*. It is during diagenesis that an indurated rock is produced from an unconsolidated, loose sediment. Diagenetic processes begin immediately after deposition and continue until metamorphism takes over; this is when reactions are the result of elevated temperatures (in excess of 150–200 °C) and/or pressures. A distinction is made between early diagenetic events, taking place from sedimentation until shallow burial, and late diagenetic events, occurring during deep burial and subsequent uplift.

Diagenetic processes, which can be introduced here but are considered further in later sections (2.9, 3.6 and 4.7), are compaction, recrystallization, dissolution, replacement, authigenesis and cementation. Compaction is both a physical and chemical process arising from the mass of the overlying sediment, which causes water to be squeezed out and grains to become closer packed. Some grains and minerals deposited in a sediment or forming a sediment are unstable and during diagenesis they may recrystallize (i.e. their crystal fabric changes but the mineralogy is unaltered) or they may undergo dissolution and/or be replaced by other minerals. The effects of dissolution and replacement are common in limestones, sandstones and evaporites. The formation of dolomite largely takes place by replacement of limestone. The precipitation of new minerals within the pore spaces of a sediment is referred to as authigenesis, and if precipitated in sufficient quantity then cementation of the sediment results. Concretions and nodules, such as commonly occur in mudrocks, form through localized mineral precipitation. The ions for cementation are derived from pore waters and grain dissolution.

In the same way that relative sea-level changes and global sea-level stand are a fundamental control of many aspects of the deposition of sedimentary rocks, they can also account for the major diagenetic events. Much early diagenesis relates to sequence boundaries, produced by second- to third-order sea-level fluctuations, but climate (arid versus humid) is important at this stage too. The nature of near-surface diagenesis very much controls the path of later (burial) diagenesis.

Diagenetic processes are important for several reasons. They can considerably modify a sediment, both in terms of its composition and texture, and in rare cases, original structures are destroyed completely. Diagenetic events also affect a sediment's porosity and permeability, properties that control a sediment's potential as a reservoir for oil, gas or water.

1.3 Methodology

The study of sedimentary rocks invariably begins in the field but after that there are several avenues which can be explored, depending on the objectives of the study and the interests of the investigator. Samples collected can be examined on a macro-, micro- and nanoscale. Sophisticated techniques and machines can be used to discover a sediment's mineralogy and geochemistry. Experiments can be devised to simulate the conditions of deposition. Data collected in the field or laboratory can be subjected to statistical tests and computer analysis. Account should be taken of any existing literature on the rocks being studied and of descriptions of similar rocks and facies from other areas, together with their probable modern analogues. With all this information at hand, the rocks under consideration can be interpreted with regard to origin, depositional process and environment, palaeogeography, diagenetic history and possible economic significance and potential.

1.3.1 In the field

The main point about fieldwork is being able to observe and record accurately what you see. With a little field experience and some background knowledge, you will soon know what to expect and what to look for in a certain type of sedimentary rock of particular

facies. It is obviously important to appreciate the significance of the various sedimentary features you see, to know which are environmentally diagnostic, for example, and also to know how they can be used to obtain maximum information: what to measure, what to photograph, what to collect. The field study of sedimentary rocks is discussed in Tucker (1996); the description and interpretation of sedimentary structures are explored in Collinson & Thompson (1989) and the statistical analysis of field data is covered in Potter & Pettijohn (1977) and Graham (1988).

The study of sedimentary rocks in the field requires the initial identification of the lithology (often with the aid of a lens) in terms of composition, grain size, texture and fossil content (see Table 1.2). These attributes can be confirmed and quantified later in the laboratory. Sedimentary structures are usually described and measured in the field because of their size. It is relatively easy to see structures in hand specimen or block, but those on the scale of a quarry or cliff face are easily overlooked. So observe on all scales. It is important to note the size and orientation of structures. Many sedimentary structures can be used for palaeocurrent analysis and these and others reflect the processes operating in the environment (see Sections 2.3, 2.4, 3.2 and 4.6). Sedimentary structures should be described within their lithological context; many are related to grain size or composition, for example. These days considerable emphasis is being placed on the larger-scale geometric relationships of sedimentary strata, seeking the onlap, offlap, downlap, etc., arrangements (see Fig. 2.86) that reflect long-term relative sea-level changes. In mountainous regions of good exposure, these seismic-scale relationships can be observed directly; in other instances they may need to be mapped out. The identification of sequence boundaries is also important, as many vertical and lateral facies patterns can be explained by a sequence stratigraphic approach, and much diagenesis ties into these boundaries too.

One of the best methods of recording sedimentary rocks is to construct a log of the section. Basically, measure the thickness of each bed or facies unit, note its composition, grain size, colour, sedimentary structures, fossils and any other features. If a palaeocurrent measurement can be taken, record this too. A *graphic log* can be drawn up in the field using an appropriate vertical scale for the sediment thickness and a horizontal scale for the sediment grain size (for examples see

Figs 2.64, 2.66, 2.68 & 2.74). Different types of shading can be used for the various lithologies and symbols and abbreviations can be used for the sedimentary structures and fossils (see Graham, 1988; Collinson & Thompson, 1989; Tucker, 1996). The value of such graphic logs lies in the immediate picture that is obtained of the vertical succession of facies. In logging a section, the lateral extent and continuity of beds must be taken into account. Many beds are actually lenticular.

Although in the field study of sedimentary rocks, it is likely that a geological map will be at hand, some detailed mapping of small areas could well be required to ascertain the relationships between facies and facies packages, and effects of local structural complications.

In many cases the interpretation of sedimentary rocks hinges on the fieldwork, so much care and attention should be paid to it. Localities need to be visited several times; it is amazing how many new things you can see at an exposure on a second or third visit.

1.3.2 In the laboratory

A great deal can be done with sedimentary rocks in the laboratory, and there are several books concerned with laboratory procedures (see Tucker, 1988). Starting with a hand specimen, cutting and polishing a surface may reveal sedimentary structures poorly displayed or invisible in the field. With limestones, etching with acid and staining a surface may further enhance the structures. With unconsolidated sediments and sedimentary rocks that are readily disaggregated, sediment grain-size can be measured through the use of sieves and sedimentation chambers (see Section 2.2.1). The heavy minerals (Section 2.5.5) can be extracted from loose sediment using heavy liquids, but this can be dangerous.

Much detailed work is undertaken on thin sections cut from sedimentary rocks or resin-impregnated unconsolidated sediments. With limestones, acetate peels are used frequently, and the staining of these and thin-sections with Alizarin Red S and potassium ferricyanide helps identify the carbonate minerals present. Stains also can be used for feldspars in terrigenous clastic sediments. There are a relatively small number of common minerals in a sedimentary rock and with a little experience it is not necessary to examine their optical properties to identify them each time. The proper-ties of the common sedimentary minerals are given in Table 1.5. The precise composition of many sedimentary rocks (the sandstones and limestones in particular), which allows them to be classified, is obtained from microscopic studies by the use of a point counter. Several hundred grains are identified as the thin-section is moved systematically across the microscope stage. Grain sizes of indurated silt- to sand-sized rocks are measured from a thin-section or peel using a calibrated eye-piece graticule. Grain shape and orientation also can be assessed. Many aspects of diagenesis in sandstones, limestones and evaporites are deduced from thin-section studies. Use of a cathodoluminoscope, which bombards a rock slice with electrons and causes luminescence, can reveal details of cements and overgrowths (see Plate 13b,c for an example and D. J. Marshall, 1988), and UV fluorescence also is useful for identifying organic matter as well as revealing 'hidden textures'. In view of the interest in porosity and reservoir potential, many sedimentary rocks are now impregnated routinely with a resin containing a dye before they are thin-sectioned (e.g. Plates 2c & 6d). For illustrations of sedimentary rocks and minerals in thin-section see Scholle (1978, 1979), Adams and MacKenzie (1998), Adams *et al.* (1984) and MacKenzie and Adams (1994).

In recent years, much sedimentological work has been carried out with the scanning electron microscope (SEM). This instrument allows examination of specimens at very high magnifications; features down to 0.1 μm can be seen. The SEM is especially useful for fine-grained sedimentary rocks, such as cherts, and for observing clay minerals and the cements of sandstones and limestones: see for example, Figs 2.54, 2.55, 4.29 & 4.31. The back-scattered mode is useful for textural studies of mudrocks, see Fig. 3.2. See Trewin (1988) for a review of the SEM in sedimentary studies.

For mineral identification in fine-grained sediments and sedimentary rocks, X-ray diffraction (XRD) is widely used. Clay minerals in mudrocks are invariably analysed in this way (see Hardy & Tucker (1988) for details on this technique).

It is becoming apparent that geochemical analyses of sedimentary rocks, especially limestones and shales, can give useful and vital information on the environment of deposition and path of diagenesis. Major and minor elements are determined mostly by atomic absorption spectrophotometry (AAS), inductively coupled plasma optical-emission spectroscopy (ICP-OES)

Table 1.5 Optical properties of common minerals in sedimentary rocks as observed with the petrological microscope

Mineral	Chemical formula	Crystal system	Colour	Cleavage	Relief
Quartz	SiO_2	trigonal	colourless	absent	very low (+)
Microcline	$KAlSi_3O_8$	triclinic	colourless	present	low (−)
Orthoclase	$K(Na)AlSi_3O_8$	monoclinic	colourless	present	low (−)
Albite	$Na(Ca)AlSi_3O_8$	triclinic	colourless	present	low (−)
Muscovite	$KAl_2(OH)_2AlSi_3O_{10}$	monoclinic	colourless	planar	moderate
Biotite	$K_2(Mg,Fe)_2(OH)_2AlSi_3O_{10}$	monoclinic	brown to green	planar	moderate
Chlorite	$Mg_5(Al,Fe)(OH)_8(AlSi)_4O_{10}$	monoclinic	green	planar	fair
Kaolinite	$Al_2O_32SiO_2.\,2H_2O$	triclinic	colourless–yellow	planar	low (+)
Illite	$KAl_2(OH)_2[AlSi_3(O,OH)_{10}]$	monoclinic	colourless–yellow	—	low (+)
Montmorillonite	$(MgCa)O.\,Al_2O_3.\,5SiO_2.\,nH_2O$	monoclinic	colourless–pink	—	low (−)
Berthierine–chamosite	$Fe_3^{2+}Al_2Si_2O_{10}.\,3H_2O$	monoclinic	green	—	moderate
Glauconite	$KMg(Fe,Al)(SiO_3)_6.\,3H_2O$	monoclinic	green	planar	moderate
Aragonite	$CaCO_3$	orthorhombic	colourless	rectilinear	moderate
Calcite	$CaCO_3$	trigonal	colourless	rhombic	low to high
Dolomite	$CaMg(CO_3)_2$	trigonal	colourless	rhombic	low to high
Siderite	$FeCO_3$	trigonal	colourless	rhombic	low to high
Gypsum	$CaSO_4.\,2H_2O$	monoclinic	colourless	planar	low
Anhydrite	$CaSO_4$	orthorhombic	colourless	rectilinear	moderate
Halite	$NaCl$	cubic	colourless	rectilinear	low
Collophane	$Ca_{10}(PO_4,CO_3)_6F_{2-3}$	a mineraloid	shades of brown	—	moderate
Pyrite	FeS_2	cubic	opaque	—	—
Hematite	Fe_2O_3	hexagonal	opaque	—	—
Magnetite	Fe_3O_4	cubic	opaque	—	—

or mass spectrometry (ICP-MS) and X-ray fluorescence (XRF). On the scale of individual grains and crystals, the electron microprobe and laser ablation with ICP-MS are used to determine trace elements on areas only a few micrometres across. A consideration of the isotopes of such elements as oxygen and carbon, measured with a mass spectrometer, is a powerful tool in the study of limestone and chert diagenesis (see Fairchild *et al.* (1988) for a review of geochemical techniques in sedimentary studies). Analysis of fluid inclusions in calcite, quartz and halite crystals also gives much important information on the temperature and salinity of pore waters from which the minerals were precipitated (see Goldstein & Reynolds, 1994).

One further laboratory approach has been to carry out experiments to determine the conditions under which sedimentary structures, grain types, minerals, etc., were formed. Perhaps the best known are those involving laboratory channels or flumes, where the effects of water flowing over sand have been monitored (Section 2.3.2), and the attempts to precipitate dolomite.

Once the data on the sedimentary rocks under investigation have been gathered, then the interpretations can begin. Information on sediment composition and microfacies can be combined with field data to deduce the environment and conditions of deposition. Petrographic studies of sandstones can give information on the geology of the source area (the provenance) and the plate-tectonic setting. Diagenetic studies can be integrated with facies and burial history to account for the patterns of cementation and dissolution, and porosity evolution.

Statistics and computers are being used increasingly

Birefringence	Other features	Form and occurrence	See Section
weak		as detrital grains (monocrystalline and polycrystalline types), cements and replacements: fibrous quartz (chalcedony), microquartz, megaquartz	2.5.2 2.9.2 9.2
weak	grid-iron twinning	as detrital crystals, also authigenic	2.5.3
weak	simple twinning (Carlsbad)	commonly altered to clays, so appearing dusty	2.9.4
weak	multiple twinning		
strong	parallel extinction	common detrital minerals occurring as flakes	2.5.4
strong	parallel extinction		3.4.3
weak	best identified through X-ray diffraction because usually so fine-grained	as detrital minerals, particularly in mudrocks, also as cement (as in sandstones) and replacements, such as of feldspars and volcanic grains	2.9.5
weak			3.4.1
strong			10.7
moderate			
weak		ooids and mud in ironstones	6.4.3
moderate		forms synsedimentary grains	6.4.3
moderate	can be distinguished by staining (Section 4.1)	form grains, matrix, cement and replacements in limestones, dolomites, sandstones, etc.	4.2, 4.3, 4.7
extreme			4.8, 2.9.3
extreme			
extreme	alters to brown colour	fine and coarse crystals in ironstones	6.4.2
weak		anhedral to euhedral crystals	5.2
strong	parallel extinction	equant to lath-shaped crystals	5.2
—	may have fluid inclusions	often coarsely crystalline	5.3
isotropic or weak	if bone—organic structure	forms ooids, pellets, bones, some shells	7.2
—	yellow in reflected light	aggregates and cubic crystals, authigenic	6.4.4
—	red–grey in reflected light	cryptocrystalline, a pigment and replacement	6.4.1, 2.9.6
—	grey–black in reflected light	cryptocrystalline, detrital	6.4.1

for the evaluation and interpretation of sedimentological and petrographic data. Mathematical geology is now an established branch of the earth sciences and reference should be made to the available textbooks on this subject. Field data also can be subjected to statistical analysis, to identify cycles in a succession of facies for example (see Graham, 1988). Quantitative modelling of sedimentary basin filling and simulation of stratigraphic sequences and cycles are also making great contributions to our understanding of the factors controlling deposition (see review by Paola, 2000).

1.3.3 In the library: sedimentological reading

However good your field and laboratory work is, it must be supported with a knowledge of the literature on the subject. Publications on the petrology of sedimentary rocks go way back into the last century, but in fact most advances have come in the past four decades.

There are many textbooks available that cover some aspects of the subject in more detail than can be given here or that deal with directly related topics. Recent texts on sedimentary processes include Pye (1994), Reading (1996), Allen (1997), Leeder (1999) and Nichols (1999). At the end of each of the other chapters in this book, a list of readily accessible textbooks and papers is given for further reading on that particular sedimentary rock type. Books dealing with more practical aspects of sedimentology/sedimentary petrology include Friedman & Johnson (1982), Lewis (1984), Lindholm (1987) and Tucker (1988). Recent books on sequence stratigraphy include Emery & Meyers (1996), Miall (1997) and Gradstein *et al.*

(1998). Basin analysis, an important branch of the earth sciences in its own right now, but very relevant to sedimentary studies and vice versa, is discussed by Busby & Ingersoll (1995), Einsele (1990), Maill (2000) and Allen & Allen (2001). There are many collections of papers on a specific topic within sedimentology and here can be cited the series of special publications of the Society of Sedimentary Geologists (formerly Society of Economic Paleontologists and Mineralogists, SEPM) and the International Association of Sedimentologists (IAS), some memoirs of the American Association of Petroleum Geologists (AAPG) and special publications of the Geological Society of London.

Most research papers, however, are published in the learned journals. Interested students should keep their eyes on the current journals for the latest information and ideas. Books soon go out of date(!). The three principal periodicals are the *Journal of Sedimentary Research* (formerly *Journal of Sedimentary Petrology*), *Sedimentology* and *Sedimentary Geology*, published by SEPM, IAS and Elsevier, respectively.

Others devoted to sediments or containing many sedimentological papers are *Geology*, *Bulletin of the Geological Society of America*, *Bulletin of the American Association of Petroleum Geologists*, *Facies*, *Journal of Geology*, *Marine Geology*, *Palaeogeography, Palaeoecology and Palaeoclimatology*, and *Palaios*. In addition there are many other journals that often contain relevant articles; a regular perusal of the current periodicals in the library will spot these as they are published. Increasing numbers of journals are now available online; check your library's website for a list of these.

Finally, there are many abstracts, indexes and bibliographies available, from which you can search for references on a particular topic within sedimentology. There are several periodicals that regularly cite all published papers. However, there are now many online databases for locating journal articles, books, theses and conference proceedings. Of particular interest to sedimentology are the Web of Science (Science Citation Index), GeoRefS and GEOBASE.

2 Siliciclastic sediments I: sandstones, conglomerates and breccias

2.1 Introduction

Siliciclastic sediments are a diverse group of rocks, ranging from fine-grained mudrocks, through sandstones to the coarser-grained conglomerates and breccias. The sediments are composed largely of grains (clasts) derived from pre-existing igneous, metamorphic and sedimentary rocks. The clastic grains are released through mechanical and chemical weathering processes, and then transported to the depositional site by a variety of mechanisms, including wind, glaciers, river currents, waves, tidal currents, debris flows and turbidity currents. Conglomerates are made mainly of pebbles and boulders, and these can be of a wide variety of rock types. Sandstones also contain rock fragments, but the majority of the grains are individual crystals, chiefly of quartz and feldspar, abraded to various degrees. The finer breakdown products of the original rocks, formed during weathering and consisting mainly of clay minerals, are predominant in mudrocks and form the matrix to some sandstones and conglomerates. In a broad sense, the composition of siliciclastic sediments is a reflection of the weathering processes, determined largely by the climate and the geology of the source area (the *provenance* of the sediment). Source areas generally are upland, mountainous regions undergoing uplift, but detritus also may be supplied from erosion in lowland and coastal areas. Sediment composition also is affected by distance of sediment transport and by diagenetic processes.

Two important features of siliciclastic sediments are their sedimentary structures and textures. Many of these are produced by the depositional processes, whereas others are post-depositional or diagenetic in origin. Many of the sedimentary structures in sandstones also occur in limestones and some of the other sedimentary rock types. For studies of sandstone composition, texture, diagenesis and porosity, thin-sections are routinely used. Staining a section can help distinguish the feldspar grains (Section 2.5.3) and

cathodoluminescence is useful too (Sections 2.5.2 and 4.1). A scheme for describing sandstones from thin-sections is given in Table 2.1, and notes for microscope work in Table 2.2.

In this chapter, the coarser siliciclastic sediments, the sandstones, conglomerates and breccias, are treated; the finer-grained siliciclastic sediments, the mudrocks, are discussed in Chapter 3.

2.2 Sediment texture

Studies of sediment texture involve considerations of grain size and grain-size parameters, grain morphology, grain-surface texture and sediment fabric. On the basis of its textural attributes, a sediment can be considered in terms of its *textural maturity*. The texture of a clastic rock is a reflection largely of the depositional process(es). Therefore many modern sediments from different environments have been studied in order to determine their textural characteristics, and this information is used as an aid to the interpretation of sedimentary rocks.

2.2.1 Grain size and grain-size parameters

The basic descriptive element of all sedimentary rocks is the grain size. A number of grain-size scales have been proposed, but one which is widely used and accepted is that of J. A. Udden, based on a constant ratio of two between successive class boundaries together with terms for the classes by C. K. Wentworth (Table 2.3). The Udden–Wentworth grain-size scale divides sediments into seven grades: clay, silt, sand, granules, pebbles, cobbles and boulders, and subdivides sand into five classes and silt into four. This scale has been modified and extended by Blair & McPherson (1999) to give more detail in the coarser grades. Pebble, cobble and boulder classes are subdivided and block, slab, monolith and megalith are introduced for very large clasts. For sediments/sedimentary rocks, the meaning of the terms sand/sandstone, silt/siltstone and clay/

Table 2.1 Scheme for petrographic description of sandstones

Hand specimen

Note the colour; grain size; grain shape; bedding, lamination and any other sedimentary structures. Any fossils present? Determine composition/mineralogy of grains and cements if possible

Thin-section

Check the macroscopic features of the thin-section by holding up to light and noting any lamination, large fossils or grains

Texture: determine the grain size, sorting, roundness of grains, grain shape, fabric (any preferred orientation of grains?) and nature of grain–grain contacts

Grains: identify grain types; determine relative proportions of quartz, feldspar, lithic grains and matrix

Matrix: check whether it is detrital; it may have formed from alteration and compaction of labile grains

Compaction: look for concavo-convex and sutured grain contacts, broken/bent mica flakes or bioclasts

Cementation: identify cements, e.g. quartz, calcite, dolomite, clays, and habits—overgrowths, pore-filling, pore-lining, etc.

Replacement/dissolution of grains: e.g. feldspar by calcite or clay; partial to complete dissolution of grains; look for oversized pores, where whole grains dissolved out

Porosity: if present determine origin and type—intergranular, dissolutional, fracture, etc.

Classification: from assessment of matrix content, is sandstone an arenite or wacke? If arenite, assess type (quartz arenite, arkose or litharenite) from grain composition. From texture, assess the maturity

Interpretation

Depositional environment: suggest from texture and composition, and any other information available, such as sedimentary structures and fossils from hand specimen and field data

Diagenesis: determine nature and order of diagenetic events and whether near-surface (pre-compaction) or burial (post-compaction) on basis of textural evidence; suggest evolution of pore fluids and destruction or creation of porosity in context of burial history

---

claystone is obvious. Gravel is applied to loose sediments of granule to boulder grade, and megagravel for coarser sediment still, although these usually have significant amounts of finer matrix. Rudite or rudaceous rock is used for indurated gravels/megagravels, and includes the conglomerates and breccias, and megaconglomerates and megabreccias. Figure 2.1 gives a classification for mixtures of sand, gravel and mud. For mudrock terminology and sand–silt–clay mixtures see Section 3.1.

Using millimetres as the units, the Udden–Wentworth scale is geometric (i.e. 1, 2, 4, 8, 16). W. C. Krumbein introduced an arithmetic scale (i.e. 1, 2, 3, 4, 5) of phi units (ϕ), where phi is the logarithmic transformation of the Udden–Wentworth scale: $\phi = -\log_2 d$, where d is the grain size in millimetres. Also used are psi units (ψ), equal to the negative of phi. For all serious work involving sediment grain size the phi scale is used because it has the advantage of making mathematical calculations much easier. For detailed work, class intervals in the sand field are taken at quarter-phi intervals.

When studying sandstones in the field, a first approximation of the grain size can be made with a hand-lens. The clast size of conglomerates and breccias can be measured directly with a tape measure. For accurate grain-size analysis, several laboratory methods are available. With poorly cemented sandstones and unconsolidated sands, sieving is the most popular technique. Medium silt to small pebbles can be accommodated in sieves and it is the practice to use a similar sieving time, of around 15 min for all samples, and similar weights, about 30 g, or more for the coarser grades. Sedimentation methods, which measure the settling velocity of grains through a column of water, can be used for clay to sand grades. For well-cemented siltstones and sandstones (and limestones), thin-sections have to be used, with several hundred grain sizes measured using an eyepiece graticule and point-counter. Reviews of these techniques are given by McManus (1988), Syvitski (1991) and Lewis & McConchie (1994). An image analysis system connected to a microscope can give a grain-size analysis and parameters almost instantly.

Once the grain-size distribution has been obtained then the sediment can be characterized by several parameters: mean grain size, mode, median grain size, sorting and skewness. A further parameter, kurtosis, has little geological significance. The parameters can be calculated from graphic presentation of the data (as in Fig. 2.3), or from a computer program.

For graphic presentation, the histogram, smoothed frequency curve and cumulative frequency curve are plotted (Figs 2.2 & 2.3). It is the practice for grain size to decrease along the abscissa (x-axis) away from the origin. The histogram and smoothed frequency curve show the frequency of grains in each size-class and usefully give an immediate impression of the grain-size distribution, particularly as to whether the distribution is unimodal or bimodal (Fig. 2.2). The cumulative

Table 2.2 Scheme for describing sandstones in thin-section

Features	Thin-section 1	Thin-section 2
Grains present and percentage Quartz (types) Feldspar (types) Lithic grains (types) Mica (types) Bioclasts (types) Others		
Texture Roundness, sorting, fabric, packing, preferred orientation of grains		
Cements Quartz, calcite, dolomite, hematite, clays, anhydrite; cement geometry and timing		
Replacements Alteration, dissolution, feldspar preservation, calcite and clays replacing grains		
Evidence of compaction Broken and squashed grains, concavo-convex and sutured contacts, stylolites		
Porosity Intergranular—reduced/enhanced, mouldic, fracture, stylolitic, etc.		
Sandstone type Arenite/wacke, quartz arenite, arkose, litharenite, greywacke—lithic/feldspathic/quartzitic		
Depositional environment Marine/non-marine, fluvial/aeolian, shallow/deep, low/high energy		
Order of diagenetic events	1: 2: 3:	

frequency curve shows the percentage frequency of grains coarser than a particular value. When plotting cumulative frequencies it is best to use semilog probability graph paper, which gives a straight line if the distribution is normal, i.e. Gaussian, as generally is the case with sediments. From the cumulative frequency plot, the percentiles of the distribution are obtained, i.e. the grain sizes that correspond to particular percentage frequencies (percentiles), so that at the nth percentile, n% of the sample is coarser than that grain size.

The grain-size parameters are defined in Table 2.4 and are used with grain size measured in phi.

The median grain size, simply the grain size at 50%, is not as useful as the mean grain size, that is an average value taking into account the grain sizes at the 16th, 50th and 84th percentiles. The mode is the phi (or millimetre) value of the mid-point of the most abundant class. Most sediments are unimodal, i.e. one class dominates, but bimodal (Fig. 2.2) and even polymodal sediments are not uncommon (matrix-rich conglomerates, for example, Fig. 2.10). Where a grain-size distribution is perfectly normal and symmetrical, then the median, mean and mode values are the same.

Trends in grain size over large areas can be used to

Table 2.3 Grain-size scale for sediments and sedimentary rocks. After Udden and Wentworth, and Blair & McPherson (1999)

Length (mm)	φ		Class	Sediment/rock name
			block	mega-conglomerate
4096	−12			
2048	−11	vc		
1024	−10	c	boulder	
512	−9	m		
256	−8	f		gravel conglomerate
128	−7	c	cobble	
64	−6	f		
32	−5	vc		
16	−4	c	pebble	
8	−3	m		
4	−2	f		
2	−1		granule	
1	0	vc		
0.50	1	c	sand	sand sandstone
0.25	2	m		
0.125	3	f		
0.063	4	vf		
0.031	5	c		
0.015	6	m	silt	silt siltstone
0.008	7	f		
0.004	8	vf		
			clay	clay claystone

infer the direction of sediment dispersal, with grain size decreasing away from the source area. Such down-current changes occur in fluvial and deltaic systems, and in turbidites in deep-sea basins. Sediment-size decrease mostly relates to selective sediment transport, rather than abrasion (Hoey & Bluck, 1999; Rice, 1999). From shorelines across shelves, the offshore decrease in grain size relates to a decrease in wave and current energy as the water depth increases.

With conglomerates, it is useful to measure the maximum clast size, along with the bed thickness, as the relationship between these two parameters does vary with the depositional process, chiefly the competence of the flow. In the fluvial environment, for example, mudflow and stream-flood conglomerates generally show a positive correlation between maximum pebble size and bed thickness, whereas braided stream conglomerates do not.

Sorting is a measure of the standard deviation, i.e. spread of the grain-size distribution. It is one of the most useful parameters because it gives an indication of the effectiveness of the depositional medium in separating grains of different classes. Terms used to describe the sorting values obtained from the Folk & Ward (1957) formula are:

φ less than 0.35	very well sorted
0.35–0.50	well sorted
0.50–0.71	moderately well sorted

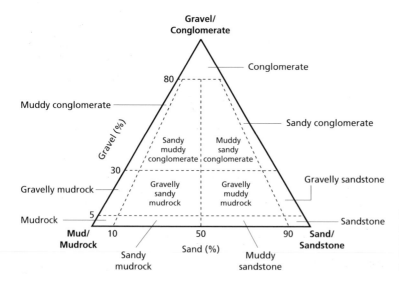

Fig. 2.1 Scheme for classifying sand–gravel–mud mixtures and the terms for sediment and rock (after Udden–Wentworth and Blair & McPherson, 1999).

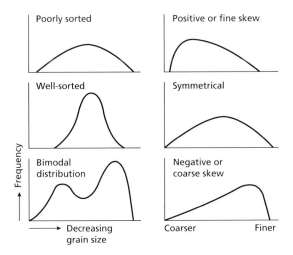

Fig. 2.2 Smoothed frequency distribution curves showing types of sorting and skewness.

Fig. 2.3 An example of the graphic presentation of grain-size data (500 grain-size measurements from a sandstone), presented as a histogram (a), and cumulative frequency curves plotted with an arithmetic scale (b) and log probability scale (c). Also given are the grain-size parameters derived graphically, using Folk & Ward's (1957) formulae of Table 2.4.

0.71–1.00	moderately sorted
1.00–2.00	poorly sorted
greater than 2.00	very poorly sorted

With thin-sections of sandstones (and limestones) there is a problem of apparent sorting, when the sediments appear to be more poorly sorted than they actually are. Reviewing published comparators, Jerram (2000) has developed additional, geologically more realistic two-dimensional and new three-dimensional visual comparators based on computer-generated three-dimensional distributions of spheres (Fig. 2.4).

Table 2.4 Formulae for the calculation of grain-size parameters from a graphic presentation of the data in a cumulative frequency plot. The percentile measure ϕ_n is the grain size in phi units at the nth percentage frequency

Parameter	Folk & Ward (1957) formula
Median	$Md = \phi_{50}$
Mean	$M = \dfrac{\phi_{16} + \phi_{50} + \phi_{84}}{3}$
Sorting	$\sigma\phi = \dfrac{\phi_{84}-\phi_{16}}{4} + \dfrac{\phi_{95}-\phi_5}{6.6}$
Skewness	$Sk = \dfrac{\phi_{16} + \phi_{84}-2\phi_{50}}{2(\phi_{84}-\phi_{16})} + \dfrac{\phi_5 + \phi_{95}-2\phi_{50}}{2(\phi_{95}-\phi_5)}$

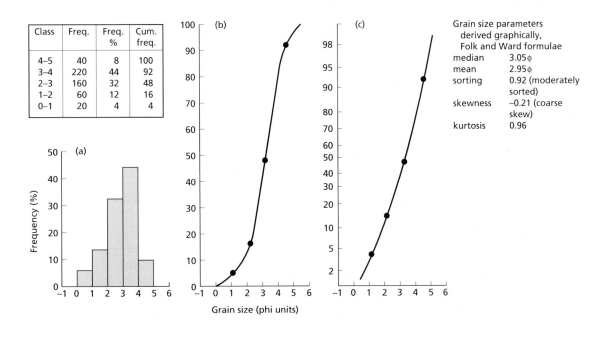

Class	Freq.	Freq.%	Cum. freq.
4–5	40	8	100
3–4	220	44	92
2–3	160	32	48
1–2	60	12	16
0–1	20	4	4

Grain size parameters derived graphically, Folk and Ward formulae

median	3.05ɸ
mean	2.95ɸ
sorting	0.92 (moderately sorted)
skewness	−0.21 (coarse skew)
kurtosis	0.96

Grain size (phi units)

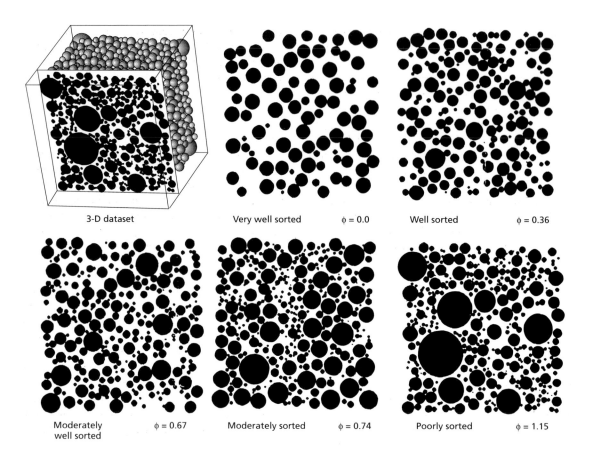

3-D dataset   **Very well sorted** $\phi = 0.0$ **Well sorted** $\phi = 0.36$

Moderately well sorted $\phi = 0.67$ **Moderately sorted** $\phi = 0.74$ **Poorly sorted** $\phi = 1.15$

Fig. 2.4 Visual comparators for sorting for thin-sections based on three-dimensional distributions of spheres that are log-normal by volume frequency (from Jerram, 2001).

Sorting of the sediment is determined by several factors. First there is the question of the sediment source: if a granite is providing the sediment, the grain sizes will be quite different from those of a sediment supplied by the reworking of a sandstone. The second factor is grain size itself: sorting is dependent on the grain size in that coarse sediments, gravels and conglomerates, and fine sediments, silts and clays, generally are more poorly sorted than sand-sized sediments, which are more easily transported and therefore sorted by wind and water. The third factor, which is where sorting is used for interpretation, is the depositional mechanism. Sediments that were deposited quickly, such as storm beds, or were deposited from viscous flows, such as mud flows, are generally poorly sorted;

sediments that have been reworked by the wind or water, the sandy deposits of deserts, beaches and shallow shelf seas, for example, are much better sorted. Generally, sorting improves along the sediment transport path. This is the case with desert sands, for example, where grain size also decreases downwind.

Skewness is a measure of the symmetry of the distribution and visually is best seen from the smoothed frequency curve (Fig. 2.2). If the distribution has a coarse 'tail', i.e. excess coarse material, then the sediment is said to be negatively skewed; if there is a fine 'tail', then the skew is positive. If the distribution is symmetrical, then there is no skew. Terms for skewness (Sk) derived from the Folk & Ward (1957) formula are:

Sk greater than +0.30	strongly fine-skewed
+0.30 to +0.10	fine-skewed
+0.10 to –0.10	near-symmetrical
–0.10 to –0.30	coarse-skewed
greater than –0.30	strongly coarse-skewed

Apart from being a useful descriptive term for a sediment sample, skewness is also a reflection of the depositional process. Beach sands, for example, tend to have a negative skew because finer components are carried off by the persistent wave action. River sands are usually positively skewed, because much silt and clay is not removed by the currents, but is trapped between larger grains. Some desert dune sands have a negative skew but this is more complicated in origin because it relates to the effect of wind on a coarse sediment. Very fine- to medium-grained sand is blown out of the source area to form the dunes. Some coarser sand also is blown into the dunefield, but much of the finer sediment (silt) is efficiently blown out of the system altogether. In general, sediment becomes more negatively skewed (and finer grained) along its sediment transport path, whereas the source sediment (lag) becomes more positively skewed and relatively coarse.

Interpretation and use of grain-size analyses

Grain-size analyses are a routine procedure in many sedimentary studies in order to characterize the sediment or rock and give information on its depositional mechanism and depositional environment.

Grain-size analyses can be used to distinguish between sediments of different environments and facies and to give information on the depositional processes and flow conditions. Many studies have attempted to distinguish between the sediments of modern depositional environments using the grain-size distribution. Scatter diagrams are usually constructed, such as sorting plotted against skewness. On this basis, it is possible to distinguish between beach, dune and river sands. Grain-size analyses of sandstones alone should not be used in environmental interpretations, but combined with studies of sedimentary structures they can be useful in facies description and analysis. Points to bear in mind are the possibilities of sand being reworked or supplied from an adjacent or pre-existing environment, when there is then a problem of inherited characteristics, and where clay-grade material is present there is the problem of its origin. The origin may well be depositional and so of consequence to the environmental interpretation, but a fine-grained matrix also can be infiltrated, a breakdown product of labile grains and a diagenetic precipitate. To complicate the interpretation of grain-size analyses in terms of depositional process, several different processes may well have operated in one environment, and similar processes do take place in different environments. See Syvitski (1991) for papers on this topic.

2.2.2 Grain morphology

Three aspects of grain morphology are the shape, sphericity and roundness. The *shape* of a grain is measured by various ratios involving the long (L), intermediate (I) and short (S) axes. Shape is most effectively plotted on a triangular diagram with the end members spheres, discs and rods, with blades being intermediate between the last two (see Fig. 2.5 and Illenberger, 1991). *Sphericity* is a measure of how closely the grain shape approaches that of a sphere. *Roundness* is concerned with the curvature of the corners of a grain and six classes from very angular to well rounded are usually distinguished (Fig. 2.6). Several formulae have been proposed for the calculation of sphericity and roundness. However, for environmental interpretations, the roundness is more significant than sphericity or shape, and for most purposes the simple descriptive terms of Fig. 2.6 for roundness are sufficient.

The morphology of a grain is dependent on many factors; initially the mineralogy, nature of the source rock and degree of weathering, then on degree of

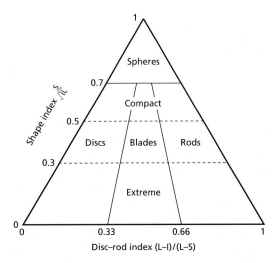

Fig. 2.5 The four classes of grain shape: spheres, discs, rods and blades, based on the shape index (a measure of the sphericity) and the disc–rod index. L, I and S represent the long, intermediate short axes of the grains, respectively (after Illenberger, 1991).

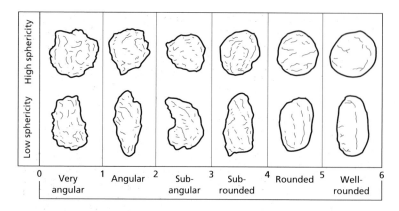

Fig. 2.6 Categories of roundness for sediment grains. For each category a grain of low and high sphericity is shown. After Pettijohn *et al.* (1987).

abrasion during transport, and later on dissolution during diagenesis. The shape of quartz grains in a sandstone does depend on their origin. If derived directly from crystalline rocks quartz tends to be highly non-spherical and angular, with embayments and fractures. If derived from a pre-existing sediment, then if that was wind deposited, the grains would be spherical and very well rounded; if from a quartz-cemented sediment, they would be moderately angular. In a very general way, the degree of roundness increases with the duration of transport and reworking. Beach and desert sands, for example, are typically better rounded than river or glacial-outwash sands. As with grain-size analyses, care must be exercised in interpreting roundness values; the roundness characteristic of grains may be inherited and intense abrasion may lead to fracturing and thereby angular grains.

The morphology of pebbles can be assessed more easily than that of sand grains. Some pebble shapes are diagnostic of a particular environment. In deserts, for example, sand blasting produces flat-sided pebbles (ventifacts).

2.2.3 Grain-surface texture

The surface of sand and coarser grains commonly has a distinct appearance. The dull, frosted nature of desert sand is an oft-cited example. With pebbles from a glacial deposit, striations are common on the surface. Crescentic impact marks occur on pebbles from beaches and river channels. Studies of the surface texture of modern sands with the scanning electron microscope have shown that there are features which are produced by the mechanisms of transport, and

some may be environmentally diagnostic. Beach sand grains show minute v-shaped percussion marks; the surfaces of desert sands possess 'upturned plates', and sands from glacial deposits have conchoidal fracture patterns and parallel striations (Fig. 2.7). The surface texture of grains in sandstones must be interpreted with great caution because diagenetic processes can considerably modify the surface, with pore-water dissolution producing a dull, etched surface. However, if used in conjunction with grain size and shape, and field sedimentology, it can provide useful supporting data. Recent reviews have been provided by Bull (1981), J. R. Marshall (1987) and Trewin (1988).

2.2.4 Grain fabric

The term fabric for grains in a sedimentary rock refers to their orientation and packing, and to the nature of the contacts between them (Fig. 2.8).

In many sandstones and conglomerates the sand grains and pebbles are aligned with their long axes in the same direction. This *preferred orientation* is a primary fabric of the rock (unless the rock has been deformed tectonically) and it is produced by the interaction of the flowing depositional medium (e.g. wind, ice, water) with the sediment.

Prolate pebbles in fluvial and other waterlain deposits can be orientated both normal and parallel to the current direction. Rolling of pebbles produces the normal-to-current orientation, whereas the parallel orientation arises from a sliding motion. In glacial sediments, the orientation of clasts is more commonly parallel to the direction of ice movement. One common fabric of oblate pebbles in waterlain deposits is

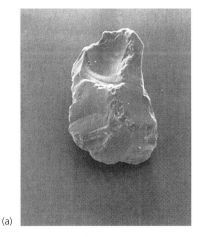
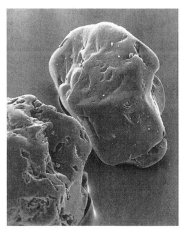
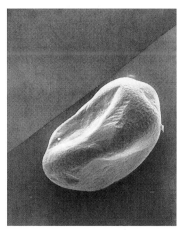

(a)   (b),(c)

Fig. 2.7 Scanning electron micrographs of quartz sand grains from three modern environments. (a) Grain from glacial outwash deposit, Ottawa, Canada, showing conchoidal fractures and angular shape. (b) Grain from high-energy beach, Sierra Leone, West Africa, showing rounded shape and smooth surface with small v-shaped percussion marks. (c) Grain from desert sand sea, Saudi Arabia, showing frosted, pock-marked surface (as a result of upturned plates, which are visible at higher magnifications) and conchoidal fractures resulting from mechanical chipping.

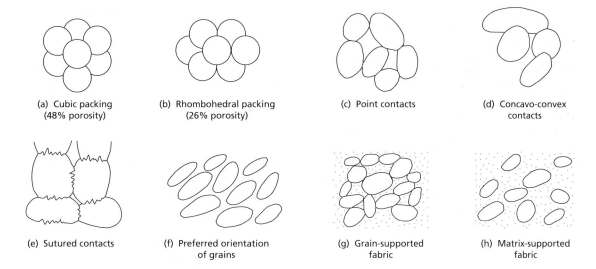

Fig. 2.8 Grain fabric in sediments: packing, contacts and orientation of grains and grain–matrix relationships.

imbrication, where the pebbles overlap each other, dipping in an upstream direction. Imbrication can thus be used in palaeoflow studies (for an example using flat limestone pebbles in the Cambrian of the Appalachians, see Whisonant, 1987). Elongate grains in sandstones also can show both parallel-to-current and normal-to-current preferred orientations, although in many sandstones the parallel orientation seems more prevalent. Grain orientation can be used as a palaeocurrent indicator, particularly if sedimentary structures are poorly developed (see Section 2.4). There are palaeomagnetic techniques to measure the anisotropy (e.g. Hounslow, 1987). Apart from detrital grains and pebbles, elongate fossils such as plant debris, belemnites and graptolites can show preferred orientations (see Fig. 2.44).

The *packing* of sediment grains is an important consideration because it affects porosity and permeability. Packing is largely dependent on the grain size, shape and sorting. Modern beach and dune sands, composed of well-sorted and rounded grains, have porosities from 25 to 65%. Where porosities are high, packing is loose and approaches the cubic packing of spheres (Fig. 2.8); lower porosities result from tighter packing in a rhombohedral arrangement. Poorly sorted sediments have a closer packing and thus a lower porosity through the greater range of grain size and the filling of pore space between large grains by finer grains. A packing density or *grain fraction* can be calculated, expressing the grain volume as a percentage of the total rock volume. Grain-fraction percentage increases with increasing overburden pressure.

The common types of *grain contact* are: point contacts where the grains are touching each other (giving the sediment a *grain-support fabric*), concavo-convex contacts where one grain has penetrated another, and sutured contacts where there is a mutual interpenetration of grains (Fig. 2.8). In addition, where there is a lot of matrix, grains may not be in contact, but 'float' in the matrix (that is, the grains are *matrix-supported*). Clast–matrix relationships separate the *orthoconglomerates* (clast-supported) from the *paraconglomerates* (matrix-supported).

The fabric of sediments is another aspect of sediment texture that can give useful information on the depositional process. With conglomerates, pebbles floating in a matrix are a common feature of debris-flow deposits (debrites, Fig. 2.9) and glacial tills and tillites (see Fig. 2.84). These contrast with conglomerates of river channel, stream flood and beach origin, where pebbles are in contact (clast-supported), and there may be little matrix (Fig. 2.10). As with many aspects of sediment texture, fabric studies should be considered along with the sedimentary structures.

2.2.5 Textural maturity

Texturally immature sediments are those with much matrix, poor sorting and angular grains; mature sediments are those where there is little matrix, moderate to good sorting and subrounded to rounded grains; texturally supermature sandstones are those with no matrix, very good sorting and well-rounded grains.

Fig. 2.9 Matrix-support fabric: pebbles 'float' in matrix. Notice also subtle synsedimentary folds. Tertiary deep-water pebbly mudstone of debris-flow origin. California, USA.

Fig. 2.10 Clast-support fabric; pebbles, mainly quartzite, are in contact and were deposited on a fan delta. Late Precambrian, Southern Norway.

Primary porosity and permeability increase with increasing textural maturity, because the more mature the sediment is, the less matrix and more pore space it possesses.

Textural maturity in sandstones is largely a reflection of the depositional process, although it can be modified by diagenetic processes (Section 2.9). Where there has been minimal current activity, the sediments generally are texturally immature; persistent current or wind activity results in a more mature sandstone.

Examples of texturally immature sediments include many fluvial and glacial deposits; supermature sediments are typified by desert, beach and shallow-marine sandstones.

2.3 Sedimentary structures

Sedimentary structures are the larger-scale features of sedimentary rocks and include the familiar cross-bedding, ripples, flute and load casts, dinosaur footprints and worm burrows. The majority of structures form by physical processes, before, during and after sedimentation, whereas others result from organic and chemical processes. Sedimentary structures, particularly those formed during sedimentation, have a variety of uses:

1 for interpreting the depositional environment in terms of processes, water depth, wind strength, etc.;

2 for determining the way-up of a rock succession in an area of complex folding;

3 for deducing the palaeocurrent pattern and palaeogeography.

Many of the structures are on the scale of tenths to tens of metres and so are studied, recorded and measured in the field. In recent years, much experimental and theoretical work has been undertaken on the development of sedimentary structures, particularly those formed by the interaction of water and sediment. These experiments have been seeking the conditions of formation of the structures and they have allowed more meaningful interpretations to be made of these structures in the geological record.

Although there is no generally accepted classification of sedimentary structures, the four main groups are (a) erosional, (b) depositional, (c) post-depositional/diagenetic and (d) biogenic structures.

Numerous photographs of sedimentary structures and descriptions are given in Allen (1982), Fritz & Moore (1988), Collinson & Thompson (1989) and Ricci-Lucchi (1995).

2.3.1 Erosional sedimentary structures

Most of the structures in this category have formed through erosion by aqueous and sediment-laden flows before deposition of the overlying bed, and by objects in transport striking the sediment surface. The most familiar are the sole structures, flute and groove casts, which occur on the underside of many beds deposited by turbidity currents, and also storm currents. Other structures of this group are scours and channels.

Flute marks have a distinctive appearance (Fig. 2.11), often described as spatulate or heel-shaped, consisting of a rounded or bulbous upstream end, which flares downstream and merges into the bedding plane. In section they are asymmetric, with the deepest part at the upstream end. They average 5–10 cm across and 10–20 cm in length, and occur in groups, all with a similar orientation and size. The formation of flutes is attributed to localized erosion by sand-laden currents, passing over a cohesive mud surface. The flute develops through a process of flow separation, whereby the current leaves the sediment surface at the upstream rim of the flute and a small eddy, rotating in a horizontal plane, is trapped within the flute. While the current is moving swiftly the eddy keeps sediment out of the flute; deceleration of the current causes sediment to be deposited, which fills the flute.

Flute marks are a characteristic structure of turbidites (see Section 2.11.7) and they give a reliable indication of the flow direction.

Groove marks are linear ridges on the undersides of sandstone beds that formed by the filling of a groove cut into the underlying mud (Fig. 2.12). Groove marks may occur singly or many may be present on one undersurface, all parallel or deviating somewhat in orientation. It is generally held that grooves are formed through a tool, a fossil, pebble or mud clast for example, which was carried along by the current, gouging the groove into the mud. In rare cases the tool has been found at the end of a groove. Groove marks

Fig. 2.11 Flute marks on sole of turbidite sandstone. Flow from bottom to top. Silurian. Southern Uplands, Scotland.

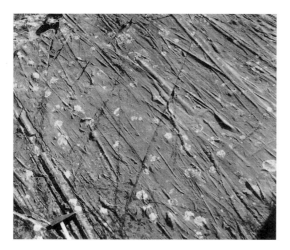

Fig. 2.12 Groove marks on sole of a turbidite sandstone (actually a greywacke). Note that the orientation of the grooves varies through about 30°. Carboniferous. Devon, England.

Fig. 2.13 Lower part of cliff shows lateral accretion surfaces in fluvial channel: notice downcutting (lower left) of channel sands into horizontally bedded floodplain mudrocks (which contain gypsum). Persistent thin sandstone in central part of cliff (arrow) is a crevasse-splay deposit. Above, another channel sand has a markedly erosive base. Triassic meandering stream facies. Arizona, USA.

are common on the soles of turbidite beds but they can form elsewhere, on floodplains when a river breaks its banks for example, and on shallow-marine clastic shelves and carbonate ramps when there are storm currents. Groove casts are a useful palaeocurrent indicator.

Gutter casts look similar to groove marks, being elongate ridges on the base of a sandstone or coarse limestone. They are U- or V-shaped in cross-section, 10–20 cm across and nearly as deep. They are straight to sinuous and may persist for many metres. These structures are common in shallow-marine sediments, where they are attributed to fluid scour, in many cases by storm currents.

Impact marks are produced by objects striking the sediment surface as they are carried along by a current. Various types have been distinguished, including prod, bounce, brush, skip and roll marks, depending on how the impact is thought to have taken place. Objects making the marks are chiefly fossils and pebbles. Impact marks are common on the bases of turbidite beds and tend to occur in more distal situations, when and where current flow has waned.

Channels and scours are found in sediments of practically all environments. Channel structures are generally on the scale of metres, in some cases kilometres, whereas scours are smaller-scale erosional features, occurring within or on the bases of beds. They can both be recognized by the cutting of bedding planes

and lamination in underlying sediments (Fig. 2.13). Scours are generally oval to elongate in plan view, with a smooth to irregular, concave-up shape in vertical section. Slightly coarser sediment or even pebbles may occur within the scours. Scour structures represent short-lived erosion events.

Channels are more organized structures than scours and many were the pathways for sediment and water transport over considerable periods of time. The larger channels can sometimes be mapped out on a regional scale, giving useful palaeogeographical information. Channels generally are filled with coarser sediments than those beneath or laterally equivalent, and commonly a thin lag deposit of pebbles and intraformational clasts is present at the base of the channel. Many channels contain cross-bedded sands from the down-current migration of dunes and sandwaves. Many channel fills are elongate, linear sand bodies, but some channels migrate laterally (meandering stream and tidal channels, for example) and these can generate a much more extensive, sharp-based sheet sand. In these channel deposits, evidence of lateral migration usually is provided by low-angle accretion surfaces, which are orientated normal to the palaeocurrent direction recorded by smaller-scale cross-stratification (see Figs 2.13 & 2.66).

Channels develop in many environments; those in fluvial (e.g. Fig. 2.13) and deltaic settings are especially well known, but they also occur in glacial, tidal-flat, shelf-margin, slope and submarine-fan locations. In some situations, marine slopes, for example, slump scars can be mistaken for channels. With most slump scars, however, the discontinuity surface is relatively smooth and concave-up, and the sediment below is similar to that above. Folds and thrusts are common in the slumped mass.

2.3.2 Depositional sedimentary structures

Sediment transport and aqueous flows

Much sediment is eroded and transported in response to unidirectional aqueous flows, such as the flow of a river, tidal or storm current. A wide range of sedimentary structures arises from water flowing over a sediment surface (discussed below). With tidal currents, the ebb and flood produces further structures (see Sections 2.11.5 and 2.11.6). Sediment is also transported by the action of waves and by the wind, to produce characteristic structures (see Sections 2.11.5 and 2.11.2, respectively). In addition, sediment is moved downslope under the direct action of gravity; most important here are the sediment gravity flows (sediment–fluid mixtures), in particular turbidity currents and debris flows. These flows and their deposits are described in Section 2.11.7. Sediment is also transported by glaciers (see Section 2.11.8).

Sediment in both water and air is transported either in suspension or as bedload. Suspended sediment is kept in suspension by the fluid turbulence. Coarser sediment generally is moved as bedload along the sediment surface by saltation (in short jumps) or by rolling and sliding. A traction carpet of moving grains develops in some situations.

For transport in suspension, the upward components of fluid turbulence must exceed a particle's fall velocity (w). This is given by Stokes' law of settling:

$$w = \Delta\rho d^2 g / 18\mu$$

where $\Delta\rho$ is the density difference between particle and fluid, g is the acceleration resulting from gravity, d is grain diameter and μ is dynamic viscosity. This relationship shows that the settling velocity is related to grain diameter (shown for grains in air in Fig. 2.33).

To describe the strength of aqueous flows the mean flow velocity ($T\bar{U}$) is used most commonly, but also the bed shear stress, denoted by (τ_0) is used, which is the average force per unit area exerted by the flow on the sediment surface. Flow depth (h) is a useful descriptive parameter, but less important than the flow strength in terms of the sedimentary structures formed. Viscosity (μ) and density (ρ) are two other important properties of a fluid and their ratio (μ/ρ) gives the kinematic viscosity (V). Two basic types of aqueous flow are *laminar flow*, where layers of flowing water move smoothly over one another with little diffusion or exchange, and *turbulent flow*, which develops at higher flow velocities and where eddies cause much diffusion; there are transverse components of motion and water moves as packets. Laminar flow is developed only in slow-moving water, but it does occur in more viscous fluids such as sediment–water mixtures, and in glaciers. Water mostly flows in a turbulent state, as does air.

Flow can be characterized in several ways. The *Reynolds number* is a ratio of inertial forces, which are related to the scale and velocity of the flow, to the viscous forces: $Re = d\bar{U}L/V$, where L is a characteristic dimension for the flow system (such as the flow depth). The inertial forces tend to promote turbulence, whereas the viscous forces tend to suppress it (higher velocity and/or greater depth give more turbulence; higher viscosity gives less turbulence). With any aqueous-flow system, the change from laminar flow to turbulent flow takes place at a specific Reynolds number. The *Froude number* is a ratio of inertial to gravity forces and is given by $Fr = \bar{U}/\sqrt{gh}$. A flow with Froude number less than unity is referred to as subcritical or tranquil flow; a Froude number greater than 1 is supercritical or rapid flow. In tranquil flow, packets of turbulent water move freely, whereas in rapid flow, the water appears streaked out and turbulence is suppressed. When a current in rapid flow is decelerating, it changes suddenly downstream to tranquil flow, and a breaking wave or hydraulic jump is generated. Erosion can take place here. With $Fr < 1$, surface waves may propagate upstream. Another way of describing a flow is in terms of its *stream power*, given by $\bar{U}\tau_0$; the sediment transport rate is a function of the stream power.

For aqueous flows, the relationship between grain size and the current velocity for sediment movement (critical erosion velocity) is given approximately by Hjülström's diagram (Fig. 2.14). This shows that fine

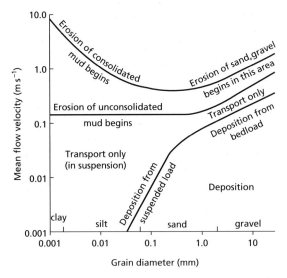

Fig. 2.14 Schematic representation of relationship between current velocity and sediment erosion, transport and deposition (Hjülstrom's diagram, deduced experimentally from flows of 1 m depth). Note that sediment may continue to be transported after the current velocity has fallen below the level at which it was initially eroded.

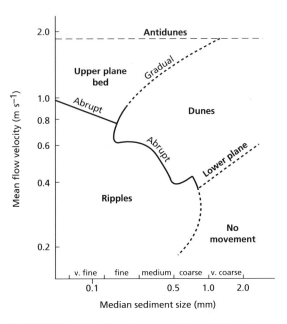

Fig. 2.15 The relationship between mean flow velocity and median grain size showing the stability fields of the various subaqueous bedforms for a flow depth of 0.25–0.4 m. After Ashley *et al.* (1990).

sand grades are more easily eroded than finer or coarser grades, but once eroded, clay and silt are held in suspension until the flow velocity is minimal. In fact, much mud is deposited as aggregates and floccules, rather than individual particles. The higher erosion velocities required for silt and clay are a reflection of the cohesive forces between particles. Detailed accounts of the mechanics of flow and movement of grains in water are given in Allen (1997) and Leeder (1999), and many other textbooks.

Once sediment is moving as a result of the effects of flowing water, then the nature of the sediment surface with its sedimentary structures, or *bedforms* as they are called, is dependent on the flow conditions. Laboratory studies have shown that there is a definite sequence of bedforms (Fig. 2.15), related to increasing flow strength and grain size. For some grain sizes and flow velocities there are overlaps of the bedform fields and transitional zones. For sediments finer than about 0.1 mm (down to about 0.03 mm) and flow depth of around 20 cm, under conditions of increasing flow strength, an initially flat bed with no sediment movement gives way to ripples and then to an upper flat bed (also called upper plane bed phase). For sands

of mean size from about 0.1 to 0.6 mm, the sequence is no movement, ripples, dunes, upper flat bed. For sands coarser than about 0.6 mm, the sequence is no movement, lower flat bed, dunes, upper flat bed. The characteristic feature of the upper flat bed is *primary current lineation*: low flow-parallel ridges, a few grains high, seen as streaks on the sediment surface. At higher flow strengths still, sinusoidal undulations again develop on the sandy bed; these are called *antidunes*, and the waves on the surface of the fast-flowing water occasionally break and move upstream.

Ripples and dunes are asymmetric bedforms that gradually move downstream as sediment is transported through erosion of the upstream-facing (stoss) side and its deposition over the bedform crest on the downstream-facing (lee) side (see Fig. 2.18). By way of contrast, the antidune bedform gradually moves upstream; sand is still going down-current of course, very rapidly, but by being eroded from the downstream-facing side of the bedform and being deposited on the upstream-facing side of the next one. Flow conditions under which ripples and dunes are formed are referred to as *lower flow regime*; upper plane bed and antidunes form in the *upper flow regime*. In addition to flow velocity and grain size, depth is another control on sandy bedforms. At very shallow depths, dunes do not form. Upper flow regime bedforms require higher velocities at greater depths, and they form more easily in the finer grain sizes. The features of these bedforms and the internal sedimentary structures they produce are discussed in the following sections.

In a number of situations, aqueous flows slow down through time, and the bedforms on the sediment surface change accordingly (e.g. upper flat bed to ripples). One particular type of flow that usually decelerates soon after it is initiated is the *sediment gravity flow*, a turbulent mixture of sediment and water flowing under the influence of gravity (see Section 2.11.7). Turbidity currents, which are important in deeper-water basins, are one common type, and similar density currents can be generated by storms on clastic shelves and carbonate ramps. A common succession of structures recorded by sediment deposited from a decelerating flow is of flat bedding (also called horizontal lamination, formed in the upper flow regime) passing up into cross-lamination (formed from ripples in the lower flow regime).

Bedding and lamination

The characteristic feature of sedimentary rocks, their stratification or bedding, is produced mostly by changes in the pattern of sedimentation, usually changes in sediment composition and/or grain size. *Bedding* generally is defined as a sedimentary layering thicker than 1 cm. Finer-scale layering, only millimetres thick, is termed *lamination* (see Table 2.5).

Table 2.5 Terminology for thickness of beds and laminae

Beds	Laminae
very thick bed	very thick lamina
— 1000 mm —	— 30 mm —
thick bed	thick lamina
— 300 mm —	— 10 mm —
medium bed	medium lamina
— 100 mm —	— 3 mm —
thin bed	thin lamina
— 10 mm —	— 1 mm —
very thin bed	very thin lamina

Lamination is commonly an internal structure of a bed (e.g. Fig. 2.16). The majority of beds have been deposited over a period of time ranging from hours or days, as in the case of turbidites and storm beds, to years, tens of years or even longer, as in the case of many marine shelf sandstones and limestones. Bedding planes themselves may represent considerable periods of time when there was little deposition. Sediment may well have been moved around during these intervals. Subsequent modifications to bedding planes result from erosion as the succeeding bed is deposited, deformation through loading and compaction, and dissolution owing to overburden pressure, especially in limestones, in which stylolites form (see Section 4.6.1). Tectonic movements also affect bedding planes.

Lamination arises from changes in grain size between laminae, size-grading within laminae, or changes in composition between laminae (e.g. Fig. 2.16). In many cases, each lamina is the result of a single depositional event. Although time-wise this is almost instantaneous, it can be much longer.

Parallel lamination, also termed planar, flat or horizontal lamination, can be produced in several ways. In fine sands and clays it commonly is formed through deposition from suspension, slow-moving sediment clouds or low-density turbidity currents. Such laminae occur in the upper parts of turbidite beds (Section 2.11.7) and in varves and rhythmically laminated deposits of glacial and non-glacial lakes (Fig. 3.4 and Section 3.5). Reversing tidal currents also can produce rhythmites, alternations of sand and mud with

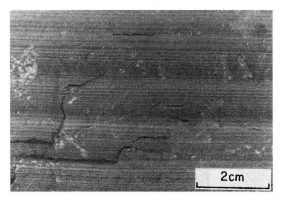

Fig. 2.16 Part of a bed showing well-developed lamination. Silt-grade quartz-rich laminae alternate with clay/organic-rich laminae in a lacustrine limestone (see Fig. 3.4). Middle Devonian. Caithness, Scotland.

thickening-up and thinning-up of the laminae reflecting lunar cycles of changing current strength (see papers in Alexander *et al.*, 1998). Laminae also can be formed by chemical precipitation, as in subaqueous evaporites (Section 5.2.3), and by phytoplankton blooms giving organic-rich layers (Section 8.8). Microbial laminae form planar stromatolites (formerly cryptalgal laminites), which are common in tidal-flat limestones (Section 4.3.3).

Parallel lamination in sand-grade sediment, also termed flat bedding, is formed chiefly by turbulent flows at high flow velocities (see Fig. 2.14). The characteristic feature of this upper plane bed phase is the presence of *parting lineation* or *primary current lineation* on the bedding-plane surface (Fig. 2.17). It consists of very low ridges, only a few grain diameters

Fig. 2.17 Parting lineation, also called primary current lineation, on the bedding surface of a parallel-laminated fluvial sandstone. Carboniferous. Northumberland, England.

high, which give the sandstone a visible fabric or streakiness. In sands with a mean grain size coarser than 0.6 mm diameter, a flat bedding can be produced by bedload transport at low flow velocities (i.e. lower plane bed phase, Fig. 2.15). Current lineation is not developed in this case.

Current ripples, dunes and cross-stratification

Current ripples and dunes are downstream-migrating bedforms produced by unidirectional aqueous flows. Their formation depends on the flow velocity, flow depth and the sediment grain size (see Fig. 2.15 and earlier section). They are common in rivers, estuaries, tidal flats, delta channels, along shorelines and on shallow-marine shelves; ripples also occur on the deep-sea floor. In the geological record, current ripples are commonly preserved intact, but the actual dune bedforms rarely so; however, the cross-stratification produced by the migration of these larger-scale bedforms is one of the most common depositional structures in sand-grade rocks.

Current ripples are small-scale bedforms with wavelengths of less than a few tens of centimetres and heights of less than several centimetres. In profile they are asymmetric with a steeper, downstream-facing lee side and a gentle upstream-facing stoss side (Fig. 2.18). Ripples can be described by the wavelength to height ratio, referred to as the *ripple index*. For current ripples this ranges between 8 and 20. *Subaqueous dunes* have wavelengths of a metre or more and heights of several tens of centimetres. Dunes have a similar triangular profile and index to that of ripples. The shape of ripples and dunes is described as two-dimensional

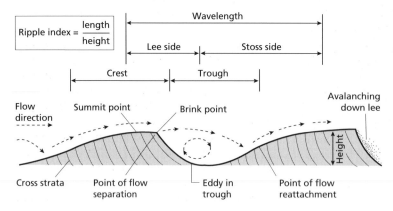

Fig. 2.18 Ripple terminology and flow pattern.

if the crests are straight, or three-dimensional, if the crests are sinuous, catenary, lunate or linguoid (Fig. 2.19). The shape of ripples and dunes is related to flow strength; with increasing flow velocity, ripples show the sequence: straight-crested, sinuous, linguoid; for dunes the sequence is: straight-crested, sinuous, catenary, lunate. Figure 2.20 shows some current ripples.

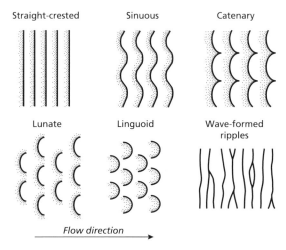

Straight-crested   Sinuous   Catenary

Lunate   Linguoid   Wave-formed ripples

Flow direction

Fig. 2.19 Terminology for the shape of the crests of ripples and dunes formed by unidirectional currents. For comparison, the typical crest pattern of wave-formed ripples is also shown.

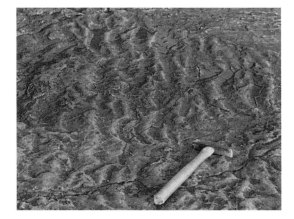

Fig. 2.20 Current ripples (flow to the right) with crests showing straight to catenary shape. The surface is also covered in desiccation cracks. A thin sheet-flood sandstone from a Triassic alluvial fan. Glamorgan, Wales.

In addition to 'subaqueous dune', other terms that have been used for these larger-scale bedforms are megaripple, sandwave and bar. However, it is now clear that they are all part of the same bedform spectrum, so that it is recommended that *dune* be used for them all (Ashley *et al.*, 1990). The term *sandwave*, however, is well entrenched in the shallow-marine clastics (and carbonates) literature and so will doubtless continue to be used. It refers to large-scale, usually asymmetric structures, which have smaller-scale dunes and ripples upon them.

Ripples and dunes migrate downstream through sediment being eroded from the stoss side and carried to the crest, from which it avalanches down the lee slope (Fig. 2.18). As a consequence of the stream flow over the bedform, flow separation takes place at the crest and an eddy develops within the trough. The backflow of the eddy may cause reworking of sediment at the toe of the lee slope, and in the case of large dunes, upstream-directed ripples (backflow ripples) may develop there. Also with large dunes, downstream-directed ripples may develop in the troughs downstream of the point of flow reattachment. Some fine sediment is carried over the crest in the separated flow and deposited out of suspension on to the lee surface.

The downstream migration of ripples and dunes under conditions of net sedimentation gives rise to *cross-stratification*, a structure in the past loosely called current bedding or false bedding (terms to avoid). The cross-strata, referred to as foresets, represent the former position of the ripple or dune lee face. Two basic types of cross-stratification, shown in Fig. 2.21, are (a) planar cross-strata, produced by two-dimensional bedforms (those with straight crests) and (b) trough cross-strata produced by three-dimensional bedforms (those with curved crests). An individual bed of cross-strata is termed a *set*; a group of similar sets is a *coset*.

In planar cross-bedding (Fig. 2.22), the *foresets* (sloping beds) dip at angles up to 30° or more, and they may have an angular or tangential basal contact with the horizontal, depending largely on the flow velocity/ sediment transport rate. If tangential, the lower part of the cross-bed is referred to as a bottom-set. Sets range from a few decimetres to a metre or more in thickness. Planar cross-bedding mostly forms tabular sets, although wedge-shaped sets also occur. In trough cross-bedding (Fig. 2.23), the scoop-shaped beds normally have tangential bases. Trough cross-bedding

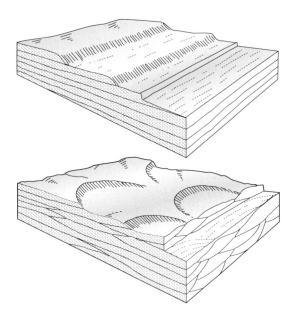

Fig. 2.22 Planar cross-bedding in a tabular set. On the bedding surface above, straight, parallel lines are seen, confirming the planar nature of the cross-beds in the third dimension. This type of cross-bedding is produced from the down-current migration of straight-crested subaqueous dunes. Carboniferous distributary channel sandstone. Northumberland, England.

Fig. 2.21 Block diagrams showing the two common types of cross-stratification: planar and trough. Above: planar cross-stratification, chiefly formed through the migration of straight-crested (i.e. two-dimensional) ripples, producing planar cross-lamination, and straight-crested dunes, producing planar cross-bedding. Below: trough cross-stratification, chiefly formed through migration of three-dimensional bedforms, especially lunate and sinuous dunes, producing trough cross-bedding (illustrated here). Linguoid ripples give a trough cross-lamination.

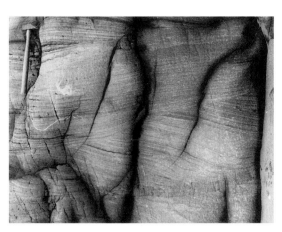

Fig. 2.23 Trough cross-bedding; several sets of cross-beds with tangential bases, formed through the migration of lunate dunes under conditions of net sedimentation in a fluvial channel. Old Red Sandstone, Devonian. Dyfed, Wales.

commonly forms through migration of sinuous and lunate dune bedforms.

In many situations, there are changes of flow velocity or depth during bedform migration so that the dunes are modified and eroded. When deposition resumes, an erosion surface will be present within the cross-strata. This is known as a *reactivation surface*. They are common in fluvial–deltaic cross-bedded sandstones, resulting from fluctuations in the stage of the river, and also in shallow-marine cross-bedded sandstones, resulting from tidal-current reversals and storm erosion. In detail, the cross-beds themselves (the foresets) may show an alternation of coarse and fine laminae (from an alternation of deposition from avalanching and from suspension fallout) and some coarse layers may be reversely graded (from grain-flow deposition, see Section 2.11.7).

Cross-bedding is produced in a *tidal regime* through migration of large dunes (sandwaves), but there are usually additional features reflecting the ebb and flood

currents. If one tidal-current direction is dominant, then unidirectional cross-bedding can form, but there may be back-flow ripples and reactivation surfaces from current reversals and storm effects. If there are bipolar tidal currents, then *herringbone cross-bedding* can form (Fig. 2.24). Mud drapes on cross-bed surfaces occur during slack-water deposition. In some tidal sandwave deposits, there are regular variations in grain size, cross-bed thickness and mud-drape occurrence (Fig. 2.25), and these 'tidal bundles' reflect spring–neap lunar cycles (reviewed by Johnson & Baldwin, 1996; Allen, 1997).

Tabular and trough cross-lamination, with set heights of less than a few centimetres, are produced mainly by straight-crested and linguoid ripples, respectively. Where there is rapid deposition, ripples

Fig. 2.24 Herringbone cross-bedding formed through tidal current reversals, in shallow-marine bioclastic limestone. Jurassic. Dorset, England.

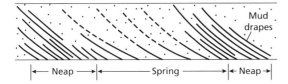

Fig. 2.25 Schematic sketch of tidal cross-bedding with tidal bundles defined by mud drapes upon foresets. Spacing of the sand–mud couplets can be indicative of spring–neap tidal cycles: thick, more sandy layers represent spring tides; thin, more muddy layers represent neap tides. The horizontal thickness of the neap–spring cross-bed unit is in the range 0.5–2 m.

build up as well as forward, so that a ripple 'climbs' up the stoss side of the one downstream. This produces *climbing-ripple cross-lamination* (also called ripple drift). If there is very rapid sedimentation from suspension, then laminae can be formed on the stoss side continuous with the foreset laminae (Fig. 2.26).

Where mud deposition is intermittent with ripple migration, thin streaks of mud occur between sets of cross-lamination and mud is concentrated in ripple troughs. This type of structure is known as *flaser bedding*, whereas the term *lenticular bedding* is applied to isolated ripples, seen in section as cross-laminated lenses, within mud or mudstone (Fig. 2.27). Both types are common in tidal-flat sediments where mud is deposited during slack-water periods; they also occur in delta front, prodelta and other situations where there are fluctuations in sediment supply and flow strength. The term *heterolithic* sediment or facies refers to mixtures of sand and mud.

One particular type of large-scale cross-bedding known as a *lateral accretion surface* (shown in Fig. 2.13) is formed by the lateral migration of point bars developed on the inside of meandering river and tidal channels. These surfaces are orientated normal to the current direction and usually separate units of medium- or small-scale cross-stratification, formed by dunes and ripples migrating downstream on the point bar.

Cross-bedding also can occur in conglomerates. Many deposits in braided-stream systems, for example, show a large-scale cross-bedding produced by the downstream movement of gravel bars. Large-scale cross-bedding is also generated by deposition in 'Gilbert-type' deltas and fan deltas (now generally called coarse-grained deltas). The former occur mostly

Fig. 2.26 Two thin siliciclastic turbidite beds showing: sharp scoured base with convoluted and loaded sandstone balls and flame structures, and ripple cross-lamination with preservation of stoss-side laminae (laminae dipping at low angle to right, current flow from right to left). Both beds show graded bedding, from fine sand at base to silt and clay at top. Carboniferous. Devon, England.

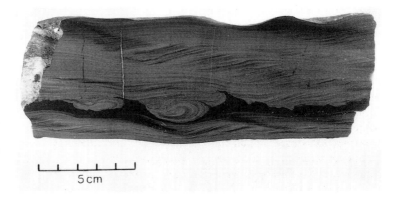

5 cm

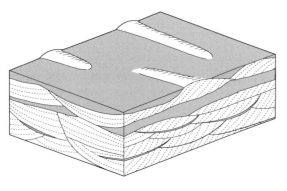
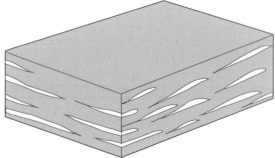

Fig. 2.27 Flaser bedding (left) and lenticular bedding (right).

in lakes (see Section 2.11.3) and the latter where alluvial fans pass straight into the sea (see Section 2.11.4).
In both situations, a steep delta front with sediment
sliding and avalanching produces steeply dipping
angle-of-repose foresets, passing down into bottom
sets of finer-grained sediments (see Fig. 2.69). Such
foresets may be 100 m in height, or more!

Antidunes and antidune bedding

Antidunes are low, undulating bedforms that develop
in the upper flow regime at high current velocities (see
Fig. 2.15). A low-angle cross-bedding, directed upstream, is developed through erosion on the downstream side of the structure, and deposition on the
upstream-facing (stoss) side. Antidune bedding is
rarely preserved but it does occur in foreshore (beach)
sands from swash–backwash deposition, in washover
fans and in base-surge tuffs (Section 10.4.2).

Wave-formed ripples and cross-lamination

Wave-formed ripples are common in many shallow-
marine to intertidal, deltaic and lacustrine sediments,
both sandstones and limestones. These ripples are
characterized by a symmetrical profile and continuous
straight crests (Fig. 2.28). However, many are asymmetric, although these tend to have more persistent
crestlines than current ripples. Wave-ripple crests are
commonly rather pointed compared with their more
rounded troughs. Crests commonly bifurcate and they
may enclose small depressions ('tadpole nests'). These

Fig. 2.28 Wave-formed ripples: ripple profiles are symmetrical
and crests bifurcate. Triassic lacustrine sandstone. Glamorgan,
Wales.

features serve to distinguish wave-formed ripples from
current ripples.

The ripple index for wave-formed ripples (6–10) is
lower than that for current ripples (8–20) in sediment
of similar grain size. The wavelength of wave-formed
ripples depends on the sediment grain size, diameter
of water-particle orbits in the waves and the water
depth. Waves are able to move sediment at depths less
than half their wavelength, a depth referred to as
wave-base.

The internal structure of wave-formed ripples
is variable and commonly the laminae are not con-
cordant with the ripple profile. Wave-ripple cross-
lamination can consist of a chevron to undulating
pattern of laminae, or unidirectional foresets.

Ripples on tidal flats usually show complicated
patterns resulting from changes in water depth,
and wind and runoff direction. Interference, double-
crested and ladderback ripples are common types (e.g.
Fig. 2.29).

Hummocky cross-stratification

Hummocky cross-stratification (HCS) is an undulating bedding that is thought to be formed by wave-generated oscillatory flows or combined flows (waves plus currents) produced by the passage of storms. It is thought to form in the area between fairweather wave-base and storm wave-base. It is characterized by gently curved, low-angle (< 10–15°) cross-lamination arranged in a convex-upward (a 'hummock') and concave-downward (a 'swale') pattern (see Figs 2.30a & 2.31, also Fig. 2.41). The spacing of the hummocks is several tens of centimetres to several metres, and in plan view they have a low domal shape. Some hummocky beds show a sequence of divisions: basal (B), planar-bedded (P), hummocky (H, the main part of the bed), flat-bedded (F), cross-laminated (X) to mudrock (M) (see Fig. 2.32), which reflects a change from a strong unidirectional current (BP) to oscillating flow (HFX) to deposition from suspension (M).

Related to HCS is *swaley cross-stratification* (SCS), where hummocks are rare (Fig. 2.30b). Flat bedding is associated with SCS, and it probably forms in shallower water than HCS, perhaps in the outer shoreface zone (see Section 2.11.5). The HCS and SCS sandstones (or limestones) form one end of a spectrum of sediments deposited by storms (Fig. 2.32). At the other end are graded beds with sole structures ('*tempestites*'). These are deposited beyond storm wave-base from waning storm currents and usually are interbedded with bioturbated shelf mudrocks. Cheel & Leckie (1993) review SCS, HCS and storm beds,

Fig. 2.29 Interference ripples formed on a Cambrian tidal flat. Anti-Atlas Mountains, Morocco.

(a) **Hummocky cross-stratification (HCS)**   (b) **Swaley cross-stratification (SCS)**

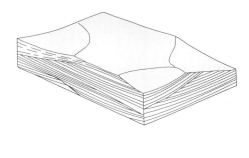

Hummock

Swale

Sharp base

Directional sole marks

Fig. 2.30 Hummocky cross-stratification (HCS) and swaley cross-stratification (SCS).

and recent examples in the literature include Midtgaard (1996) and Datta *et al.* (1999).

Similar to storm deposits are sediments deposited by tsunamis—long-wavelength waves produced primarily from submarine faulting (and meteorite impacts). *Tsunamites* will have erosive bases, probably with gutter casts, and HCS–SCS; they may be associated with seismites, sediments with structures indicating seismic activity (see Section 2.3.3 and Shiki *et al.* 2000).

Fig. 2.31 Hummocky cross-stratification in shallow-marine sandstone. Scale is 30 cm long. Cretaceous. Alberta, Canada. Photograph courtesy of Guy Plint.

Wind ripples, dunes, draas and aeolian cross-bedding

As with sediment transport by water, there is a critical erosion velocity for sediment entrainment by wind. Very fine sand is the first to move, at a velocity of only $15\,cm\,s^{-1}$ (Fig. 2.33). Fine sediment is transported in suspension and coarser sediment by saltation. Very coarse sand moves mostly by surface creep, brought about by the bombardment of smaller saltating grains. Ripples soon develop as the sand is moved, and at higher wind velocities, a flat surface is generated. However, in a desert the larger-scale bedforms, dunes and draas, are the major structures that give rise to thick aeolian sandstones. The physics of grain movement by wind is reviewed by Nickling (1994), Allen (1997) and Leeder (1999).

Wind ripples generally are straight-crested asymmetrical forms with some crest bifurcation. Ripple wavelength and height depend on the grain size and wind strength, and in particular on the length of the saltation path of the moving sand grains. Compared with current and wave ripples, the ripple index is generally much higher.

The larger-scale bedforms, dunes and draas, have wavelengths of tens to hundreds of metres and hundreds of metres to several kilometres, respectively. Many draas are in effect compound dunes, having small superimposed dunes upon them, with crests tangential to the main draa crest. In some deserts it

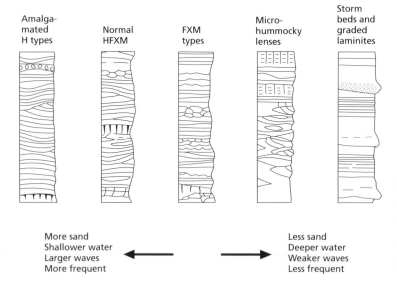

More sand
Shallower water
Larger waves
More frequent

Less sand
Deeper water
Weaker waves
Less frequent

Fig. 2.32 The spectrum of storm deposits from HCS-dominated sands through to graded storm beds ('tempestites'). H, hummocky cross-stratification; F, flat bedding; X, cross-lamination; M, mud. Logs are schematic but the thickness of strata shown is of the order of 0.5 m in each case.

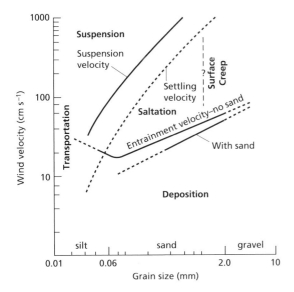

Fig. 2.33 The relationship between sediment grain size and velocity for transport and deposition by wind. Sediment is transported by surface creep (coarse grain sizes), by saltation (sand) and in suspension (silt and fine sand). Note that the entrainment velocity increases with increasing grain size, but it is slightly lower when sand is already being carried by the wind. Also see that the settling velocity is higher for larger grain sizes (Stokes' Law) and that the suspension velocity is higher than settling velocity for a given size.

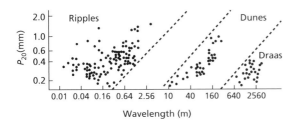

Fig. 2.34 Wavelength of aeolian bedforms against grain size (P_{20}: the coarse 20 percentile) showing the three separate fields: ripples, dunes and draas. Based on data from the Sahara.

Fig. 2.35 Aeolian cross-bedding. Notice the large scale (set heights here are 5–10 m), the wholly sandstone nature of the succession (cliff height is about 200 m), the shape of the cross-beds (steep to tangential) and the direction of dune migration (all to the south, left). Navajo Sandstone, Jurassic. Utah, USA.

appears that there is a basic dynamic difference between dunes and draas, as shown by the grain-size–wavelength relationship (Fig. 2.34), but also important are wind strength and pattern, sediment supply and the nature of the underlying surface. On the basis of orientation with respect to the dominant wind direction, the large aeolian bedforms are divided into transverse, longitudinal and oblique types. The familiar dunes are the barchan, a lunate-shaped dune generally forming where sand supply is limited, and the seif dune, an elongate ridge, near-parallel to the prevailing wind direction. Other dune types are parabolic, star-shaped and transverse (normal to prevailing wind). Seif draas are major ridges of sand with smaller, commonly barchanoid dunes upon them.

In an analogous manner to subaqueous dunes, sand is moved up wind-facing slopes (stoss side) to the crest of the dune, from where it periodically avalanches down the lee slopes, when the angle of repose (around 35°) is exceeded to generate cross-beds (Fig. 2.35). This process is termed sand flow or grain flow, and it generates laminae which may be reversely graded. As a result of flow separation at the dune crest, finer sand is carried in suspension in the jet of air over the top of the crestline and deposited directly on the lee slope. Laminae produced by this grain fall are thus usually finer grained, thinner and perhaps normally graded, compared with grain-flow laminae. One further lamina type, 'climbing-translatent', is formed by the migration of wind ripples. This typically is reversely graded, but as ripples occur mostly on the windward side of the dunes, rather than on the slip faces, this lamination has a low preservation potential.

On average, the angles of dip of aeolian cross-beds, 25–35°, are steeper than those of subaqueous origin.

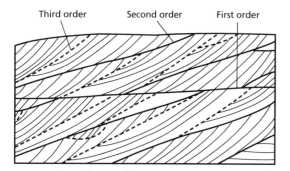

Fig. 2.36 Three orders of bounding surface in aeolian cross-strata. Such features also occur in subaqueously deposited complex sand bodies.

The set thickness in aeolian cross-beds, commonly several metres thick, but reaching 30 m or more, generally is greater than that of subaqueous dunes and sandwaves.

Movement of barchan dunes under conditions of net sedimentation gives rise to a large-scale trough cross-bedding, dipping downwind. Simple seif dunes, on the other hand, generally elongate parallel to the wind direction, have an internal structure of cross-beds dipping in opposite directions, although commonly one direction will be dominant. Seif draas, with their parasitic dunes, give rise to a very complex cross-bed/palaeowind pattern.

Within aeolian sandstones, there are commonly different orders of *bounding surface* (Fig. 2.36). First-order surfaces cut through the whole sand body. They are usually flat-lying and may be regularly spaced. They represent major events in the development of the dune/draa field, such as a rise of the water table or migration of a draa across the area. Second-order bounding surfaces are low-angle truncation surfaces, probably formed from the migration of a dune upon a draa. Third-order surfaces are less prominent, dipping erosion surfaces cutting the cross-bedding, formed through a change in the angle of the dune slip-surface. These are reactivation surfaces and can result from a severe sand storm. See Pye & Lancaster (1993) and Kocurek (1996) for further information on aeolian sedimentary structures, and Section 2.11.2 for aeolian sandstones and desert deposits.

Graded bedding

This feature relates to the grain-size changes upward through a bed and mostly develops in response to changes in flow conditions during sedimentation. Many different types of graded bedding have been described but two types of *normal grading* are (a) a gradual decrease in the average grain size up through the bed (content grading), and (b) a gradual upward decrease in size of the coarsest grains (coarse-tail grading). Multiple grading, where there are several graded subunits in a bed, also occurs. Coarse- to fine-graded bedding is generated through deposition from decelerating sediment-laden currents; thus this normal graded bedding is a characteristic feature of turbidites deposited from waning density currents (see Figs 2.26 & 2.80). On clastic shelves and carbonate ramps, graded beds ('tempestites') are deposited from waning storm currents (see Fig. 2.32). Decelerating flows giving graded beds also occur on river floodplains, from crevassing.

Reverse or *inverse grading*, where grain size increases up through a bed, is not common. On a small scale, it occurs in beach lamination, through accelerating wave backwash, and it occurs in some cross-beds as a result of grain flow, where the dispersive pressure of colliding grains pushes coarser grains to the zone of least shear stress near the top of the flow. In coarse-grained, deeper-water gravels and sands, reverse grading is present in the lower part of some beds, as a result of deposition from a traction carpet at the base of a high-density turbidity current (see Section 2.11.7). Reverse grading is present in some volcaniclastic deposits (Sections 10.3 and 10.4).

Massive bedding refers to beds without any apparent internal structure. Although many sedimentary rocks can appear massive or structureless, closer inspection may show that this is not the case. If the massive bedding is real, it could have formed through rapid deposition, a 'dumping' of sediment as from high-density, sediment gravity flows (Section 2.11.7). Alternatively, original internal depositional structures could have been destroyed by bioturbation, dewatering or recrystallization-replacement.

Mudcracks: desiccation and syneresis

Mudcracks are common in fine-grained sediments, and, if formed through desiccation, they indicate subaerial exposure. Desiccation cracks are typically polygonal (present in Fig. 2.20). They are common in tidal-flat facies, and their reworking leads to the gener-

Fig. 2.37 Subaqueous shrinkage (syneresis) cracks. Note spindle and trilete shapes. Huronian, Early Proterozoic. Ontario, Canada.

ation of intraclasts. Shrinkage cracks also may develop subaqueously as a result of *syneresis*: contraction of the sediment through loss of pore water. It is thought that slight changes in water chemistry are responsible for this, inducing volume changes in certain clay minerals. Osmotic effects also may be involved. Syneresis cracks usually do not form a complete polygonal pattern, but are trilete and spindle shaped (Fig. 2.37). They are common in lacustrine sediments. Both desiccation cracks and syneresis cracks may be filled with sand when deposition resumes. Later compaction may distort these crack fills.

Shrinkage cracks also may develop *within* muddy strata as a result of compaction, perhaps with seismic shocks acting as a trigger (Tanner, 1998). Diastasis cracks form in stiff mud interbedded with loose sand (clastic or carbonate) through their differential mechanical behaviour on compaction (Cowan & James, 1992). 'Molar-tooth structures', crumpled veins filled with microspar, which are common in Precambrian muddy calcareous sediments, are a matter of controversy, but probably are the result of seismic shocks causing shrinkage, dewatering and fissuring of colloidal sea-floor mud (James *et al.*, 1998; Pratt, 1998).

Rain spots are occasionally found in mudrocks of continental and shoreline environments.

2.3.3 Post-depositional sedimentary structures

In this group are included slides and slumps, convolute bedding, load casts, sandstone dykes and dewatering structures.

Slides and slumps

These terms refer to downslope mass movement of sediment upon a glide plane, with a slide involving *en bloc* movement with little internal deformation of the sediment, and a slump involving significant internal deformation. Folds and thrusts and wholesale brecciation occur within slumped masses. Slide and slump blocks range from metres to several kilometres in size. The majority are initiated by earthquake shocks and contemporaneous fault movements, but some may be associated with relative falls in sea-level (megabreccias). They are mostly found in deep-water slope and basin-margin deposits (see Section 2.11.7), and may involve shallow-water shelf-margin deposits (such as blocks of reef rock) or the slope sediments themselves. Slump and slide scars may be mistaken for the bases of channels, and slump folds may be misinterpreted as tectonic. Slump folds can be used to determine palaeoslope.

Convolute bedding develops in cross- and planar-laminated strata and consists of regular to irregular folds and contortions (Fig. 2.38). It is usually only the uppermost part of a bed that is affected, and the convolutions may be planed off, demonstrating their synsedimentary nature. The origin of convolute lamination is not fully understood, but probable causes are differential liquefaction and lateral–vertical intrastratal flow (dewatering processes, see below), and shearing of the sediment surface by currents. Convolute lamination is common in turbidite beds, but it also occurs in fluvial, tidal-flat and other sediments.

Related to convolute bedding is *overturned cross-bedding* where the upper part of a bed of foresets is turned over, normally in a consistent downstream direction (Fig. 2.39). This has been attributed to a shear force exerted by the current depositing the cross-strata, probably when sand was weakened by partial liquefaction following rapid deposition. Such structures are common in fluvial sandstones.

Load casts are a common sole structure, seen as bulbous, downward-directed protuberances of a sandstone bed into underlying sediment, normally a mudrock (Fig. 2.40). Load casts show considerable variation in shape and size, and one common feature is the squeezing of mud up into the sand to form flame structures (seen in Fig. 2.26). Lobes of sediment may become detached from the sand horizon to form load balls. The structures are formed as a result of

Fig. 2.38 Convoluted and deformed lamination in a thin, siliciclastic turbidite. Ordovician. Ayrshire, Scotland.

Fig. 2.39 Planar cross-bedding showing overturning in downstream direction (to the right). Carboniferous braided-stream sandstone. Northumberland, England.

Fig. 2.40 Large load casts on the underside of fluvial sandstone; the mudrock of the underlying bed is visible between the loads. Late Precambrian. Southern Norway.

a vertical density contrast of more dense sand overlying less dense mud, so that the sand sinks down into the mud.

Related to load casts are *ball-and-pillow structures* where a sand bed lying within mudstone has broken up into pillow-shaped masses, still partly connected or free floating in the mud (Fig. 2.41). It is thought that with many ball-and-pillow structures an earthquake shock was the triggering mechanism. High sedimentation rates also favour the development of this and the other deformation structures.

A number of soft-sediment deformation structures result from *dewatering*, the sudden loss of pore water causing the sediment to lose its strength. The two chief processes are fluidization, whereby upward-moving water produces a fluid drag on the grains, and liquefaction, whereby particles are shaken loose from each

Fig. 2.41 Sandstone pillows formed by loading into mud (now fissile shale), truncated by erosion before deposition of succeeding sandstone. The latter shows HCS–SCS bedding. Carboniferous marine-shelf mudrock overlain by storm sandstones. Northumberland, England.

other through some applied stress, such as that associated with an earthquake. Dewatering can give rise to a whole range of water-escape structures. These include disruptions and contortions of bedding (as noted above), and sandstone dykes cutting across primary structures and, if reaching the surface, forming a sand volcano. Mud volcanoes (monroes to some!) also may form. One particular water-escape structure in sandstones is *dish-and-pillar structure*. This consists of thin, concave-up laminae of clay (the 'dishes'), which may be turned up vertically at their ends as 'pillars' of sand cut through. This structure is often difficult to see, occurring in sandstones that otherwise look massive. Dish-and-pillar structures are thought to be the result of upward, as well as some lateral, movement of pore water. They are common in rapidly deposited, thick turbidite sandstones, but do occur in other facies. Rossetti (1999) described a range of soft-sediment deformation structures related to palaeoseismicity and Owen (1996) devised experiments to help understand the conditions of formation of such structures. The term *seismite* is used for sedimentary rocks containing deformation structures produced by seismic events. See papers in Shiki *et al.* (2000).

Some post-depositional structures in sandstones are the result of chemical processes operating during burial diagenesis. *Stylolites*, through-going suture planes resulting from pressure dissolution in lithified rocks, do occur in sandstones (Section 2.9.1), but are more common in limestones (see Section 4.7.5). Differential

cementation in sandstones can give rise to large nodular structures, sometimes called *doggers* (nodules are described in Section 3.2).

2.3.4 Biogenic sedimentary structures

The sedimentary structures formed by organisms range from trace fossils (ichnofossils) with distinctive features that can be attributed to a particular organism and/or activity and are given Latin names, through to vague bioturbation structures produced by animals churning up and burrowing through the sediment. Bioturbation disrupts and even destroys primary bedding and lamination. It may produce a nodularity in the sediment, with subtle grain-size differences between burrow and surrounding sediment. A colour mottling can be produced by burrowing organisms. Later diagenesis may enhance burrow structures, by preferentially dolomitizing or silicifying them; many flint nodules in chalk are actually silicified burrows. In more muddy sediments, burrows are flattened during burial compaction. Also the result of biological processes are *stromatolites*, produced by microbial mats. These usually occur in limestones and are described in Section 4.3.3.

With many trace fossils, it is not known which type of organism was responsible, but the behaviour of the animal can be deduced. Similar trace fossils can be produced by quite different animals if they had similar modes of life. Trace fossils may be the only evidence of organisms living in a sediment if body fossils are not preserved. Trace fossils can be divided into two groups: those formed on the sediment surface (tracks and trails) by epibenthic organisms, and those formed within the sediment (burrows) by the endobenthic organisms. Generally, tracks and trails are seen only on bedding planes, whereas burrows are seen both on horizontal and vertical sections through beds.

Six categories of trace fossil can be distinguished on the basis of the animal's activity: on the sediment surface—resting, crawling and grazing traces—and within the sediment—dwelling, feeding and escape burrows (see Figs 2.42 & 2.43). Some trace fossils are formed through a combination of activities.

Resting traces are formed by vagile epibenthic animals. They are impressions showing the broad shape of the animal. Impressions of starfish are a distinctive variety of resting trace occasionally found in the geological record; more common are the resting

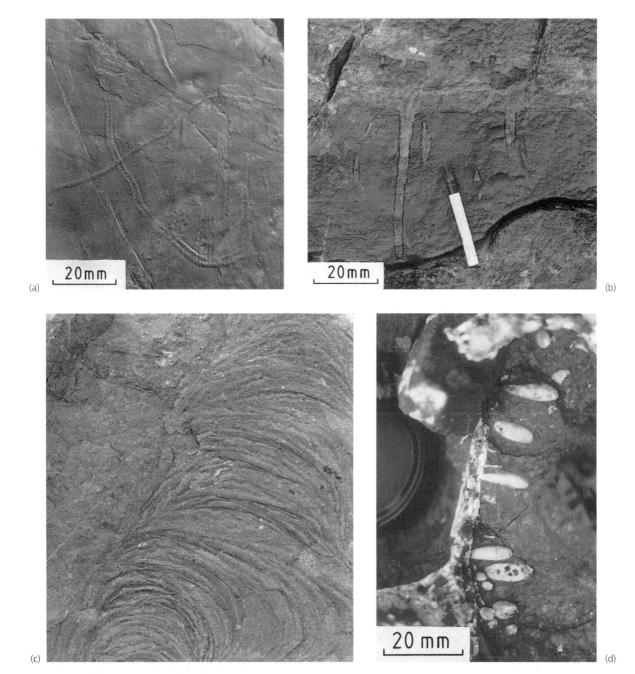

(a)

(b)

(c)

(d)

Fig. 2.42 Examples of trace fossils. (a) Crawling trails on base of turbidite (deep-water) sandstone. Cambrian. Southern France. (b) Simple vertical dwelling burrows, *Skolithos* type, filled with oolite. Carboniferous shallow-water limestone. Gower, Wales. (c) Feeding-burrow system (*Zoophycos*) from Carboniferous shallow-water limestone. Northumberland, England. (d) Borings of the bivalve *Lithophaga* in Triassic dolomite. Miocene. Mallorca, Spain.

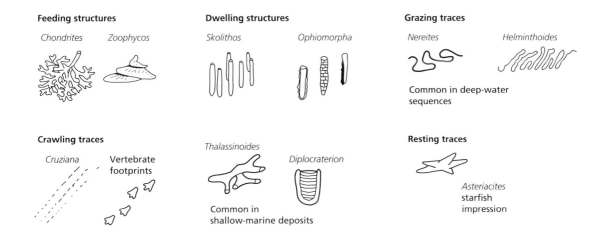

Fig. 2.43 Illustrations of the common types of trace fossil.

traces of bivalves (*Pelecypodichnus*) and trilobites (*Rusophycos*).

Crawling (or locomotory) traces are made by any mobile animal, from a trilobite to a dinosaur. They are typical of predators, scavengers and some deposit feeders. The trails and trackways of moving animals are usually linear or sinuous (Fig. 2.42a), and uncomplicated by comparison with grazing and feeding traces. *Cruziana*, the trilobite foraging trail, is common on the surface or sole of Lower Palaeozoic sandstone beds. Vertebrate footprints are relatively common throughout the Mesozoic and Cenozoic.

Grazing traces are produced mainly by mobile, deposit-feeding epibenthic organisms that feed at or near the sediment surface. The traces are epistratal, i.e. located on the bedding surface, and consist of curved, coiled or radiating furrows formed by the organisms systematically ingesting the sediment for food. Examples include *Helminthoides* and *Nereites*, which are typical of deeper-water sediments.

Dwelling burrows are mainly formed by sessile and semisessile endobenthic animals, particularly suspension feeders, predators and scavengers. The burrows may be simple vertical tubes, such as *Skolithos* (Fig. 2.42b), U-shaped or more complicated burrow systems. Some U-shaped burrows, such as *Diplocraterion*, give rise to a web-like series of concave-up laminae (termed *spreite*) formed through upward or downward movement of the animal in response to sedimentation or erosion. Many dwelling burrows are lined by pellets or mud, which serve to distinguish them from feeding structures. Examples include *Ophiomorpha* and *Thalassinoides*, crustacean burrows of intertidal–shallow-subtidal environments, and *Skolithos*, a simple vertical tube common in intertidal zones.

Feeding structures are intrastratal, i.e. formed below the sediment surface, and are made by endobenthic deposit feeders, living within a burrow system. The trace fossil typically consists of a network of filled burrows, branching or unbranched. *Chondrites* and *Zoophycos* (Fig. 2.42c) are feeding structures found in many Phanerozoic shelf sediments. Although forming within the sediment, they are often seen on bedding surfaces. *Escape structures* are produced when burrowing organisms are smothered with sediment and they burrow up to regain their preferred position relative to the sediment–water interface. In so doing, they leave behind characteristic vertical chevron structures cutting and deflecting downward the sedimentary lamination.

Borings are another type of trace fossil, although more common in limestones. They differ from burrows by being made into a hard substrate, such as a hardground surface (Section 4.6.1), pebble, reworked concretion or carbonate skeleton. In the field, the borings of lithophagid bivalves (Fig. 2.42d; flask-shaped, up to 2 cm long) and clionid sponges (beaded, 5 mm across) are readily identified.

The importance of trace fossils lies in the information they can give on the depositional environment, because in broad terms certain trace fossils or suites of trace fossil are characteristic of a particular environment and, in certain cases, of a specific depth range.

Table 2.6 Trace-fossil assemblages, depositional environments, trace-fossil forms and typical ichnofossils

Trypanites assemblage	*Glossifungites* assemblage	*Skolithos* assemblage	*Cruziana* assemblage	*Zoophycos* assemblage	*Nereites* assemblage
Rocky coast, +1–5 m depth	Semiconsolidated substrate along shoreline, 0–10 m depth	Sandy shoreline, foreshore and shoreface, 0–10 m depth	Sublittoral zone, offshore inner shelf, c. 10–100 m depth	Bathyal zone, outer shelf and slope, c. 100–2000 m depth	Abyssal zone, deep-sea floor, c. 1000–5000 m depth
Borings into bedrock and boulders	Sharp-edged burrows, vertical, U- or flask-shaped	Vertical burrows, U-shaped, or pellet-lined simple	Wide variety: surface tracks and trails, complex burrows	Limited number on surface and shallow subsurface	Regular patterns on sediment surface
Borings by clionid sponges, lithophagid bivalves and worms	*Glossifungites Ophiomorpha*	*Skolithos, Rhizocorallium, Diplocraterion*	*Zoophycos, Chondrites, Planolites, Thalassinoides*	*Zoophycos Helminthoides*	*Nereites, Palaeodictyon*

Trace fossils have to be studied with the sedimentology, however, because there are many instances of shallow-water-type trace fossils occurring in deep-water facies, and vice versa. Organisms respond to the local conditions of sedimentation and at times these may not be very different in deep and shallow water. Trace fossils commonly reflect the substrate, rate of sedimentation and degree of current reworking, as well as water depth. Sediments can be subdivided using their trace-fossil assemblages into 'ichnofacies'. Six common *trace-fossil assemblages* are recognized from coastal to deep-water environments, named after a characteristic trace fossil (see Table 2.6). Water depth and substrate are the major controls.

Trace fossils are discussed at length in Goldring (1991), Maples & West (1992), Donovan (1994) Bromley (1996) and Brenchley & Harper (1998).

2.4 Palaeocurrent analysis

Many erosional and depositional sedimentary structures can be used to deduce the direction or trend of the currents that formed them. A knowledge of the palaeocurrents gives vital information on palaeogeography, sand-body geometry and sediment provenance, and in addition certain palaeocurrent patterns are restricted to a particular depositional environment.

Some sedimentary structures are vectorial, i.e. they give the direction in which the current was flowing; cross-stratification, asymmetric ripples, flute marks

Fig. 2.44 Preferred orientations of graptolites on bedding surface of hemipelagic mudrock showing that there are significant currents on the deep-sea floor. Ordovician. Wales.

and imbrication are structures of this type. Other structures give the trend or line of direction of the current; these include parting lineation, preferred orientation of grains, pebbles and fossils (e.g. Fig. 2.44), symmetrical ripples and groove marks.

Palaeocurrent data are collected in the field by simply measuring the orientation of the structure. However, if the strata are tilted to any extent (greater than 15°) the beds must be brought back to the horizontal by using a stereogram. This necessitates measurement of the dip and azimuth of the structure and dip and strike of the beds. Although the more data one has the more accurate is the palaeocurrent picture, some 20–30 readings for a locality or small area can be sufficient if dealing with a unimodal current system; for a bimodal or polymodal current system, more readings are desirable. To find the mean palaeocurrent direction, readings cannot simply be totalled and averaged because the data are directional. To obviate this, vector means can be calculated or constructed graphically, but only if the distribution is unimodal. Statistical tests can be applied to palaeocurrent measurements to allow more precise comparisons between different sets of data. Palaeocurrent measurements normally are presented as a rose diagram, as in Fig. 2.45, where conventionally readings are plotted in a 'current-to' sense. If a trend only can be measured then the resulting rose diagram is symmetrical. There are four basic types of palaeocurrent pattern (Fig. 2.45): unimodal, bimodal–bipolar, bimodal–oblique and polymodal.

Palaeocurrent interpretation needs to be combined with studies and interpretations of the facies for maximum information. Measurements from fluvial, deltaic and shallow-marine deposits are most representative of the dominant flow direction(s) when cross-bedding is used. Smaller-scale structures, such as ripples and cross-lamination, usually reflect secondary flows and so may deviate from the direction of interest. With turbidites the flow direction is best given by the sole struc-

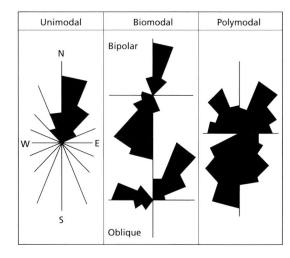

Fig. 2.45 The four common palaeocurrent patterns.

Table 2.7 Depositional environments, directional structures and typical dispersal pattern

Environment	Directional structures	Dispersal pattern
Aeolian	Large-scale cross-bedding	Unimodal if formed by barchan-type dunes and then indicates palaeowind direction; bimodal if formed by simple seif dunes and polymodal if complex seifs; may relate to trade winds; independent of regional palaeoslope.
Fluvial	Best to use cross-bedding from dune/sandwave bedforms; also parting lineation, ripples, scours, imbrication, regional grain-size changes	Palaeocurrents reflect palaeoslope and indicate provenance direction; unimodal pattern with small dispersion if low-sinuosity rivers; unimodal with larger scatter if high-sinuosity river or alluvial fan
Deltaic	Best to use cross-bedding from dunes/sandwave bedforms; also channels, parting lineation, ripples	Typically unimodal pattern directed offshore, although marine processes (tidal and storm currents and waves) can complicate palaeocurrent pattern
Shallow-marine shelf	Best to use cross-bedding from dunes/sandwave bedforms; also ripples and scours	Pattern can be complex and difficult to interpret; bimodal pattern through tidal current reversals although tidal currents may be parallel or normal to shoreline; can be unimodal if one tidal current dominates; polymodal and random patterns also occur; complicated by wave and storm effects
Turbidite basin	Best to use sole structures, especially flutes; also parting lineation, cross-lamination, ripples, grain orientation, slump folds	Unimodal pattern common from turbidites, although may be downslope or along basin axis, or radial if on submarine fan. Contourites give palaeocurrent pattern parallel to the strike of the slope

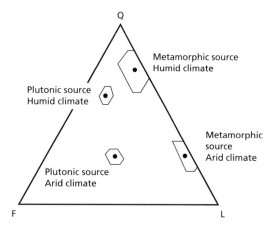

Fig. 2.46 Average compositions of medium sand-size fraction of first-cycle stream sediment derived from plutonic igneous and metamorphic sources under different climatic conditions. Q, quartz; F, feldspar; L, lithics.

tures. A turbidity current is apt to wander when it is slowing down so that measurements from parting lineation, grain orientation and cross-lamination within the bed often differ from those of sole structures. The palaeocurrent patterns of the principal depositional environments are given in Table 2.7. Palaeocurrents are discussed at length in Potter & Pettijohn (1977) and their measurements, statistical evaluation and interpretation are also covered by Graham (1988) and Tucker (1996).

2.5 Detrital components of siliciclastic sediments

Sandstones, conglomerates and breccias consist of: *detrital grains*, which form the framework of the sediments, fine-grained *matrix* located between grains, and *authigenic minerals* and *cements* precipitated after deposition of the sediment, during diagenesis (Section 2.9).

Practically any of the naturally occurring minerals and fragments of any known rock type can occur as grains in a clastic sediment. However, certain minerals and rock types are much more stable than others in the sedimentary environment, so that in fact the number of common grain types is relatively small. The abundance of a particular mineral in a sedimentary rock is dependent on its availability, i.e. source-area geology, but also on its mechanical and chemical stability and resistance to weathering and alteration.

With regard to *chemical stability*, minerals can be arranged into a series from the most to the least stable:

quartz, zircon, tourmaline
chert
muscovite
microcline
orthoclase
plagioclase
hornblende, biotite
pyroxene
olivine

Dissolution of minerals takes place at the site of weathering, particularly if the weathering processes are more chemical than physical. The prevailing climate is important; mineral dissolution is more prevalent in hot and humid regions than in hot and semi-arid or polar regions. This is well shown in Fig. 2.46, where there is a clear difference in the composition of sands derived from similar volcanic and plutonic rocks but under different climatic regimes. In addition to climate, relief in the source area is very important too. With low relief and weathering under a hot and humid climate, quartz probably will form the only grains that survive for erosion and transportation. In regions of very intense weathering, quartz too will dissolve, and laterites and bauxites (residual deposits), then form. In an area of high relief, however, some unstable grains will always be liberated for erosion and later deposition, even if the weathering is extreme, because there will be some fresh rocks somewhere. Dissolution of mineral grains during transport by water appears to be insignificant, but it is very important during diagenesis, namely the process of *intrastratal dissolution*.

The *mechanical stability* of a mineral depends on the presence of cleavage planes and the mineral's hardness. Quartz, being relatively hard with no cleavage, is mechanically very stable and can survive considerable abrasion during transportation. On the other hand, feldspars with their strong cleavage, and many rock fragments with their generally weak intercrystalline or intergranular bonds, are more easily broken during sediment transport. Such unstable grains are called *labile grains*.

The detrital particles in siliciclastic rocks can be divided into six categories: (a) rock fragments, (b) quartz, (c) feldspar, (d) micas and clays, (e) heavy minerals and (f) other constituents. On the basis of their detrital components, sediments can be considered usefully in terms of their compositional maturity.

2.5.1 Rock fragments

These dominate conglomerates and tend to be the coarser particles in sandstones. As rock fragments become smaller they tend to break up into their constituent minerals and grains. The composition of the rock fragments depends basically on source-rock geology and durability of particles during transportation. In sandstones the lithic fragments are commonly of (a) fine-grained sedimentary (Ls) and metasedimentary (Lm) rocks such as mudstone, shale and siltstone, and slate, pelite and mica schist, respectively (e.g. Plate 1a); (b) siliceous sedimentary rocks such as chert (usually classified as a polycrystalline quartz grain (Qp); and (c) igneous, in particular volcanic rocks (Lv). Plutonic igneous rocks tend to be more coarsely crystalline and so for sandstones have broken into individual crystals, rather than remained as composite grains. Volcanic grains can be recognized from the presence of feldspar laths in a fine groundmass (microlitic) and from microgranular and vitric textures (Plate 1b). Felsitic rock fragments can be difficult to distinguish from chert grains, but the latter do not usually contain any feldspar. Rock fragments are very useful in studies of the provenance of a sandstone, but intrabasinal lithics, which are commonly grains of mud and carbonate, are usually excluded when percentages of grain types are calculated.

In provenance studies, it is very important to study sandstones of a similar grain size. The percentage of rock fragments increases with increasing grain size, and misleading conclusions would be reached if this factor were not taken into account. In fact, in the 'Gazzi–Dickinson' method, sand-sized crystals and grains within a larger rock fragment are assigned to the category of the crystal or grain, rather than to the rock fragment class. This method is thought to be better for relating sandstone composition to tectonic setting, as well as avoiding the problem of grain size.

Compaction and diagenetic alteration may render fragments of shale, slate and igneous rock indistinguishable from a fine-grained muddy matrix. Igneous rock fragments may be replaced by chlorite and zeolites. Many rock fragments are unstable (labile grains) but they are characteristic of certain sandstone types, in particular the litharenites and greywackes (Section 2.6).

Within *conglomerates and breccias*, two types of clast occur: extraformational and intraformational clasts. Intraformational clasts, formed within the area of sedimentation, are chiefly lumps of mud, derived from erosional reworking of previously deposited muds. Extraformational clasts are derived from outside the area of sedimentation and can be of almost any rock type, even very unstable varieties if the transport path is short. Conglomerates or breccias may contain a great variety of clasts, particularly if the material is far-travelled. Such deposits are termed *polymictic* and examples include some glacial deposits and fluvial gravels. Conglomerates containing only one type of clast are referred to as *monomictic* (or oligomictic), and a local source usually is implied.

Rock fragments in sandstones give very specific information on the *provenance* of a deposit if they can be tied down to a particular source formation. Rock fragments generally are derived more from supracrustal rocks undergoing rapid uplift and erosion. Mountain belts and volcanic areas supply large quantities, whereas continental/granitic basement does not. The types of lithic grain do relate to plate-tectonic setting of the provenance terrane and adjoining sedimentary basin (see Section 2.8). Calculating the percentage of sedimentary rock fragments (Ls) and low-grade (Lm_1, slate+quartzite) and high-grade (Lm_2, phyllite+schist+quartz/mica/albite aggregates) metamorphic grains up through a succession may show trends related to source-area uplift (e.g. Fig. 2.47).

2.5.2 Quartz

The most common mineral in sandstone is quartz, the most stable of all minerals under sedimentary conditions. The average sandstone contains some 65% quartz, but some are practically 100%. Many quartz grains in Mesozoic and Cenozoic sandstones are in their second or third cycle of sedimentation. The majority of quartz grains are derived from plutonic granitoid rocks, acid gneisses and schists. Various types of quartz can be distinguished: monocrystalline quartz grains (Qm) are composed of a single crystal and polycrystalline grains (Qp) of two or more crystals (see Plate 1c). Polycrystalline grains can be divided into those with two to three crystals (Qp_{2-3}) and those with more ($Qp_{>3}$). It is also possible to examine the crystal

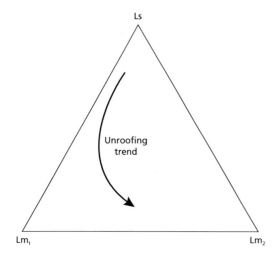

Fig. 2.47 The trend in lithic grains (Ls, sedimentary; Lm$_1$, low-grade metamorphic; Lm$_2$, medium-grade metamorphic) in sandstones derived from the unroofing of a sedimentary–metasedimentary complex of an arc–continent collision belt.

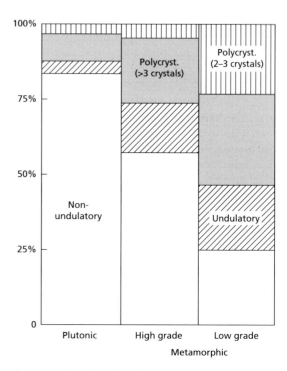

Fig. 2.48 Relative abundance of detrital monocrystalline and polycrystalline quartz grains in Holocene sands derived from known plutonic and metamorphic sources. After Basu *et al.* (1975).

boundaries in a Qp grain, for sutured, straight and irregular contacts. Chert fragments can be classified as Qp. In calculating percentages of the various grain types in a sandstone, Qp grains are usually included with lithics (L), see Section 2.7.

Further subdivision of monocrystalline quartz grains can be made on the basis of extinction and inclusions. Crystals may possess unit extinction, where the whole crystal is extinguished uniformly under crossed polars (Qmnu — monocrystalline, non-undulatory), or undulose extinction, where extinction is not uniform but sweeps across the crystal as it is rotated over more than 5° (Qmu — monocrystalline, undulatory) (see Plates 1c & 5a). Undulose extinction usually is a reflection of strain in the crystal lattice. Inclusions are common in quartz grains and are either vacuoles filled with fluid (appearing brown or black in transmitted light, silver in reflected light) or minute crystals of other minerals, especially rutile, mica, chlorite, magnetite and tourmaline.

With many quartz grains it is not possible to assign a precise origin, although there are differences in the proportions of the various quartz types in the common source rocks (Fig. 2.48). Some features of the quartz grains are thought to be more diagnostic of one provenance than another. Quartz grains derived from volcanic igneous rocks, for example, are typically

monocrystalline with unit extinction and no inclusions. Quartz from hydrothermal veins may be monocrystalline or coarsely polycrystalline but they characteristically possess numerous fluid-filled vacuoles. Pebbles of vein quartz, identified by their milky-white colour imparted by the fluid inclusions, are common in extraformational conglomerates. Polycrystalline quartz grains from a metamorphic source typically possess many crystals and they are usually elongate, with a preferred crystallographic orientation. Undulose extinction, once thought to indicate a metamorphic source, is quite common in igneous quartz crystals. Taken together, close examination of extinction properties and polycrystallinity can separate sediments derived from plutonic and low-grade and high-grade metamorphic terranes.

Monocrystalline quartz grains with undulose extinction and polycrystalline quartz grains are both less stable than unstrained monocrystalline quartz. As a result of this, they are selectively removed during weathering, transportation and diagenesis, and sandstones contain a higher monocrystalline/polycrystalline quartz ratio and unstrained/strained quartz ratio than the source igneous or metamorphic rock. This ratio can be used as a maturity index. A high ratio of polycrystalline to monocrystalline quartz suggests derivation from a metamorphic region without extensive weathering and/or transportation. Recycling of quartz grains from an older sandstone leads to further loss of the less stable quartz grain types. Recycled monocrystalline quartz grains can be recognized by the abrasion of overgrowths that were precipitated around the grains in the original sandstone.

The technique of cathodoluminescence can distinguish between quartz grains of different origins, and it also shows up the difference between grain and overgrowth (see the review by Matter & Ramseyer, 1985).

2.5.3 Feldspars

The feldspar content in sandstones averages between 10 and 15%, but in the arkoses (Section 2.7.2) it commonly reaches 50%. The mechanical stability of the feldspars is lower than that of quartz, because feldspars are softer and have a stronger cleavage. This leads to disintegration of feldspar crystals during transportation and in turbulent environments, so that on a broad scale, for example, fluvial sediments contain more feldspar than beach, shallow-marine and aeolian sandstones. The chemical stability of feldspars is also lower because of the ease with which they are hydrolysed. Chemical alteration typically involves replacement by clay minerals such as sericite (a variety of muscovite), kaolinite and illite (Section 2.8.5). Incipient alteration gives the feldspars a dusty appearance or there are bright specks (see Plates 1d & 3c,d). Complete replacement produces clay-mineral pseudomorphs after feldspar. Feldspar alteration takes place at the site of weathering, if it is dominantly chemical rather than physical weathering, and during diagenesis, either on burial or subsequent uplift. Diagenetic replacement of feldspar by calcite also is common. Feldspars also are subject to dissolution, and partial to total loss of feldspar can take place during diagenesis (see Plates 2c,d & 4c,d). Sandstones near the surface and those beneath unconformities in particular, may have grain-sized pores where feldspar has been dissolved out through contact with meteoric water.

Of the different types of feldspar, the potash feldspars (Fk), orthoclase and microcline, are much more common in sandstones than the plagioclases (Fp). There are two reasons for this. K-feldspar has a greater chemical stability than plagioclase, and so the latter is altered preferentially in the source area. In addition, K-feldspar is much more common in continental basement rocks (granites and acid gneisses), which are the provenance of many sandstones in the geological record. Plagioclase is more common in sandstones derived from uplifted oceanic and island-arc terranes, which generally are less important source areas.

Microcline is readily identified in thin-section from the grid-iron (cross-hatch) twinning pattern (see Plates 1d & 3c,d), but there is often a problem in recognizing *orthoclase*. The crystals usually break along the twin planes so that the simple Carlsbad twinning is rarely seen in an orthoclase in a sandstone. With its first-order low birefringence, orthoclase will often look like quartz. Apart from optical tests, orthoclase usually can be recognized by its dusty and partly altered appearance (to kaolinite or sericite), especially relative to the very clear and clean-looking quartz grains (see Plates 1d & 3c,d). Cleavage may be visible too in the feldspar (seen in Plate 5a,b). The orthoclase content is often underestimated in sandstone modal analysis. *Plagioclase* usually can be identified by its polysynthetic, multiple twinning. Albite (Na^+-rich) is much more common than anorthite (Ca^{2+}-rich), mostly because it is a more important mineral in source rocks and it is more stable in the sedimentary environment, but also because anorthite is altered to albite during diagenesis (see Section 2.9.4). In all serious studies of sandstones a stain is applied which gives the various feldspars (K-, Na- and Ca-feldspar) a different colour (see Miller (1988a) for details).

Feldspar grains are derived from the same crystalline rocks as quartz. These chiefly are granites and gneisses, where potash feldspar dominates over sodic plagioclase. Textures in feldspar crystals may contain clues to their origin. Various types of zoning are frequently seen, and this is more common in feldspars from a volcanic source. Pyroclastic feldspars tend to be anhedral crystals, which frequently are broken. Perthites are the result of slow cooling and so are more typical of plutonic source rocks. The majority of feldspars in sediments are first cycle.

Apart from suitable source rocks, the feldspar content of a sediment is controlled largely by the rate of erosion and the climate. A humid climate in the source area promotes feldspar destruction because of the dominantly chemical weathering, whereas in an arid area, fresh feldspars survive the dominantly physical weathering. Rapid erosion, however, as happens in an area of high relief, will produce some feldspar grains in spite of a humid climate. Such feldspars typically will vary from fresh to altered. Microprobe studies of plagioclase grains in modern deep-sea sands have shown that the various tectonic settings, passive margin, island-arc, back-arc, strike-slip, etc., can be differentiated on the basis of K^+, Na^+ and Ca^{2+} contents, although there is considerable overlap in the compositional fields (Maynard, 1984).

2.5.4 Micas and clay minerals

These sheet silicates (phyllosilicates) are particularly common in the matrix of sandstones and coarser clastics, and are the main component of mudrocks (Chapter 3).

Biotite and *muscovite* occur as detrital flakes (see Plates 2a,b & 5c), which may be concentrated along partings, laminae and bedding planes. Because of their platy nature they are easily washed out of coarser sediments and so tend to accumulate with finer sands and silts; they are also easily removed from wind-blown sediments. Muscovite and biotite are derived from many igneous rocks but especially from metamorphic schists and phyllites. Although biotite is more common in the source rocks than muscovite, the latter is more stable chemically and so is far more common.

Biotite and muscovite are easily identified from their platy nature and parallel extinction. Biotite has a brown–green pleochroism, which masks the interference colours, and muscovite is colourless in plane-polarized light, but has bright second-order colours under crossed polars.

Clay minerals in sandstones are both detrital and authigenic. Detrital clay-mineral types cannot be identified with the petrological microscope, but some authigenic clays can (see Section 2.9.5). All the chief clay-mineral groups are represented in sandstones: kaolinite, illite, chlorite, smectite and mixed-layer clay. Detrital clays reflect the source-area geology, climate and weathering processes (Section 3.5). Soon after deposition, clays may mechanically infiltrate a sandstone. During diagenesis, clay minerals may be altered to other clays (Section 3.6), or they may form at the expense of other grains, in particular, the feldspars, which commonly are replaced by kaolinite. Chlorite replaces labile rock fragments such as volcanic grains. In some sandstones, much of the matrix is actually formed by the compaction and alteration of unstable grains to form a secondary matrix or *pseudomatrix* (see Section 2.7.4). Moraes & De Ros (1992) distinguished depositional, infiltrated and authigenic clays in Jurassic fluvial sandstone reservoirs from Brazil.

2.5.5 Heavy minerals

These accessory grains are present in concentrations of less than 1%. They are chiefly silicates and oxides, many of which are very resistant to chemical weathering and mechanical abrasion. The common, non-opaque heavy-mineral grains are apatite, epidote, garnet, rutile, staurolite, tourmaline and zircon (the last is seen in Plate 3a). Ilmenite and magnetite are two common opaque detrital minerals. The specific gravity of heavy minerals, greater than 2.9, is higher than that of quartz and feldspar at 2.6. In view of this and their low concentration in the sediment, heavy-mineral grains are separated from the crushed rock or loose sediment by using a heavy liquid such as acetylene tetrabromoethane, which has a specific gravity of 2.9. Heavy minerals can be identified by their optical properties, in some cases through the use of refractive index oils, and by geochemical means. A colour atlas of heavy minerals is provided by Mange & Maurer (1992).

The study of heavy minerals can give useful information on provenance and events in the source area. Certain heavy minerals, such as garnet, epidote and staurolite, are derived from metamorphic terranes, whereas others, rutile, apatite and tourmaline for instance, indicate igneous source rocks (see Table 2.8). Major changes in the source-area geology, such as the uplift and unroofing of a granite, may be recorded in the heavy-mineral assemblage of sandstones deposited in the region (see Uddin & Lundberg (1998) for an example). Heavy-mineral suites can be used to identify petrographic provinces within a formation, where, for example, sediment was supplied by two or more rivers draining areas of different geology.

The heavy-mineral assemblage in source-area rocks is affected by weathering in the same way as other

Table 2.8 Characteristic heavy minerals of different source rocks

Source rock	Characteristic heavy minerals
Acid igneous	apatite, biotite, hornblende, magnetite, zircon
Basic igneous	augite, hypersthene, ilmenite, rutile
Pegmatite	fluorite, garnet, tourmaline
High-grade metamorphic	epidote, garnet, kyanite, sillimanite, staurolite
Low-grade metamorphic	biotite, tourmaline
Sedimentary	reworked tourmaline and zircon (rounded)

minerals. Many will be dissolved out through intense weathering but as long as there is some relief and a sufficient rate of erosion, some less-stable grains will be supplied to the nearby sedimentary basin. By and large, most heavy minerals are sufficiently strong mechanically to resist loss by abrasion during transport, although some will be lost during temporary alluvial storage in floodplains.

As a result of their higher specific gravities, heavy-mineral grains tend to be smaller than the quartz grains with which they occur. This is the concept of *hydraulic equivalence*. In some instances the heavy minerals are concentrated in particular laminae or beds (*hydraulic sorting*). This is a common feature of beaches and other locations where persistent winnowing takes place. Economic deposits, termed *placers*, may be formed in this manner. The monazite beach sands of Brazil, gold in fluvial and beach sediments of the Yukon and Alaska, and ilmenite beach sands of Queensland are well-known modern examples (see papers in *J. Geol. Soc. London*, **142**(5), 1985). In the gold deposits of Witwatersrand, South Africa, the gold is located principally in early Proterozoic river-channel conglomerates through synsedimentary concentration of detrital gold (Els, 1998).

Heavy minerals can be dissolved out during diagenesis through intrastratal dissolution. This can have a major effect on the heavy-mineral assemblage. The general order of decreasing resistance to dissolution is as follows, but it does depend on the pore-water pH, Eh and other factors: rutile, tourmaline, zircon, garnet, apatite, staurolite, kyanite, epidote, amphibole, andalusite, sphene, pyroxene, sillimanite, olivine.

There are many documented cases where the heavy-mineral assemblage shows an increasing diversity down from an unconformity, as a result of dissolution by descending pore waters. Higher temperature and pressure in the burial diagenetic realm also cause heavy-mineral dissolution. This is clearly shown where sandstones sealed from circulating pore-fluids at an early stage, such as through calcite cementation or oil infiltration, have a much more diverse heavy-mineral assemblage, including unstable grains, than adjacent, more porous sandstones. The dissolution of grains with increasing depth is well documented by Milliken (1989) from Plio-Pleistocene sandstones in the Gulf Coast region, USA. Older sandstones tend to have a less diverse heavy-mineral suite as a result of intrastratal dissolution.

Chemical analyses of heavy minerals, especially pyroxene, amphibole, epidote and garnet, can be useful in provenance studies because certain trace element contents (e.g. TiO_2, MnO and Na_2O) and REEs (rare-earth elements) and platinum group elements (PGEs) reflect particular plate-tectonic settings. Garnets are particularly useful in this respect, because they are so stable. Morton (1987) used this approach to document different source areas for sands in the Jurassic Brent Group of the northern North Sea and in the Palaeocene Forties Formation of the central North Sea. Fission-track dating of zircons and apatites can be used to identify provenance terranes and the timing of uplift. Papers on heavy minerals are contained in Bahlburg & Floyd (1999) and the processes controlling the composition of heavy-mineral assemblages in sandstones are reviewed by Morton & Hallsworth (1999).

2.5.6 Other detrital components

Carbonate grains in sandstones and coarser clastics are mostly fossils or fragments of them, and non-skeletal grains such as ooids, peloids and intraclasts (see Chapter 4). Detrital grains of limestone and dolomite do occur, but unless there is an abundant supply, they are a very minor component. The importance of fossils in a sandstone or conglomerate lies in the stratigraphic and environmental information they may give. Other mineral grains in siliciclastics are skeletal phosphate (bone fragments), glauconite and berthierine–chamosite. There generally is little disseminated organic matter in sandstones, although carbonaceous plant fragments do occur in many sandstones.

2.5.7 Compositional maturity

A compositionally immature sandstone contains many labile grains, i.e. unstable rock fragments and minerals, and much feldspar. Where rock fragments are of a more stable variety and there is some feldspar and much quartz, then the sediment is referred to as mature. For a sandstone composed almost entirely of quartz grains the term supermature is applied. Compositional maturity can be expressed by the ratio of quartz + chert grains to feldspars + rock fragments. This compositional maturity index is useful if comparisons between different sandstones are required.

Compositional maturity basically reflects the weathering processes in the source area and the degree and extent of reworking and transportation. Typically, compositionally immature sediments are located close to their source area or they have been rapidly transported and deposited with little reworking from a source area of limited physical and chemical weathering. Examples include many near-source fluvial sediments, glacial and glacio-fluvial deposits. At the other extreme, supermature sediments are the end-product of intense weathering, where all unstable grains have been removed, or they are the result of intense abrasion and sediment reworking. The concepts of compositional maturity are considerably modified where (a) the source area itself consists of mature sediments, when the weathering mantle, soil and fluvial sediments derived therefrom also will be mature, and (b) sediment is supplied directly to a beach and nearshore area from adjacent igneous–metamorphic rocks, when immature sandstones will result.

2.6 Classification of siliciclastic sediments

2.6.1 Classification of sandstones

The classification of a sandstone is based on microscopic studies and requires an assessment of the percentages of the various grain types present (shown in Table 2.9). Several hundred grains (400 recommended) are identified by point-counting a slide, and to compare different sandstones, samples of a similar grain size should be used. A rough estimate of this modal composition can be obtained by comparing the sandstone field-of-view with the diagrams of Fig. 2.49, which show the appearance of different percentages of grains in circular and square fields. A very rough esti-

Table 2.9 Classification of sand-grain types

Quartzose grains (Qt = Qm + Qp)
   Qt = total quartz grains
   Qm = monocrystalline quartz
   Qp = polycrystalline quartz

Feldspar grains (F = P + K)
   F = total feldspar grains
   P = plagioclase grains
   K = potassium feldspar grains

Lithic fragments (Lt = Qp + Lvm + Lsm)
   Lt = total lithic fragments (L + Qp)
   L = total unstable lithic fragments (Lvm + Lsm)

Lv/Lvm = volcanic/metavolcanic lithic fragments

Ls/Lsm = sedimentary/metasedimentary lithic fragments

mate only of a sandstone's composition can be made in the field through close scrutiny with a hand-lens.

There are several classification schemes available and most use a triangular diagram with end members of quartz (Q), feldspar (F) and rock fragments (L). The triangle is divided into various fields, and rocks with an appropriate modal analysis are given a particular name. In recent years, the study of sandstone provenance has been taken much further, so that a sandstone composition can be tied more precisely to its source area and tectonic setting. This involves counting all the different types of quartz (Qmnu, Qmu, Qp_{2-3}, $Qp_{>3}$), feldspar (Fk, Fp) and lithic grains (Ls, Lm, Lv) (see Table 2.9) and plotting the results in different ways on several triangular diagrams (see Section 2.8).

In the widely used simple classification (Fig. 2.50), sandstones are divided into two major groups based on texture, that is, whether the sandstones are composed of grains only, the *arenites*, or contain more than 15% matrix, forming the *wackes*.

For the arenites, the term *quartz arenite* is applied to those with 95% or more quartz grains, a rock type formerly referred to as orthoquartzite (quartzite is the low-grade metamorphic equivalent). *Arkosic arenite* refers to an arenite with more than 25% feldspar, which exceeds the rock-fragments content, and *litharenite* is applied where the rock-fragment content exceeds 25% and is greater than feldspar. The arkosic arenites can be divided into *arkoses* and *lithic arkoses*. Two rock types transitional with quartz arenite are *subarkose* and *sublitharenite*. Specific names applied

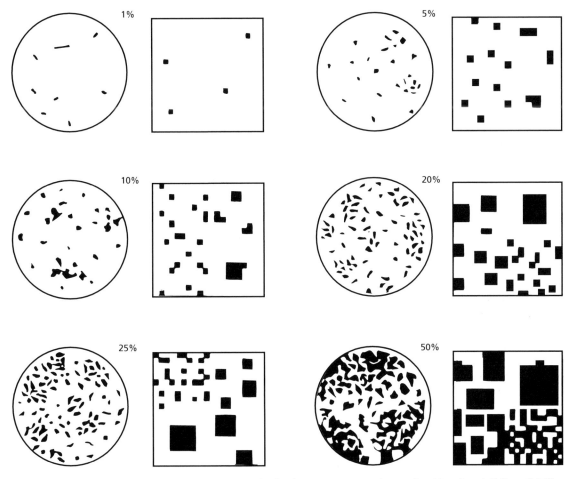

Fig. 2.49 Percentage estimation comparison charts, conventional and computer-generated. Reproduced from Terry & Chilingar (1955) and Folk *et al.* (1970) with permission of the Society of Economic Paleontologists and Mineralogists.

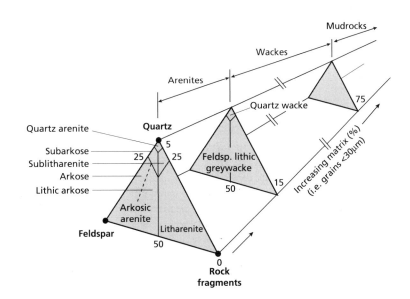

Fig. 2.50 Classification of sandstones. After Pettijohn *et al.* (1987).

to litharenites are *phyllarenite*, where the rock fragments are chiefly of shale or slate, and *calclithite*, where the rock fragments are of limestone.

The wackes are the transitional group between arenites and mudrocks. The most familiar is the *greywacke* and two types are distinguished: feldspathic and lithic greywacke. The term arkosic wacke is used for arkoses with a significant proportion of matrix. *Quartz wacke*, not a common rock type, is dominantly quartz plus some matrix.

This classification is concerned primarily with the mineralogy of the sediment and presence or absence of a matrix. It is independent of the depositional environment, although some lithologies are more common in certain environments. The nature of the cement in arenites is not taken into account. With regard to fine-grained interstitial material (matrix), a basic feature of the wackes, there is often a problem of origin. Some, perhaps most, is deposited along with the sediment grains. A part of the matrix, however, could be authigenic, a cement, and some a diagenetic alteration product of unstable grains (pseudomatrix). The latter is particularly the case with greywackes (Section 2.7.4).

In addition to siliciclastic sandstones, there are many *hybrid sandstones*. These contain a non-clastic component derived from within the basin of deposition. The three main types are calcarenaceous, glauconitic and phosphatic sandstones. In glauconitic sandstones, the glauconite occurs as sand-sized pellets (see Section 6.4.4). With phosphatic sandstones, the phosphate may be present as coprolites, faecal pellets and bone fragments (Section 7.4).

Calcarenaceous sandstones contain up to 50% $CaCO_3$ as carbonate grains. The latter are chiefly ooids, commonly with quartz nuclei, and skeletal fragments (bioclasts). Calcarenaceous sandstones occur in carbonate-producing areas where there is a large influx of terrigenous clastics. They will pass laterally into limestones or into purer sandstones towards the source of the siliciclastic sediment. Siliciclastics cemented by calcite have been referred to as calcareous sandstones. A scheme for classifying these mixed sediments has been devised by Mount (1985). For a collection of papers on these calcarenaceous sandstones see Doyle & Roberts (1988).

2.6.2 Classification of conglomerates and breccias

There are many ways of dividing up coarse clastic rocks, which are also called *rudites*. In the first place, on the basis of pebble shape, they can be split into *conglomerates*, with subrounded to well-rounded clasts, and *breccias*, with angular and subangular clasts. A division in terms of composition has been mentioned previously: polymictic conglomerates (or breccias) contain a variety of pebbles and boulders, whereas mono- or oligomictic conglomerates consist of one pebble type only. A division into intraformational and extraformational conglomerates divides those composed of clasts derived from within the basin of deposition, from those where pebbles are derived from beyond the area of sedimentation (see Section 2.5.1). On sediment fabric, there are two types: clast-supported (orthoconglomerates) and matrix-supported (paraconglomerates) (see Section 2.2.4). The latter are also called *diamictites*. Terms for sandy and muddy conglomerates are given in Fig. 2.1.

Particular types of breccia include slump breccia, consisting of broken and brecciated beds derived from downslope slumping (Section 2.3.3), collapse breccia, resulting from the dissolution of evaporites and the collapse of overlying strata (Section 5.5), and palaeokarstic breccia, produced by dissolution of limestone, formation of caves, fracture and collapse. Carbonate megabreccias, with very large clasts (>4 m diameter), may have a sequence stratigraphic significance and relate to a relative sea-level fall.

2.7 Petrography and origin of principal sandstone types

The four common types of sandstone are quartz arenite, arkose, litharenite and greywacke. They are typical of certain depositional environments, but because of the provenance control on sandstone composition (described in Section 2.8), they are not restricted to a particular depositional setting. They reflect the geology of the source area to a greater or lesser extent, depending on weathering and relief.

2.7.1 Quartz arenites

Sandstones with at least 95% quartz grains are the most compositionally mature of all sandstones (see Section 2.5.7). In addition, they usually consist of well-rounded and well-sorted grains so that textural maturity is also very high. Cements are typically quartz overgrowths, but calcite also is common, normally a poikilotopic cement (a large crystal including several grains). Monocrystalline unit-extinguishing quartz grains dominate, because the less stable, undulatory and polycrystalline grains have been eliminated. Common heavy minerals are rutile, tourmaline, zircon and ilmenite, again the most resistant types. Very pure quartz sands with little cement are referred to as glass sands. Quartz arenites commonly show the effects of pressure dissolution, with sutured grain contacts and through-going stylolites. Quartz arenites are illustrated in Plates 1c, 2c,d, 3a, 4c,d & 5a.

In many cases quartz arenites are the products of extended periods of sediment reworking, so that all grains other than quartz have been broken down by mechanical abrasion. Climate in the source area can also play a major role in producing quartz arenites. A warm humid climate will lead to the removal of many unstable grains, and if this is coupled with low relief and slow sedimentation rates, the quartz will dominate the detritus. Many quartz grains in these arenites could be second cycle, derived from pre-existing sediments. Evidence for this is provided by grains with abraded overgrowths from an earlier diagenetic cycle. Quartz arenites produced by persistent wave or current reworking were deposited on stable cratons and passive margins. They are well developed in the Cambro-Ordovician of the Iapetus continental margins (northwest Europe and eastern USA) and the Cordilleran shelf of western USA. Many aeolian sandstones also have very high quartz contents.

Quartz arenites also can be formed *in situ* by extreme chemical weathering. The ganisters of Carboniferous coal measures successions are palaeosoils formed through leaching of sandy sediment by organic acids descending from peat, which of course later forms coal (Percival, 1986). *Silcretes* are another type of quartzose soil (see Section 9.5).

2.7.2 Arkoses

Rocks generally accepted as arkoses contain more than 25% feldspar, much quartz and some rock fragments. Detrital micas are also present and some fine-grained matrix. The feldspar is chiefly potassium feldspar and much of this is microcline. The feldspar usually is fresh, although some may be altered to kaolinite and sericite. Polycrystalline quartz and quartz/feldspar rock fragments are common. Arkoses are typically red or pink, owing to the colour of the

feldspar and also the presence of finely disseminated hematite, as many arkoses occur in red-bed successions (Section 2.9.6). Arkoses are derived from granites and gneisses and vary from *in situ* weathering products, which have experienced little movement, to stratified and cross-bedded sandstones, where there has been substantial sediment transport. The texture of an arkose is typically poorly sorted to well sorted, with very angular to subrounded grains, the precise texture dependent on the degree of reworking. Grain-supported arkoses are cemented by calcite or quartz, whereas others are cemented by a matrix, usually containing much kaolinite. Some feldspars will have syntaxial overgrowths. The chemistry of arkoses is fairly uniform (Table 2.10). They are rich in Al_2O_3 and K_2O and there is an excess of Fe_2O_3 over FeO. Arkoses are illustrated in Plates 1d & 3c,d.

Arkoses are clearly derived from feldspar-rich rocks, particularly K-feldspar-bearing granites and gneisses. However, apart from an appropriate provenance geology, climate and source-area relief are also important factors. Under humid conditions, feldspars weather to clay minerals, so that semi-arid and glacial climates favour arkose formation. If erosion is very rapid, however, particularly the case where the source area has a high relief, then arkosic detritus can be produced in spite of intense chemical weathering. Many arkoses were deposited in fluvial environments. The Torridonian Sandstone of the late Precambrian in Northwest Scotland is a classic arkose, derived from the Lewisian Gneiss. Arkoses also occur in the Old Red Sandstone (Devonian) in Scotland and in the Triassic fluvial facies of the Newark Group of eastern USA.

Table 2.10 Average chemical composition of the three common sandstone types: quartz arenite, arkose and greywacke (from Pettijohn *et al.* 1987)

	Quartz arenite	Arkose	Greywacke
SiO_2	95.4	77.1	66.7
Al_2O_3	1.1	8.7	13.5
Fe_2O_3	0.4	1.5	1.6
FeO	0.2	0.7	3.5
MgO	0.1	0.5	2.1
CaO	1.6	2.7	2.5
Na_2O	0.1	1.5	2.9
K_2O	0.2	2.8	2.0
TiO_2	0.2	0.3	0.6

2.7.3 Litharenites

These sandstones are characterized by a rock-fragment content that is in excess of feldspar. They range widely in composition, both in terms of grain types and chemistry, depending largely on the types of rock fragment present. These are chiefly fragments of mudrock and their low-grade metamorphic equivalents, and volcanic grains; other components are flakes of mica, some feldspar and much quartz of course. There is little primary matrix; otherwise they are similar to greywackes in composition and in fact they have been referred to as subgreywackes. Cements usually are either calcite or quartz, and authigenic clays are common. Although they have a variable chemistry, litharenites generally possess a high Al_2O_3 content from the dominant clay-/mica-rich rock fragments, and low Na_2O and MgO. Litharenites are illustrated in Plates 1a & 3b.

Litharenites account for some 20–25% of all sandstones. Their immature composition implies high rates of sediment production from supracrustal sources followed by short to moderate transport distances. Many fluvial and deltaic sandstones are litharenites, such as the Devonian (Old Red Sandstone) and Carboniferous sandstones of Britain and eastern USA, derived from uplift and erosion of folded Lower Palaeozoic strata. Many of the Tertiary sandstones of the US Gulf Coast, such as the Frio and Wilcox, are litharenites, as are Jurassic–Cretaceous–Tertiary sandstones of the US Western Interior, derived from the Cordillera. Lithic sandstones are common in the Alpine Molasse Basin, a foreland basin formed by thrust loading after continental collision. The nature of rock fragments in litharenites may change through time reflecting uplift in the source area, and the availability of different rock types (see Sections 2.5.1 and 2.8).

2.7.4 Greywackes

The characteristic feature of greywackes is the fine-grained matrix, which consists of an intergrowth of chlorite, sericite and silt-sized grains of quartz and feldspar. Of the sand fraction, quartz dominates over rock fragments and feldspar. Many different rock fragments are usually present, but in many cases, fine-grained sedimentary and metasedimentary rock types dominate. Igneous rock fragments are common in some greywackes, especially grains of more acidic and andesitic extrusives. Feldspar grains are chiefly sodic plagioclase and these usually are fresh in appearance. As indicated by the name, greywackes are dark grey or black rocks, usually well indurated; they can look like dolerite. Greywackes are illustrated in Plates 2b & 4a,b.

The origin of the matrix has been referred to as the 'greywacke problem'. There are two possibilities: fine-grained sediment deposited along with the sand fraction, and diagenetic alteration of unstable lithic grains to produce a pseudomatrix. The evidence cited against a primary origin is the fact that modern, deep-sea turbidite sands do not contain much mud and that during transportation and deposition by turbidity currents, one would expect a cleaner separation of mud and sand. Evidence for a diagenetic origin of the matrix through replacement of the labile components is provided where there are early calcite cements and then rock fragments are well preserved. The consensus of opinion is that much of the matrix is of diagenetic origin (i.e. a pseudomatrix), but some part is still likely to be fine-grained detritus.

On the whole, greywackes have a uniform chemical composition, which contrasts with that of arkoses (Table 2.10). Greywackes have high contents of Al_2O_3, total Fe ($FeO + Fe_2O_3$), MgO and Na_2O. The high MgO and FeO values are reflections of the chloritic matrix and the high Na_2O results from the dominant albitic plagioclase. However, albitization of more intermediate plagioclases and K-feldspar does occur (see Section 2.9.4). Greywackes differ from arkoses in the dominance of FeO over Fe_2O_3, MgO over CaO and Na_2O over K_2O.

Many greywackes were deposited by turbidity currents in basins of various types, usually off continental margins, in back-arc and fore-arc basins, and in association with volcanics. These greywackes can show all the typical turbidite features (Section 2.11.7). Classic greywackes occur in the Lower Palaeozoic of Wales, Ireland and the Southern Uplands of Scotland, and in the Devonian–Carboniferous of Southwest England and Germany. Greywacke sandstones also form many of the Cretaceous–Tertiary flysch successions in the Alps. In North America, they occur in the Mesozoic of California (Franciscan Formation), Washington and Alaska, and in the Palaeozoic of Newfoundland and the Appalachians (e.g. Bock et al., 1998). Although many greywackes are uniform in composition, in detail they do vary, especially in the nature of the rock fragments, so that distinct *petrofacies* often can be recognized. These variations can have important implications with regard to the tectonic situation at the time of deposition, because many greywackes were synorogenic. Many were deposited during periods of active vertical and horizontal plate movements, often coupled with island-arc volcanism.

2.8 Sandstone composition, provenance and tectonic setting

In recent years, there has been much effort to relate the detrital composition of a sandstone to the tectonic setting of its provenance region. The detrital modes of both modern and ancient sands have been used in these studies, which often have been supplemented with chemical analyses of grains, including age-dating of zircons and rock fragments (see papers in Johnsson & Basu, 1993). In a simple quartz–feldspar–lithics plot of modern deep-sea sands, Yerino & Maynard (1984)

showed that the five major tectonic settings could be distinguished, but with much overlap (Fig. 2.51). In the work of Dickinson (1985) on ancient sands, four major provenance terranes were distinguished: stable craton, basement uplift, magmatic arc and recycled orogen. Stable cratons and basement uplifts form the continental blocks, i.e. tectonically consolidated regions of amalgamated ancient orogenic belts, which have been eroded to deep levels. Magmatic arcs include the continental and island arcs associated with subduction, and these are areas of volcanics, plutonic rocks and metamorphosed sediments. Recycled orogens are uplifted and deformed supracrustal rocks, which form mountain belts, and they mostly consist of sediments, but include volcanics and metasediments. Detritus from the various provenance terranes generally has a particular composition and the debris is deposited in associated sedimentary basins, which occur in a limited number of plate-tectonic settings (Table 2.11).

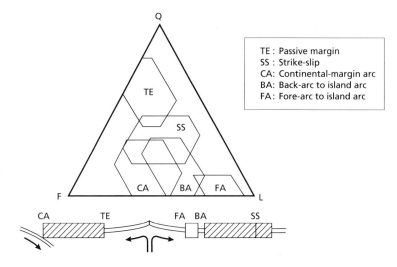

Fig. 2.51 Composition of modern deep-sea sands from trailing-edge (TE, also called passive margin), strike-slip (SS), continental-margin arc (CA), back-arc to island-arc (BA) and fore-arc to island-arc (FA) tectonic settings. After Yerino & Maynard (1984).

Table 2.11 The major provenance terranes, their tectonic setting and typical sand compositions. After Dickinson (1985)

Provenance type	Tectonic setting	Derivative sand composition
Stable craton	continental interior or passive margin	quartzose sands (Qt-rich) with high Qm/Qp and Fk/Fp ratios
Basement uplift	rift shoulder or transform rupture	quartzo-feldspathic (Qm–F) sands low in Lt with Qm/F and Fk/Fp ratios similar to bedrock
Magmatic arc	island arc or continental arc	feldspatho-lithic (F–L) volcaniclastic sands with high P/K and Lv/Ls ratios grading to quartzo-feldspathic (Qm–F) batholith-derived sands
Recycled orogen	subduction complex or fold–thrust belt	quartzo-lithic (Qt–Lt) sands low in F and Lv with variable Qm/Qp and Qp/Ls ratios

From a modal analysis of a sandstone, the percentages of various combinations of grains are plotted on triangular diagrams, and these are used to differentiate the different provenance terranes (see Fig. 2.52). The categories of grain determined (Qt, Qm, Qp; F, Fp, Fk; L, Lv, Ls, Lt) are shown in Table 2.9. A triangular plot of Qt–F–L takes all the quartz grains together (Qm + Qp) and so places emphasis on the maturity of the sediment. Plots of Qm–F–Lt include Qp with the lithic grains and so give weight to the source rock. Plots of Qp–Lv–Ls consider just the rock fragments and those of Qm–Fp–Fk involve only the single mineral grains (these last two not shown here). Care must be exercised where there is more than 10% pseudomatrix (Sections 2.5.4 and 2.7.4) in the sandstone (Cox & Lowe, 1996). The use of these diagrams allows sandstones from the four major terranes to be discriminated (Fig. 2.52).

Stable cratons of low relief generally produce quartzose sands from the granite–gneiss basement and recycling of earlier sedimentary strata. They are deposited on the cratons or transported to passive continental margins. Basement uplifts are areas of high relief along rifts and strike-slip zones, and the dominantly quartzo-feldspathic, lithic-poor sands are deposited in extensional and pull-apart basins.

Magmatic arcs produce sands with high contents of volcanic rock fragments, and as they are dissected down to their more plutonic roots, quartzo-feldspathic debris is generated. A volcanic to plutonic trend may thus result. The sands are deposited in fore-arc and interarc basins. The volcanic grains commonly will have andesitic compositions and usually they are microlitic. After diagenesis, greywacke-type sandstones may be produced.

Detritus derived from the recycling of orogenic belts is very varied in composition, reflecting the different types of orogen (broadly either continent–continent or continent–ocean collision). Sediment from a recycled orogen may fill adjacent foreland basins and remnant ocean basins or be transported in major river systems to more distant basins in unrelated tectonic settings. Lithic grains dominate in many recycled-orogen sandstones, and in those derived from continental collision mountain belts (such as the Alps and Himalayas), quartz plus sedimentary rock fragments dominate, and then the metamorphosed equivalents of the latter as deeper levels of the orogen are uplifted. These sands thus tend to be more quartzo-lithic, with few feldspar and volcanic grains (a high Ls/Lv ratio). Detritus from an uplifted subduction complex in a continent–ocean

Fig. 2.52 Triangular diagrams showing average compositions of sand derived from different provenance terranes. After Dickinson (1985), based on several sources.

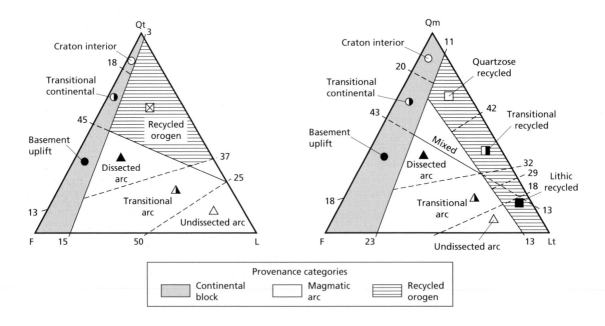

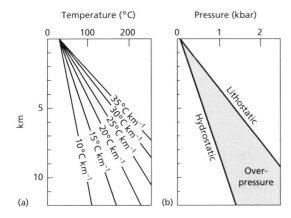

Fig. 2.53 (a) Increase in temperature with increasing depth for different geothermal gradients. (b) Increase in hydrostatic and lithostatic (overburden) pressure with increasing depth.

orogen, by way of contrast, will have a high igneous rock-fragment content, as well as fine-grained sedimentary rock fragments such as chert. Feldspars will be more abundant too.

Studies of sandstone petrofacies within a basin can be used to unravel the geological history of the provenance terrane. Examples have already been given where uplift in a source area reveals deeper levels to erosion, so that the composition of the detritus gradually changes. One well-documented study is of the petrofacies of the Cretaceous sandstones filling the fore-arc basin in the Great Valley of California, derived from uplift of the magmatic arc of the Sierra Nevada (Ingersoll, 1983). The sandstones are more quartzo-feldspathic and less lithic upwards, and potash feldspar increases relative to plagioclase, as volcanics in the arc were eroded and then more plutonic rocks were exposed. Studies of petrofacies and palaeocurrents in Neogene sediments in Alaska showed that there was a complete reversal of the drainage system and that a change in palaeoclimate had an effect on sandstone composition too (Ridgway *et al.*, 1999). Also see papers in Zuffa (1985), Morton *et al.* (1991), Johnsson & Basu (1993) and Bahlburg & Floyd (1999). Zuffa *et al.* (1995) considered arenite composition in a sequence stratigraphic context and the role of relative sea-level change.

2.9 Sandstone diagenesis

As noted in Chapter 1, diagenesis has been divided into

two broad stages: early diagenesis, for processes taking place from deposition and into the shallow-burial realm, and late diagenesis for those processes affecting the sediments at deeper levels, and on uplift. The terms eogenesis, mesogenesis and telogenesis have been used for early and burial diagenesis and diagenesis-on-uplift, respectively.

Two important diagenetic processes that are dependent largely on depth of burial are compaction and pressure dissolution. Chemical processes of diagenesis include the precipitation of minerals, leading to the cementation of the sediments, the dissolution of unstable grains, and the replacement of grains by other minerals. Chemical processes take place in the medium of water, so that salinity, pH (a measure of the hydrogen ion concentration) and Eh (redox potential) and the ability of water to move through the sediment (dependent on the porosity and permeability) are of critical importance to diagenesis. In the early stages of diagenesis, lasting for some 1000–1 000 000 years and affecting sediments to depths of around 1–100 m, pore waters are related to the depositional environment: seawater in the marine environment, and fresh meteoric waters in the majority of continental sedimentary environments. In many cases these pore waters are soon modified by the breakdown of organic matter and microbial activity. With many marine sediments, for example, the initial stages of diagenesis take place in oxidizing pore waters, which with increasing depth below the sediment–water interface become reducing as oxygen is used up in bacterial processes.

During deep burial, pore waters are modified further by reactions with clay minerals, dissolution of unstable grains, precipitation of authigenic minerals and mixing with waters from other sources. Burial diagenesis operates over many tens of millions of years and affects sediments to depths of around 10 000 m, where temperatures are in the region of 100–200°C (Fig. 2.53). Beyond this, processes of burial metamorphism take over. In general, pore waters in deeply buried sediments are saline, neutral and alkaline. In comparison with sea water, these formation waters, as they are called, have lower Na^+, Mg^{2+}, SO_4^{2-} and K^+ values relative to chlorine, but higher Ca^{2+}, Sr^{2+} and silica.

Processes taking place in sedimentary rocks on uplift typically involve fresh groundwaters with low Eh and acid pH. The extent of telogenetic processes depends largely on the porosity and permeability, which may well have been largely occluded during burial diagenesis.

There are many factors affecting sandstone diagenesis. The depositional environment, climate, composition and texture of the sediment are initial controls and then pore-fluid migrations, the burial history and other factors affect the course of diagenesis. The principal processes are compaction and pressure dissolution, silica and calcite cementation, clay-mineral and feldspar authigenesis, and the formation of hematite coatings and impregnations. Reviews of sandstone diagenesis are contained in Burley *et al.* (1985) and Morad *et al.* (2000).

2.9.1 Compaction and pressure dissolution

In the initial stages, compaction involves dewatering and a closer packing of grains. When initially deposited, well-sorted and rounded grains may have a loose, cubic packing with more than 50% porosity (see Fig. 2.8). Compaction quickly reduces the porosity considerably as packing becomes more rhombohedral. Further compaction through overburden pressure results in local fracturing and bending of weak grains (as in Plates 2a, 3b & 5c). Soft grains, such as muddy lithic grains, may be squashed and deformed at this stage and then lose their identity to produce a pseudomatrix. The *grain fraction* (see Section 2.2.4) gives an indication of the extent of compaction. Near-surface, it is of the order of 50–60%, but it reduces to 90% or more in many sandstones with burial. Sandstones of the Alaska North Slope show a 25% increase in grain fraction by 800 m of burial through simple grain rearrangement, and then a slower increase to 90% with further burial to 4 km through plastic deformation (Smosna, 1989). The near regular change in grain fraction (and porosity, see Fig. 2.60) with increasing depth of burial means that the degree of compaction can be represented by mathematical equations and so be predicted (this is the concept of 'compaction law'; see Smosna, 1989).

Of especial importance during burial is the dissolution of grains at points of contact. This produces sutured contacts between the grains if they are of a similar solubility/hardness, and concavo-convex contacts where one grain dissolves preferentially (see Fig. 2.8). Pressure dissolution at grain contacts is minimal where the sediment is cemented early, before deep burial, or where there is much matrix, because in these cases the load is spread and the contact pressure reduced. Pressure dissolution can take place once a rock is fully cemented to produce irregular or sutured planes, known as *stylolites*. These cross-cut grains and cements, and a thin layer of insoluble material is concentrated along the stylolites (for a stylolite in dolomite see Plate 13a). The importance of grain-contact pressure dissolution and stylolitization is that it could be the process by which material is supplied for cementation of the sediment. Sutured contacts and stylolites are prominent in some quartz arenites, where it appears that pressure dissolution is more intense in finer-grained sandstones and those with illite rim cements (Houseknecht, 1988).

2.9.2 Silica cementation

One of the most common types of silica cement is the quartz overgrowth (see Plates 2c,d & 4c,d). Silica cement is precipitated around the quartz grain and in optical continuity with it, so that the grain and cement extinguish together under crossed polarizers. In many cases the shape of the original grain is delineated by a thin iron oxide or clay coating between the overgrowth and the grain (a dust-line). A thicker clay rim around the quartz grain, however, inhibits precipitation of a syntaxial overgrowth. In some cases the boundary between the grain and overgrowth cement cannot be discerned with the light microscope (i.e. there are no dust-lines); the quartz grains then appear angular and tightly interlocking. The use of cathodoluminescence (Section 1.3.2) will often bring out the difference between grain and overgrowth.

The origin of the silica for this cementation frequently has been attributed to pressure dissolution. Pore solutions become enriched in silica, which is then reprecipitated as overgrowths when supersaturation is achieved. Silica-rich solutions may migrate significant distances from the sites of pressure dissolution. Other possible sources of quartz are dissolution of silica dust, other silicates and biogenic silica, and ground water. Silica dust could be derived from grain abrasion, especially if the rock is an aeolian sandstone. Dissolution of feldspars, amphiboles and pyroxenes would provide silica, as would the mineral transformations montmorillonite to illite and feldspar to kaolinite. In marine sediments, pore waters often contain significant concentrations of silica derived from the dissolution of diatoms, radiolaria and sponge spicules. These siliceous skeletons are composed of metastable, amorphous opaline silica, which has a higher solubility than quartz (see Section 9.3 and Fig. 9.8). Because of this solubility difference and the increased solubility of opaline silica with increasing depth of burial, biogenic silica could be an important source. Groundwater can become supersaturated with respect to quartz and if this could be moved through a quartz sand in sufficient volume, then cementation by overgrowth development could take place. Calculations from known rates of groundwater movement indicate that a fully cemented sandstone could be produced in some 200 Myr. Studies of fluid inclusions in quartz overgrowths (e.g. Walderhaug, 1994; Primmer *et al.*, 1997) generally indicate precipitation in the range 75° to 150°C.

Apart from quartz overgrowths, silica is also present as a cement in the form of microquartz, megaquartz, chalcedonic quartz and opaline silica (see Section 9.2). In some cases, the presence of opal is related to the decomposition of volcanic particles. Silica is precipitated in some soils to form silcretes (see Section 9.5).

One important feature arising from the early quartz cementation of sandstones is that they are then able to withstand better the effects of compaction and pressure dissolution during later burial. In this way a moderate porosity can be preserved, which may be filled later with oil or gas. Papers on quartz cementation in sandstones are contained in Worden & Morad (2000).

2.9.3 Carbonate cementation

Calcite is one of the most common cements in sandstones, but other carbonate cements of more local importance are dolomite and siderite. The cement may vary from an even-to-patchy distribution, to local segregations and concretions. The latter have been called *doggers*. The two main types of calcite cement are poikilotopic crystals and drusy calcite spar. *Poikilotopic* crystals are large single crystals, up to several centimetres across, which envelop many sand grains (see Plate 5a,b). Drusy calcite mosaics consist of equant crystals, which fill the pores between grains and typically show an increase in crystal size towards the centre of the original cavity (see Section 4.7.2).

As a result of calcite precipitation there commonly is a displacement of grains so that they appear to 'float' in the cement. Calcite also may be precipitated in cracks in grains and so force them to split (as in Plate 5b). This is common with micas, but also occurs with quartz and feldspar grains. Apart from filling pores, calcite and the other carbonates may also replace grains. Quartz grains cemented by calcite commonly are corroded and etched at their margins. Feldspar grains can be replaced, with incipient alteration taking place along twin and cleavage planes.

Calcite cements are common in grain-supported sandstones, such as quartz arenites, arkoses and litharenites. Calcite is commonly the first cement. In fact, it is being precipitated in some modern river and desert sands and soils. *Calcrete* is a specific type of calcareous soil that can form in sands (see Section 4.10.1). The early precipitation of calcite inhibits later quartz overgrowth formation and feldspar alteration to clays but it can result in total loss of porosity and permeability. In other sandstones, calcite is a later precipitate, post-dating quartz overgrowths and authigenic kaolinite. Precipitation of $CaCO_3$, taking place when the solubility product is exceeded, often occurs through an increase in the activity of the carbonate ion. In the very shallow subsurface, this may happen through evaporation of vadose or near-surface, phreatic groundwater. Organic-matter-degradation reactions also may be involved. At depths, carbonate precipitation can be brought about by an increase in the pH and/or temperature. The source of the $CaCO_3$ may be the pore water itself, but in marine sandstones, much is probably derived from dissolution of carbonate skeletal grains. Morad (1998a) has reviewed carbonate cements in sandstones and papers on this topic are included in Morad (1998b).

Dolomite cements vary from pore-filling microcrystalline rhombs to coarse anhedral mosaics and large poikilotopic crystals. Perhaps the most common are well-formed millimetre-sized rhombohedra. They commonly are iron-rich (ferroan), indicating precipitation in reducing conditions. Early dolomite precipitation may be related to near-surface evaporation. Magnesium for later dolomite precipitation may be derived from clays or dissolution of magnesium-rich silicates (see Section 4.8). Siderite cements ($FeCO_3$) do occur in some sandstones, typically as microcrystalline and fibroradiate mosaics (see Section 6.3 for conditions of formation of siderite). Ankerite is common in some deeply buried reservoir sandstones in the North Sea and formed at the time of hydrocarbon maturation and clay-mineral transformation in adjacent mudrocks (Hendry *et al.*, 2000).

2.9.4 Feldspar authigenesis

Although in many sandstones feldspars are altered to clay minerals or replaced by calcite, in some instances feldspar overgrowths do form on detrital feldspar grains. They are most common on potash feldspars, but they also occur on detrital albite grains. For authigenic feldspar, alkaline pore waters rich in Na^+ or K^+, Al^{3+} and Si^{4+} are necessary. These elements are derived largely from hydrolysis and dissolution of less stable grains within the sediment, especially volcanic grains (e.g. De Ros *et al.*, 1994). Authigenic K-feldspars characterize the Cambro-Ordovician sandstones of the Iapetus shelf regions and in this case the K^+, Si^{4+} and Al^{3+} ions are thought to have been supplied by decomposition of tephra (volcanic ash). In the deeper burial environment, albitization of detrital plagioclase and K-feldspars takes place, with the required Na^+ probably coming from the breakdown of smectite to illite (Saigal *et al.*, 1988).

2.9.5 Clay mineral and zeolite authigenesis

The precipitation of clay minerals and zeolites in a sandstone is very significant as it can have a great effect on its permeability and porosity, and this may seriously reduce its reservoir potential. Clay may also filter into sandstone, carried down by pore waters from muddy interbeds. Extensive infiltration drastically alters the texture of the sediment and decreases the original textural and compositional maturity.

Illite and kaolinite are the most common authigenic clays in sandstones, but montmorillonite, mixed-layer illite–montmorillonite and mixed-layer montmorillonite–chlorite also occur. Authigenic clay minerals occur as pore-filling cements and clay rims up to $50\,\mu m$ thick around grains. The attenuation and absence of rims near and at grain contacts demonstrates their diagenetic origin. The precipitation of clay rims usually is an early or the first diagenetic event, often pre-dating quartz overgrowths or calcite cementation. In some Cenozoic desert sediments this clay coating mostly consists of mixed-layer illite–montmorillonite, thought to have been precipitated from solutions that leached unstable minerals in the shallow-burial environment. The clay rim can become impregnated with

hematite (next section) or altered to other clays during diagenesis. Where clay rims are thick, they may inhibit later cementation and so preserve porosity. Illite in clay rims shows a variety of growth forms including flakes, fibres and whiskers (Fig. 2.54).

Within pores between grains, authigenic kaolinite characteristically forms 'books' or 'concertinas' of stacked pseudo-hexagonal plates (Fig. 2.55 and Plates

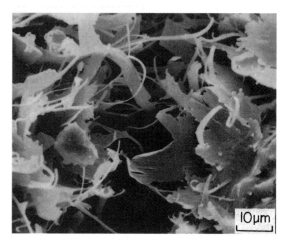

Fig. 2.54 Scanning electron micrograph of authigenic illite in the form of radially arranged flakes and whiskers growing into pore space between two sand grains (left and right of picture). Rotliegend desert sandstone, Lower Permian. Northern Germany.

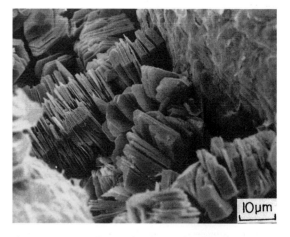

Fig. 2.55 Scanning electron micrograph of authigenic kaolinite, consisting of stacked pseudohexagonal platy crystals, between rounded sand grains. Rotliegend desert sandstone. Lower Permian. Northern Germany.

2a & 5c). At depth, alkaline pore fluids and higher temperatures result in the replacement of kaolinite by illite, as has happened in the Permian Rotliegendes of the North Sea (Lanson *et al.*, 1996). The typical kaolinite 'book' texture may be retained (pseudomorphed) by the illite.

For clay mineral authigenesis, neutral to alkaline pore fluids are required for illite, together with sufficient K^+, Si^{4+} and Al^{3+} (see Fig. 2.58, below). Kaolinite requires more acid pore waters and low K^+ (Fig. 2.58) and these can be produced by flushing of the sandstone by fresh water, either during an early burial stage if the sediments are continental, or if marine, during uplift after a burial phase. Many North Sea reservoir sandstones contain early kaolinite cement derived from meteoric flushing on uplift (Bjørlykke & Aagaard, 1992). Kaolinite precipitation within marine sediments also may result from decomposing organic matter setting up a low pH. The ions for kaolinite and illite precipitation are derived largely from the alteration of labile detrital minerals, in particular feldspars. Chlorite precipitates from more alkaline pore fluids containing little K^+ but sufficient Mg^{2+} (Fig. 2.58).

The type of clay mineral present is important for the reservoir potential of a sandstone (see papers in Houseknecht & Pittman, 1992). The pore-filling kaolinite reduces the porosity of a sandstone, but has little effect on the permeability, whereas the pore-lining illite reduces the permeability considerably by blocking pore throats, but it has little effect on porosity. This relationship is well shown in a poroperm plot for the gas-bearing lower Permian Rotliegendes (aeolian sandstones) of the southern North Sea (Fig. 2.56). Illite is a common late diagenetic cement in the Jurassic Brent Sands of the North Sea, where it probably precipitated from warm formation waters expelled out of hydrocarbon source rocks (see Glasmann *et al.*, 1989). Chlorite generally is precipitated at higher temperatures, as in the Spiro Sandstone gas reservoir in Oklahoma (Spötl *et al.*, 1994).

Apart from clay-mineral precipitation, much replacement of silicate minerals by clay occurs. The replacement may be irregular, peripheral or along fracture and cleavage planes. Feldspar grains and igneous and metamorphic rock fragments are commonly partially to completely replaced by clays. Detrital clays themselves may recrystallize and be replaced by other clays. The replacing clay minerals are commonly mixed-layer illite–montmorillonite and kaolin-

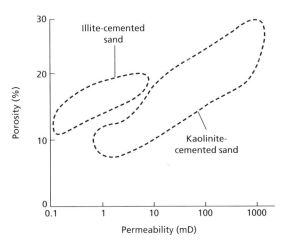

Fig. 2.56 Porosity–permeability plot for kaolinite- and illite-cemented aeolian sandstones in the Permian Rotliegendes. Southern North Sea. After Stalder (1973).

ite, although in time these are replaced by illite and chlorite (Section 3.6). One important effect of this replacement is the increase in the amount of interstitial matrix. This considerably affects the textural maturity of the sediment and gives rise to the greywacke-type texture (Section 2.7.4). The clay rarely preserves the shape of the grains it has replaced because compaction as a result of overburden pressure causes the clay to become squeezed between more rigid grains.

It is virtually impossible to identify detrital clay minerals in a thin-section of a muddy sediment, but clay cements in a sandstone can be recognized relatively easily. Kaolinite has low, first-order interference colours (shades of grey), a pore-filling habit and book texture (as in Plates 2a & 5c). Illite has a higher birefringence and normally is pore-lining.

Sandstones with a high content of volcanic rock fragments and feldspars, such as those derived from island arcs and continental arcs, are subject to much alteration during diagenesis, and apart from clay minerals, *zeolites* can be produced in abundance. Clinoptilolite is the typical low-temperature zeolite and this is systematically converted to laumontite at higher temperatures ($>100°C$)(Primmer *et al.*, 1997). The pore-filling laumontite is particularly common in Tertiary sandstones of fore-arc and back-arc basins in the northeast Pacific and California. Chlorite and smectite are usually associated. These minerals come from alteration of glass and feldspars, and non-ferroan calcite is a common burial cement in these rocks too. Pore-filling laumontite virtually destroys any reservoir potential of a sandstone so that it is often regarded as 'economic basement' when encountered in hydrocarbon exploration.

2.9.6 Hematite cementation and pigmentation: red beds

Many terrigenous clastic sediments are coloured red through the presence of hematite. In many cases these rocks were deposited in continental environments (deserts, rivers, floodplains, alluvial fans, etc.) and the term 'red beds' has been applied to them. The hematite typically occurs as a very thin coating around grains (see Plates 2c, 3c & 4c), but it also stains red infiltrated or authigenic clay minerals and feldspar. It also develops within biotite cleavage planes and in some cases replaces the biotite. The hematite is chiefly amorphous or consists of micron-size crystals. These features of the hematite, together with the absence of hematite coatings at grain contacts (as in Plate 3c) indicate a diagenetic origin.

There has been much discussion on the source and origin of the hematite pigment in red beds (see the review of Mücke, 1994). One view advocates a detrital origin, that amorphous iron compounds, formed through moist tropical lateritic weathering in upland areas, are transported and deposited along with the sediments and then converted to hematite. Many detrital sand grains have a yellow–brown stain of iron hydroxide, gained in the source area, and this also can age to hematite after deposition. From studies of recent to Tertiary desert (arid climate) sands, an alternative origin involves a purely diagenetic mechanism, whereby the iron is supplied by intrastratal dissolution of detrital silicates such as hornblende, augite, olivine, chlorite, biotite and magnetite. If the diagenetic environment is oxidizing, then the iron is reprecipitated as hematite or rather a hydrated iron-oxide precursor, which converts to hematite on ageing. The length of time involved in the ageing process is of the order of a million years; modern desert alluvium is a yellowish colour and the red colour gradually develops as the sediments become older. Only a small quantity of iron, 0.1%, is sufficient to impart a bright red colour to the sediments. The hematite develops above the water table, as well as below if the ground water is alkaline and oxidizing. If reducing conditions prevail, the iron

is present in the more soluble ferrous state and if incorporated into clays, rather than carried away in solution, it will impart a green colour to the sediments. The chemistry of iron mineral formation is discussed in Section 6.1.

Although in the majority of red beds, an intrastratal–diagenetic origin of the hematite is thought most likely, it is still difficult to exclude a detrital component, especially in matrix-rich sandstones.

Secondary alteration of the red colour takes place where reducing solutions penetrate into the sediments. This may be along more porous horizons or tectonic fractures. The colour of the sedimentary rock is then changed to green, grey or even white, if leaching is intense.

Other diagenetic minerals of only local importance in sandstones are sulphates and sulphides. Gypsum and anhydrite occur as cements in sandstones where there are evaporite beds in the succession; otherwise they are rare. Sulphate cements usually are not preserved in sandstones at outcrop. Celestite ($SrSO_4$) and barite ($BaSO_4$) rarely occur as cements. Pyrite occurs in many sandstones as small cubes, clusters and framboids (see Section 6.5), but it is only an accessory diagenetic mineral. It is most common in marine sandstones, the sulphide coming from the bacterial reduction of sulphate, which is present in sea water. Pyrite is commonly altered to goethite/limonite on surface weathering.

2.9.7 Diagenetic environments and sequences

Within any sandstone formation, the sequence of diagenetic events can be simple, involving only one mineral precipitate, or highly complex, involving many stages of precipitation, replacement and dissolution. Factors controlling the path of diagenesis in sandstones are first the depositional environment and climate, sediment composition and texture, and then later the pore-water chemistry, depth of burial and timing of uplift. Three major categories of near-surface (eogenetic) diagenetic environment can be distinguished, each with different pore-water chemistries and associated processes: the marine, the non-marine hot and humid, and the non-marine semi-arid/arid (see reviews of Burley *et al.*, 1985; Morad *et al.*, 2000).

In the marine environment, most of the sand grains are stable, so that diagenetic reactions begin when marine pore waters are modified and less stable material begins to react. Where there is organic matter, bacterial oxidation releases bicarbonate, which can lead to dissolution of small grains. When oxygen is used up, sulphate-reducing bacteria produce more bicarbonate ions and reduce the pH further by releasing H_2S. Under these circumstances, pyrite may be precipitated if there is Fe^{2+} available (as on detrital clays), and calcite and dolomite may precipitate as local cements. Some dissolution of metastable silicates may lead to clay minerals, and quartz and feldspar overgrowths being precipitated from the modified seawater. In many marine sandstones, the typical order of early diagenetic events is clay authigenesis associated with quartz and feldspar overgrowths, followed by carbonate cement. Also of marine origin and present in some sandstones are the green phyllosilicates glauconite and berthierine (chamosite) (see Section 6.3).

In the hot and humid non-marine, near-surface, eogenetic environment, pore waters usually become acidic through bacterial breakdown of organic matter. However, as there is little K^+, Mg^{2+} and SO_4^{2-} in fresh water, early diagenetic precipitates within quartz-rich sandstones are usually quartz overgrowths and kaolinite, and feldspar is dissolved. Quartz is less soluble at lower pH (see Fig. 9.8), so that extensive precipitation may occur in a near-surface sand to give *silcrete* (see Summerfield, 1983). In sands with mafic grains, dissolution will liberate Fe^{2+} and Mg^{2+}, and siderite and chlorite may precipitate if pore waters are anoxic. Where the sand contains volcanic grains, then K^+, Ca^{2+} and Mg^{2+} may become available for smectite and zeolite authigenesis.

In the hot and arid, non-marine, near-surface environment, pore waters are generally oxidizing and chemical leaching is less extreme. Two major eogenetic processes operating in sands under this climate are the formation of *red beds* and the development of *calcretes*. Red beds are typical for semi-arid regions and many form through the release of iron from mafic minerals during early burial and its precipitation around grains as a hydrated iron oxide, which ages to hematite (see Section 2.9.6). Calcretes are calcareous soils, which have many distinctive textures (see Sections 2.9.3 and 4.10.1). Other processes are feldspar dissolution and reprecipitation as overgrowths, gypsum cementation (e.g. desert roses) and alteration of volcanic grains to zeolites, such as clinoptilolite.

In the mesogenetic (burial) environment, sand-stones are subject to increasing burial and higher overburden pressure and temperature, and pore waters become more saline. Many grains are now unstable and begin to dissolve and some kinetic obstacles to chemical reactions are overcome. With increasing burial, there are changes in clay mineralogy (see Section 3.6), with smectites being altered to illite via mixed-layer clays. This is important for sandstone cementation because it releases SiO_2, Ca^{2+}, Na^+, Fe^{2+} and Mg^{2+}. Pressure dissolution also supplies ions for cementation and authigenesis, although early cementation may inhibit this. Quartz overgrowths, dolomite, ankerite and chlorite can be precipitated and plagioclase may be albitized. Volcanic grains are altered to zeolites, especially laumontite. The generation of hydrocarbons also takes place during deeper burial (1.5–3 km depth, see Section 8.10.2). This can lead to the formation of acidic pore waters, which are able to leach grains and carbonate cements, producing secondary porosity (see papers in Pittman & Lewan, 1995).

Another factor in the deep subsurface is the development of high pore-fluid pressures. Where there are low-permeability sediments in a succession (such as mudrocks), then the pore-fluid pressure in adjacent sandstones may be substantially higher than the hydrostatic pressure for that depth (see Figs 2.53 & 2.57). The maximum possible is the *lithostatic pressure*, where the pore fluid is under a pressure equal to the weight of rock above. Where sandstones are overpressured, then diagenetic reactions are retarded and grain–grain stress is reduced. It is possible for primary porosity to be preserved under these circumstances.

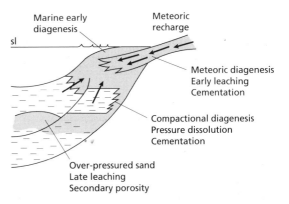

Fig. 2.57 Sketch illustrating main siliciclastic diagenetic environments. After Bjørlykke (1988).

When sandstones are uplifted into the *telogenetic environment*, climate is again important. If semi-arid, then oxidation of sulphides and iron carbonates will lead to the formation of hydrated iron oxides (goethite–limonite), which may age to hematite. In Britain, Carboniferous sandstones immediately beneath Permian and Triassic desert sediments are reddened through this weathering process. Under a more humid climate, leaching of feldspars, carbonates and heavy minerals (see Section 2.5.5) will take place near-surface and this can raise the porosity significantly.

The relative timing of diagenetic events in sandstones is important in terms of the introduction of hydrocarbons. If a sandstone's porosity is occluded by early cementation, then it cannot act as an oil reservoir. Diagenetic processes take place in an aqueous medium so that the influx of oil terminates diagenesis and prevents further reactions. The effects of this are shown in the Jurassic Brent Sand of the North Sea, where kaolinite is preserved in oil-bearing strata but has been converted to illite in areas where the sandstone pores were filled with water.

In studying the diagenesis of sandstones, attempts are now made to reconstruct the larger-scale, basin-wide patterns of pore-water movement through the sediment (see papers in Crossey *et al.*, 1996; Montañez *et al.*, 1997). There are two major processes involved.

1 Recharge from continental upland areas allows meteoric waters to penetrate to great depths in a sedimentary succession; in some instances they even emerge on the sea floor.

2 Expulsion of waters out of compacting mudrocks can lead to much lateral and vertical movement of formation waters (see Fig. 2.57); thrust loading in foreland basins is especially significant.

The early diagenesis of sandstones sometimes can be related to the development of major sequence boundaries and flooding surfaces produced by third-order relative sea-level changes, and the diagenesis as a whole of a sandstone in a sedimentary basin can be tied into the plate-tectonic setting. Sandstone diagenesis in a sequence stratigraphic context is discussed in Ryu & Niem (1999) and Morad *et al.* (2000). As an example, from the Book Cliffs, Utah, Taylor *et al.* (2000) described ferroan dolomite concretions precipitated from meteoric fluids entering marine sediments during a relative sea-level fall, and laterally extensive carbonate-cemented horizons beneath major marine-

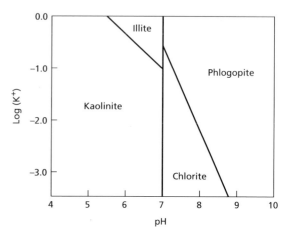

Fig. 2.58 Stability diagram for system $K_2O-MgO-Al_2O_3-SiO_2-H_2O$ at 25°C and 1 atmosphere with FeO, SiO_2 and H_2O in excess, showing the stability fields of kaolinite, chlorite, illite and phlogopite (a mica found in metamorphosed limestones). After Garrels & Christ (1965).

flooding horizons initiated during a pause in deposition during a sea-level rise.

In recent years, studies of sandstone diagenesis have become much more geochemical and concepts of thermodynamics and kinetics have been applied to help understand the complex mineral reactions taking place. Diagenetic phases are now examined routinely for their cathodoluminescence and the scanning electron microscope is used to see clay and grain fabrics. Cements and clays are analysed for their stable isotope compositions, and electron probe and XRD are used to give trace-element and mineral data. Fluid inclusions in cements are studied for the information they give on pore-fluid salinity and temperature of precipitation. The application of thermodynamic stability models allows the mineral assemblages observed to be accounted for and constraints to be placed on the chemistry of the pore fluids. As a simple example, Fig. 2.58 shows the stability fields for kaolinite, illite and chlorite with reference to pore-water pH and potassium content.

2.10 Porosity and permeability

Two important aspects of sedimentary rocks are their porosity and permeability. Porosity is a measure of the pore space and two types are defined:

absolute porosity Pt =

$$\frac{\text{(bulk volume – solid volume)}}{\text{bulk volume}} \times 100$$

effective porosity Pe =

$$\frac{\text{interconnected pore volume}}{\text{bulk volume}} \times 100$$

Absolute porosity refers to the total void space, but as some of this will be within grains, effective porosity is more important. It is the latter that determines the reservoir properties of a rock, together with permeability, the ability of sediment to transmit fluids. Porosity is a basic feature of a sediment or rock, whereas permeability depends on the effective porosity, the shape and size of the pores and pore interconnections (throats), and on the properties of the fluid itself, i.e. capillary force, viscosity and pressure gradient. From Darcy's law, permeability (K) depends on the rate (Q) at which fluid flows through a unit cross-section of rock: $Q = (K/\mu)(\mathrm{d}p/\mathrm{d}l)$, where μ is the fluid viscosity and $\mathrm{d}p/\mathrm{d}l$ is the pressure gradient in the direction of flow. In general, for sandstones, the permeability increases as the porosity increases (e.g. Fig. 2.59).

With porosity, two major types are primary and secondary porosity. Primary porosity is developed as the sediment is deposited and includes inter- and intra-particle porosity. Secondary porosity develops during diagenesis by dissolution and dolomitization (the latter in limestones, Section 4.9), and through tectonic movements producing fractures in the rock.

Primary porosity in sandstones is principally inter-particle porosity, dependent on the textural maturity of the sediment (Section 2.2.5), controlled largely by depositional processes and environments, and to a lesser extent on compositional maturity (Section 2.5.7). In general, the primary porosity increases as the grain size increases, the sediment is better sorted and more loosely packed, the grains become better rounded, and the clay content decreases. The clean, well-sorted, loosely packed sands of beaches and aeolian dunes can have porosities in excess of 50%, and they have high permeabilities too. In contrast to most sandstones, finer-grained sediments, such as siltstones and chalks (see Section 4.9), have high effective porosities, but their permeabilities are usually low, as capillary forces prevent the easy flow of fluids through the small pore throats.

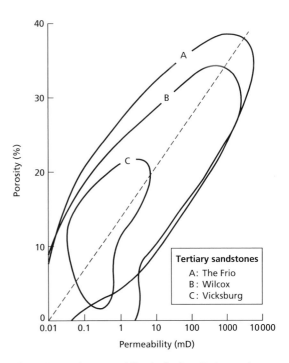

Fig. 2.59 Porosity–permeability plot for three Tertiary sandstones of the Gulf Coast subsurface, the Frio (a), Wilcox (b) and Vicksburg (c), showing the general increase in permeability with increasing porosity. After Loucks *et al.* (1984).

After deposition, most sandstones show a gradual decrease of porosity and permeability with increasing depth. However, the composition of the sand is now the major factor in porosity evolution. Frameworks of contrasting compositions behave very differently during diagenesis, and display different rates of porosity reduction with depth of burial. This effect is shown in Fig. 2.60, based in part on studies of Tertiary sandstones from the Gulf Coast of the USA where porosity reduction for quartzose sandstones is much less than 5% km^{-1}, whereas it is about 5% km^{-1} for feldspathic sandstones and much more than 5% km^{-1} for lithic sandstones. Two major diagenetic processes in porosity reduction are compaction and cementation. Compaction takes place from a few metres below the sediment surface, and results in a closer packing of grains and eventually, at depths of hundreds to thousands of metres, to pressure dissolution and interpenetration of grains (Section 2.9.1). Lithic grains are particularly susceptible to deformation and crushing by increasing overburden. Cementation, however, is

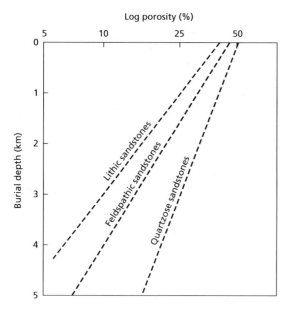

Log porosity (%)

Fig. 2.60 Porosity–depth relationship for sandstones of different composition. After Dickinson (1985), based on several sources.

the principal process of porosity loss in sandstones. Silica, calcite and clay can all be precipitated as cements (Sections 2.9.2–2.9.5), lining and filling pores and decreasing both porosity and permeability. Most non-carbonate petroleum reservoirs occur in sandstones that have been cemented only partially and so retain much of their depositional porosity. The less compositionally mature sandstones show much porosity loss through clay and zeolite replacement of labile grains. The formation of these minerals involves mostly hydration reactions of glass and feldspar, with the result that the new minerals occupy more volume than the original grains and so they fill pores too.

Porosity can be created during burial by the dissolution of grains and cements. The acidic pore waters for doing this can be produced by thermal decarboxylation of organic matter in mudrocks adjacent to sandstones, and in some cases it relates to the generation and migration of hydrocarbons (papers in Pittman & Lewan, 1995). Secondary porosity can be recognized by (a) partially dissolved grains, (b) undissolved clay rims around former grains and (c) oversized pores: i.e. large pores of the size and shape of grains. To observe these features, samples are best impregnated with a resin containing a dye before sectioning. Grains com-

Table 2.12 Factors favouring preservation and reduction of primary porosity, and formation of secondary porosity

Factors favouring the preservation of porosity	Factors contributing to reduction of porosity	Factors favouring the formation of secondary porosity
high primary grain strength	high percentage of ductile grains, e.g. volcanic and basic igneous rocks	early flushing of meteoric water causing leaching of feldspar and carbonates
low content of highly soluble minerals or amorphous silica	high content of more soluble grains, e.g. volcanic glass, biogenic silica and soluble minerals	a high content of humic kerogen in adjacent carbonate-free shale, producing large amounts of CO_2
precipitation of limited early cement, supporting the grain framework	low (hydrostatic) pore pressure (good updip communication of pore water), increasing net stress, mechanical compaction, and pressure dissolution	transformation of kaolinite, smectite and mixed-layered minerals to illite, which may also cause the leaching of feldspar
precipitation of early grain-coating cement, hindering quartz overgrowth		a high pore-fluid pressure may form secondary porosity by hydraulic fracturing
at depth, overpressure, reducing grain-to-grain stress		
early introduction of hydrocarbons, reducing precipitation from pore water		

monly dissolved out are feldspar, mafic minerals, volcanic origin and carbonate (bioclasts), and cements of carbonate, zeolite, clay and sulphate may go too. The factors favouring porosity preservation and destruction are summarized in Table 2.12.

Porosities necessary for good petroleum reservoirs are 20–35%. Reservoir rocks retaining primary intergranular porosity include the gas-bearing Permian Rotliegendes of the southern North Sea, although clay cements reduce this in many horizons (see Fig. 2.56). The oil-bearing Jurassic Brent Sands of the northern North Sea have porosities that are very much controlled by the depositional environment, but secondary porosity is also important. Porosity also is very much facies dependent in the Permian Ivishak Formation of the Prudhoe Bay field of the Alaska North Slope, but again secondary porosity is widespread owing to the burial dissolution of early carbonate cements (Melvin & Knight, 1984). The Upper Cretaceous sandstone reservoirs of the US Western Interior (e.g. the Shannon, Sussex and Terry) also have significant porosity generation at depth (Pittman, 1988). In the much-studied, oil-bearing Tertiary sandstones of the US Gulf Coast, the general porosity decrease with increasing depth is very much controlled by the original sediment composition (as in Fig. 2.60), but grain–cement dissolution at 2–3 km depth has raised the porosity, and so the reservoir potential, at many horizons (Loucks *et al.*, 1984). Feldspar authigenesis is a major process in these sandstones too (Milliken, 1988). Case histories of sandstone hydrocarbon reservoirs are contained in Berg (1986), Tillman & Weber (1987), Barwis *et al.* (1990) and Kupecz *et al.* (1997).

2.11 Depositional environments of sandstones and coarser clastics

The principal depositional sites of the coarser siliciclastic sediments are (a) fluvial environments, (b) deserts, (c) lakes, (d) deltas, (e) marine shorelines — barriers, beaches, tidal flats and estuaries, (f) shallow-marine shelves and epeiric seas, (g) continental margins and deep-water basins, and (h) glacial environments.

2.11.1 Fluvial environments and facies

Fluvial environments are complex systems of erosion, sediment transport and deposition which give rise to a great variety of landforms. At the present time fluvial systems range from alluvial fans, through braided and low-sinuosity stream networks to meandering (high-sinuosity) rivers. Their sediments range from the coarsest conglomerates through sandstones to mudrocks. In general, fluvial sandstones are usually sharp-based and cross-bedded (as in Fig. 2.13), with some flat bedding and cross-lamination. They may be lenticular (the fills of channels) or laterally more persistent (from channel or point-bar migration). Texturally and compositionally, fluvial sandstones are generally immature to mature, although this depends on the sediment provenance, climate and transport distance. Sediment grain size generally decreases downstream, and sorting improves. Many fluvial sandstones are arkoses and litharenites; those derived from reworking of older sandstones are more quartzose. Fluvial conglomerates are typically lenticular, commonly with a crude cross-bedding; many are polymictic, with both extra- and intraformational clasts, and they usually have a pebble-support fabric. Maximum clast-size–bed-thickness relationships can separate different conglomerate types (see Section 2.2.1). For describing fluvial sediments, a scheme of lithofacies codes is frequently used as a shorthand (see Fig. 2.64, below, for the codes and examples of their use).

Fossils are not common in river sediments and consist mostly of plant material and skeletal fragments of freshwater and terrestrial animals. Fluvial sediments deposited under semi-arid climates are usually red in colour from the early diagenetic formation of hematite (Section 2.9.6). Soil horizons are common in fluvial successions: calcretes if a semi-arid climate and low water table (Section 4.10.1), and vertisols with rootlets, siderite nodules and even coals, if humid conditions prevailed and the water table was high (Sections 3.7.1 and 8.7).

There are four basic types of river-channel pattern (Fig. 2.61): meandering, straight, braided and anastomosing. Meandering channels have high sinuosity and a single channel. Straight channels may have side bars and a sinuous thalweg at low discharge. Braided streams have low sinuosity but the channel is divided into subchannels by braid bars at low discharge. Anastomosing streams have several major channels, with high or low sinuosity, that are separated by permanent alluvial islands, usually vegetated. There are many factors that determine the type of stream. Of importance are the sediment grain size, the regional gradient and

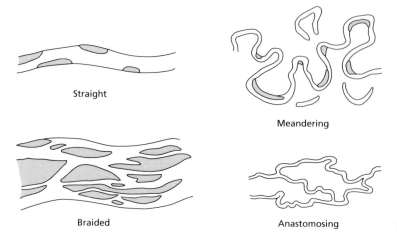

Straight

Meandering

Braided

Anastomosing

Fig. 2.61 Principal river-channel patterns.

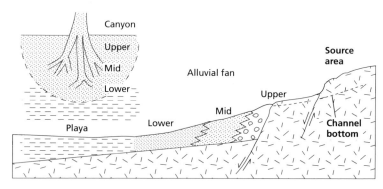

Canyon

Upper

Mid

Lower

Playa

Lower

Mid

Upper

Alluvial fan

Source area

Channel bottom

Fig. 2.62 Schematic facies model for an arid-zone alluvial fan: coarse gravels and sands of stream flood and debris-flow origin in the upper fan pass down to pebbly sands and finer sediments of the mid- to lower fan, deposited by stream and sheet floods, which in turn grade into playa muds.

the climate, which determines discharge and its seasonal variations. With braided and meandering streams, for example, meandering is favoured where the gradients are low, the discharge does not fluctuate much throughout the year, and there is a high suspended sediment–bedload ratio. Climatic changes and tectonic movements, such as source-area uplift, can cause one river type to evolve into another.

Four categories of fluvial system can be distinguished: alluvial fan, braided stream, meandering stream and anastomosing stream.

Alluvial fan facies

Alluvial fans are particularly common in semi-arid regions where there is infrequent heavy rainfall, but they also occur in humid settings. They are aprons of sediment occurring adjacent to upland areas, especially those bounded by faults, with the fan apex located at the mouth of a canyon or wadi (Fig. 2.62). Alluvial fans build out on to playas, lakes, floodplains of more permanent rivers, coastal plains and in some places directly into the sea to form *fan deltas* (see Section 2.11.4). Typical fans in a semi-arid location have a radius of 5–15 km, with their size depending on the area of the catchment basin. The surface of a fan is dissected by a network of channels radiating out from the fan apex. Fans may coalesce at the foot of a mountain front to form an apron (a bajada). Compared with other fluvial sediments, those of alluvial fans are generally coarser, consisting of much gravel and coarse sand, because of the short distance of transport. Fan sediments are compositionally immature, usually having a great range of pebble/grain types, reflecting the source-area geology. Sediment grain size generally decreases markedly down fan. Upper-, mid- and lower-fan facies can be distinguished (Fig. 2.62).

Deposition takes place from debris flows, stream

floods and sheet floods. Debris flows (or mud flows) are high-density, high-viscosity flows consisting of much fine sediment, together with clasts up to boulder size. Debris-flow deposits are laterally extensive beds without erosive bases and with a matrix-support fabric (Section 2.2.4). Stream floods are low-viscosity flows, usually confined to channels. The deposits are cross-bedded, pebbly sands and lenticular gravels with a pebble-support fabric and possibly with imbrication (Section 2.2.4). They occur mostly in the upper- to mid-fan region. Sheet floods, shallow but extensive surface flows, deposit thin, sharp-based, laterally continuous beds of sand and gravel, which may show graded bedding, and planar and cross-stratification. They are more common on the lower parts of fans.

Alluvial fans give rise to wedge-shaped clastic bodies that are commonly associated with fault-bounded basins, such as grabens, half-grabens and pull-apart basins. Vertically, successions may show a broad coarsening and thickening upward indicating active fan progradation through hinterland uplift, or fining and thinning upward through fan retreat (retrogradation). Palaeocurrent patterns from alluvial fans are usually radial.

There are many examples of alluvial-fan facies in the geological record; they occur in the Devonian of northwest Europe and Canada, the Triassic of Scotland and Permo-Pennsylvanian of Colorado. Alluvial fans are not common hydrocarbon reservoirs, generally because potential source rocks are rarely deposited close by. However, they do have good reservoir qualities and there are a number of cases where they contain oil (e.g. the Lower Cretaceous of the Sirte Basin, Libya and the Ordovician of the Oman (Heward, 1989)). Alluvial fans are reviewed in Rachocki & Church (1990) and Blair & McPherson (1994); Blair (1999a,b) has described waterflow-dominated and debris-flow dominated fans from eastern California.

Braided-stream facies

The channels of braided streams are usually broad and shallow, and where sand-grade sediment dominates they are floored by dunes. The sand bars and large dunes that divide the stream into smaller channels are straight-crested to linguoid in plan, and are exposed at moderate to low discharges. Large braid bars are complex structures with smaller dune and ripple fields developed upon them. During high stage, the bars and dunes migrate downstream and the positions of the channels themselves may alter. In gravelly braided streams, the channels consist of flat-bedded imbricated gravels, and tabular bars have low slip-faces at their downstream ends. They usually have thin sheets of sand with ripples and small dunes on their upper surfaces (bar-top sands). Fine-grained floodplain sediments are poorly developed, only occurring locally in abandoned channels. Depending largely on tectonic factors, some braided streams migrate laterally, whereas others aggrade.

Braided-stream deposits are dominated by channel and bar facies with tabular, planar cross-bedding formed from downstream bar and dune (two-dimensional) migration and some trough cross-bedding from three-dimensional dunes (Fig. 2.63). Internal erosion surfaces and reactivation surfaces are common, and channel fills may show some fining upwards of grain size. On a large scale, braided streams give rise to multistorey sand bodies with an elongate to sheet geometry depending on the amount of lateral migration. Mudrock interbeds are absent or of no great thickness. Palaeocurrents are unimodal, generally with a lower dispersion than meandering stream sandstones. The internal structure of modern braided channel bars (and other fluvial deposits) can be revealed by ground penetrating radar (e.g. Bridge *et al.*,

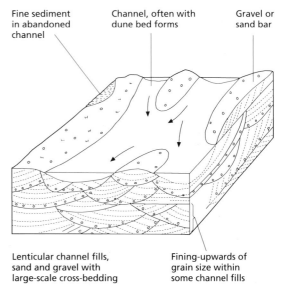

Fine sediment in abandoned channel

Channel, often with dune bed forms

Gravel or sand bar

Lenticular channel fills, sand and gravel with large-scale cross-bedding

Fining-upwards of grain size within some channel fills

Fig. 2.63 Sketch showing part of a braided-stream deposit.

1998), providing useful data for interpreting their ancient equivalents.

Modern (and ancient) braided streams are very variable. Three examples from North American rivers are shown in Fig. 2.64, as variations on the general facies model shown in Fig. 2.63. The Scott type is a very proximal gravelly braided stream with deposits consisting of massive and horizontally bedded gravels in wide, shallow channels. Cross-bedding may occur in the gravels and lenses of sand. Little order is shown by these deposits; no fining-upward units occur, for example. The Bijou Creek type is an ephemeral sandy stream where discharge is very variable. Sheet floods dominate so that the deposits consist of packets of flat-bedded and cross-bedded sand, usually of considerable lateral extent. The South Saskatchewan type is a large, complex, sandy braided stream, with bars showing a variety of bedforms and floodplains developed adjacent to the stream. The sediments are a complex of trough cross-bedded and planar cross-bedded sands. Some fining of sand occurs up through a succession and floodplain sediments may cap the unit. These braided-stream deposits may be confused with meandering-stream deposits (see below) but the former are characterized by more planar cross-bed sets, more irregularities in grain-size trends, and less fine floodplain sediment.

There are numerous descriptions of braided-stream facies in the geological record. They occur in the Old Red (Devonian) and New Red (Triassic) Sandstones of Britain, in the Devonian of eastern North America and in the Triassic of eastern Australia. The Permian Ivishak Formation of Prudhoe Bay (Alaska) is an example of a petroleum reservoir in a braided-stream sand body. In this instance, a southward prograding wedge of marine shales and sandstones, overlain by deltaics, is capped by the oil-bearing braided-stream sandstones. Cretaceous marine shales provide a seal and also may have been a source. Papers on braided streams are contained in Best & Bristow (1993).

Meandering stream facies

Meandering streams possess distinct channel and

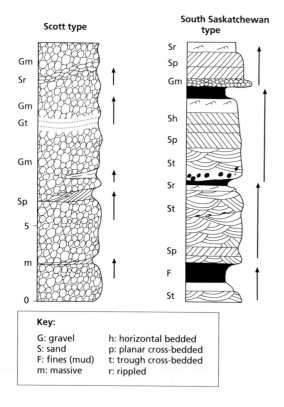

Key:

G: gravel	h: horizontal bedded
S: sand	p: planar cross-bedded
F: fines (mud)	t: trough cross-bedded
m: massive	r: rippled

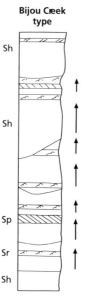

Fig. 2.64 The variation in braided-stream deposits as illustrated by three modern North American rivers: Scott, South Saskatchewan and Bijou Creek. Also on the logs are lithofacies codes for fluvial sediments.

overbank subenvironments (Figs 2.65 & 2.66). The channel itself usually has large dune structures on its floor and these also occur on the lower part of the point bars. They give rise to trough cross-beds. Flat-bedded sands of upper flow regime may be deposited on the point bar, and in the upper part, finer sands with ripples are common, giving rise to cross-laminated sands, with mud drapes and lenses. Lateral migration of a meandering stream through bank erosion and point-bar deposition occurs within a quite narrow zone to generate an alluvial ridge, which may rise several metres above the adjacent floodplain. Levees are developed along parts of the channel bank and consist of fine sands and silts deposited when the river is in high stage. Overbank flooding takes suspended sediment on to the floodplain. Crevasse channels may be cut through the main channel bank so that coarser sediments are taken on to the floodplain as crevasse splays.

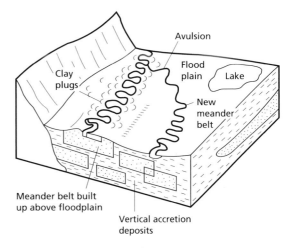

Fig. 2.65 Schematic facies model for a meandering stream system.

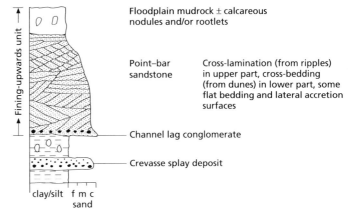

Fig. 2.66 The subenvironments of a meandering stream together with a graphic log of the sedimentary unit produced through lateral migration of such a stream. Fluvial fining-upward units are usually between about 2 and 20 m thick.

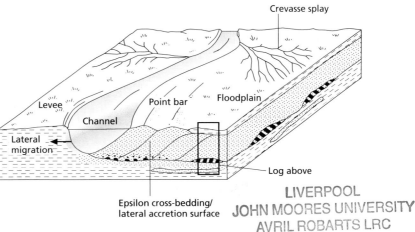

Thin-bedded sandstones interbedded with floodplain silts are produced. Floodplains may be sites of soil formation, marshes and swamps, lakes and salt precipitation, if the climatic conditions are appropriate.

Meandering streams frequently take short cuts so that meander loops are abandoned. These form oxbow lakes that are gradually filled with fine sediments. These lenses of mud help to confine the river to its alluvial ridge, but periodically a river will break out and begin a new alluvial ridge on the floodplain. This process is known as avulsion.

Meandering stream sedimentation generates a fining-upward unit through point-bar migration (Fig. 2.66). Occasional major floods will smooth off the point bar to generate a large-scale, low-angle surface, referred to as a lateral accretion surface, or epsilon cross-bedding. It is a structure which shows that meandering has taken place and is identified by its dip direction, normal to palaeoflow. Meandering stream deposits generally consist of thick floodplain deposits and linear channel point-bar sandstones (Fig. 2.65). Depending on the amount of sand in the system and aggradation rate, these linear sand bodies may be interconnected. The floodplain sediment acts as a permeability barrier and thus may separate potential reservoir sand bodies. Meandering stream deposits typically consist of many fining-upward sandstones overlain by floodplain silts (Fig. 2.66). Examples are well known from the Old Red Sandstone of Britain and the Devonian of the Catskills, New York. Meandering stream successions that are petroleum reservoirs occur in the Wilcox Group of the Tertiary of Texas. The Statfjord Formation of the northern North Sea is a major oil reservoir in fluvial-channel sands.

Anastomosing stream facies

This stream type is unable to migrate much laterally because of vegetated alluvial islands. The channel sediments therefore tend to aggrade and quite thick sand bodies can be generated. Cross-bedded sands are closely associated with finer-grained sediments of the alluvial islands, which may possess soils and peats. Modern anastomosing streams occur in the foothills of the Canadian Rocky Mountains and oil- and gas-bearing Cretaceous equivalents occur in the subsurface of Saskatchewan.

Books and compilations of papers on fluvial facies include Best & Bristow (1993), Fielding (1993), Marzo & Puigdefabregas (1993), North & Prosser (1993), Miall (1996) and Smith & Rogers (1999), and reviews are provided by Collinson (1996) and Miall (1996). Papers on floodplains are contained in Marriott & Alexander (1999). Blum & Tornqvist (2000) reviewed the response of fluvial systems to climate and sea-level change.

2.11.2 Desert environments and aeolian sand facies

Deserts are areas of intense aridity, generally located in the subtropical belts (latitudes of 20–30°), although local topography and degree of continentality affect their development. Apart from areas of wind-blown sand, alluvial fans, ephemeral streams, salt lakes and playas occur in a desert region, and there are vast areas of bare rock. Aeolian desert sands vary from a thin, impersistent cover (sand sheet) to ridges many tens of kilometres long to extensive and thick sand seas. Aeolian sands also accumulate in higher latitudes from deflation of glacial-outwash plains, the so-called 'cold deserts' of periglacial regions. They were developed extensively during the Pleistocene Ice Age and they have been described in association with more ancient glaciations (e.g. in the late Precambrian of Mali; Deynoux *et al.*, 1989).

Desert dune sands are typified by a grain size of fine to coarse sand (0.1–1.0 mm), good sorting and a negative skew (Section 2.2.1). Grain size decreases and sorting improves downwind. Some desert sands, particularly those of interdune areas, have a bimodal grain-size distribution from the preferential removal of the very fine sand fraction, the easiest to be eroded and transported by the wind (see Fig. 2.33). Grains are well rounded (see Plates 2c, 4c & 5a) with a frosted surface, produced by the frequent grain collisions (Section 2.2.3; Fig. 2.7c). The high textural maturity of aeolian sands is matched by a compositional maturity; they are mature to supermature, many being quartz arenites. Many desert sandstones are red through hematite pigmentation (Section 2.9.6) and fossils are absent apart from local vertebrate bones and footprints. Commonly associated with aeolian sandstones are wind-faceted pebbles (ventifacts), waterlain deposits resulting from sheet and stream floods, and desert-lake sediments.

The characteristic sedimentary structure of aeolian sands is large-scale, high-angle cross-bedding, seen in

Fig. 2.35 and described in Section 2.3.2. Thick sandstone successions with this structure to the exclusion of others, and without finer-grained interbeds, are likely to be of aeolian origin. On a larger scale, aeolian sands occur as thick sand blankets, which represent the deposits of sand seas (ergs), and as linear bodies (ridges), which represent seif draas. Erg deposits are well developed in the Jurassic of the western USA, such as the Wingate Sandstone of northeast Arizona, the Entrada of Utah–Colorado and the Navajo of Utah. The Permian Rotliegendes of the southern North Sea is also a sand-sea deposit, but with fluvial and playa intercalations (Sweet, 1999), whereas its lateral equivalent in northeast England, the Yellow Sands, is a seif–draa complex (a series of some 14 parallel sand ridges, up to 60 m high, separated by very thin, gravelly interdune deposits). An extensive lower Cretaceous sand sea in Namibia was buried relatively rapidly by the Etendeka flood basalts, and so has well-preserved aeolian landforms (Jerram *et al.*, 2000; Mountney & Howell, 2000). Aeolian sandstones are usually highly porous (see Fig. 2.56) and so make excellent reservoirs, for water as well as hydrocarbons.

Reviews and compilations of papers on desert sediments include Pye & Tsoar (1990), Brookfield (1992), Cooke *et al.* (1993), North & Prosser (1993), Pye & Lancaster (1993), Nickling (1994), Pye (1994), Kocurek (1996) and Goudie (1999).

2.11.3 Lacustrine environments and sandy facies

Sands and coarser sediments are being deposited along lake shorelines, in deltas where rivers drain into lakes, and on deep lake-basin floors. Coarse siliciclastic sediments are best developed in hydrologically *open lakes* (ones with an outlet and so relatively stable water-level and shoreline), rather than hydrologically *closed lakes*, which tend to be sites of evaporite and limestone deposition. Compared with their marine counterparts, beach sands and gravels of the lake shoreline are generally less well sorted and rounded because the level of wave activity is much less (reduced fetch) and there are no tides. The sandy facies of lake beaches, barriers and spits are very similar to low- to moderate-energy marine coasts. Wave-formed ripples (e.g. Fig. 2.28), polygonal desiccation cracks and syneresis cracks (e.g. Fig. 2.37) are common in finer-grained lake shoreline and nearshore sediments.

Sediment distribution in a lake delta is broadly simi-

lar to that of marine coastal deltas, although the delta sequences produced are usually on a much smaller scale. Many lake deltas are of the Gilbert-type; that is, they have a steep slope (delta front) where avalanching of the mostly coarse sediment produces a single cross-bed set (foreset). Top-set (horizontal) and bottom-set (tangential) facies may be well developed (see Section on fan deltas, 2.11.4, and Fig. 2.69).

On the deep floors of lakes, sediment-laden river underflows cut sublacustrine channels and deposit graded beds with scoured bases, in a similar manner to turbidity currents of marine basins. Finely laminated sediments (rhythmites) are deposited in lake-basin centres from dilute density currents and settling out of clay from suspension (see Section 3.7.2; Figs 2.16 & 3.4). The recognition of siliciclastic lake sediments is dependent mainly on the absence of a marine fauna and an association with certain minerals or rock types that are restricted to, or more common in, lakes (see Sections 3.7.2, 4.10.1 and 5.4.2).

Ancient lacustrine shoreline siliciclastics occur in the Devonian Orcadian lake basin of northeast Scotland and in Permo-Triassic lakes in the Karoo Basin, Natal. Reviews of lake sediments are provided by Talbot & Allen (1996) and compilations of papers are included in Anadon *et al.* (1991). Coarse-grained lacustrine deltaic sediments are described in Nemec & Steel (1988) and Colella & Prior (1990).

2.11.4 Deltaic environments and facies

Deltas are complex environments with characteristics that are determined by the nature of the river system supplying the sediment, coastal processes and climate. Deltas can be divided into several parts (Fig. 2.67). The delta plain or delta top refers to the area landward of the shoreline, and an upper delta plain, dominated by river processes, is distinguished from a lower delta plain where there is some marine influence, mainly in tidal inundation. Subenvironments of the delta plain are the distributary channels, interdistributary bays, floodplains, marshes and swamps. The delta front includes the subaqueous mouth bars and distal bars in front of the distributary channels, and the prodelta in the deeper offshore region.

The upper delta plain is the area where fluvial, lacustrine and swamp sediments occur. The nature of the deposits depends on the type of river present and climate. Both braided and meandering streams can

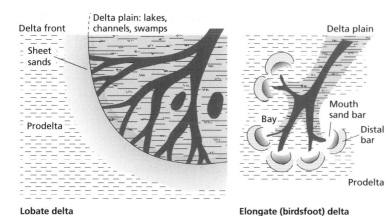

Fig. 2.67 The subenvironments of a lobate and elongate (bird'sfoot) delta. Progradation of a lobate delta gives rise to a laterally extensive delta-front, sheet sand body, whereas a linear sand body is generated by an elongate delta.

occur, although the latter are more common on upper delta plains. Branching of the main river channel may occur to give smaller channels, separated by floodplain.

Also present are shallow lakes mostly filled with fresh water, although it may be hypersaline if the climate is arid. The lakes are sites of mud deposition derived from overbank flooding of the distributary channels. Anoxic sediments may accumulate if the lakes are sufficiently deep for water stratification to develop. Frequently a river will break its banks by crevassing and a small delta will build into the lake. A coarsening-upward unit develops of silt and clay passing up into sand, and this thin, lacustrine deltaic unit would be cut through by the feeder channel. Eventually a lake may be filled and then swamp vegetation will grow there, unless the climate is arid.

On the *lower delta plain*, channels become more numerous as they divide into smaller distributaries. Levees are well developed and interdistributary bays occur at the shoreline between channels. Marshes and swamps are usually extensive, especially in a humid climate. Bays between channels are large in some deltas and these are gradually filled by sediments crevassing from major distributaries. Bay sediments, bioturbated muds with marine or brackish faunas, are overlain by coarser sediments as the bay is gradually filled. Marsh growth over the former bay gives rise to soils and peat. Sediment is now no longer supplied to the area, and through compaction-induced subsidence, marine waters eventually inundate the area to regenerate the shallow-marine interdistributary bay again. The process of bay filling can be repeated, and

this may lead eventually to a sequence of small-scale, coarsening-upward, bay-fill cycles in this lower delta-plain area.

Distributaries may aggrade or meander to deposit sand bodies on the lower delta plain. These will usually consist of sorted sand in a fining-upward unit with cross-bedding. Distributaries may be abandoned by channel migration or avulsion and they then become lakes that are filled by fine sediments and plant debris.

The delta front is the region where the sediment carried by distributaries is deposited. At the distributary channel mouth, the flow expands, mixes with sea water and deposits its bedload, mostly sand, at the mouth bar. Dunes, giving cross-bedding, and ripples, giving cross-lamination and flaser bedding (Fig. 2.27), occur on the mouth bar, but modification takes place if wave action or tidal currents are operative (see later sections). Fine sediment accumulates farther offshore in a distal bar, where fine sand and mud give laminated and lenticular bedding (Fig. 2.27). Bioturbation is common here. Material in suspension (clay) is carried farther offshore still, often as a sediment plume, to be deposited on the delta slope (prodelta). Organic-rich, laminated and bioturbated muds accumulate here. During major floods, silt and sand is carried out to the distal bar and on to the delta slope.

The delta front is the area of progradation. Deposition on the mouth and distal bars results in a seaward building of the delta front so that coarser sediments of the mouth bar come to overlie finer sediments of the distal bar and prodelta. The thick, coarsening-upward unit so produced is the characteristic feature of deltas

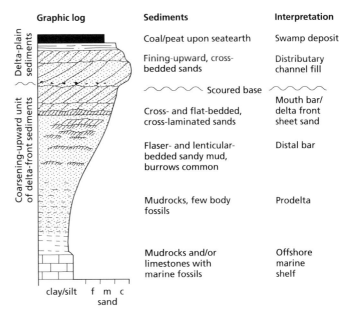

Fig. 2.68 Sketch graphic log and interpretation of typical succession produced by delta progradation. Many variations can occur at the top of the sequence where the delta-plain sediments occur, particularly if interdistributary bays are well developed and then small-scale coarsening-upward units may form. Here the simple case of a delta distributary-channel sand and swamp deposit is shown. The thickness of such a unit varies considerably, but many are between 20 and 40 m.

(Fig. 2.68). The distributary channel itself is also building seawards, and cuts into the mouth bar. Coarse, cross-bedded sands, which may fine upwards, are typical for the channel deposits.

Rapid sedimentation along the delta front and prodelta gives rise to much sediment instability. Debris flows and turbidity currents, slide blocks and slump masses all occur in this area. Prodelta mud may flow under the overburden of mouth-bar/distal-bar sands, and islands can be generated by this diapirism. Turbidity currents generated off deltas may feed fans and ramps in deep water (Section 2.11.7).

Also common in the delta-front/prodelta region are growth faults. These contemporaneous structures are major shear planes, moving continuously with deposition. With time, offsets of marker beds increase with depth, and sediment thickness increases abruptly across the fault. The fault angle decreases with depth, and in plan view it is typically concave away from the delta. Commonly associated with growth faults are rollover folds or reverse-drag structures. Growth faults and associated structures are an important component of some reservoirs in delta facies (e.g. the Niger Delta petroleum province).

Sands and sandstones of delta channels and mouth bars are typically moderately to well sorted, with rounded grains; they have high primary porosities and permeabilities. Prodelta muds and delta-plain muds

and peats can provide permeability barriers. Deltaic sandstones generally are derived from supracrustal rocks so that many are quartzose and lithic arenites. The clay mineralogy of deltaic mudrocks will reflect the source-area geology and climate (see Section 3.5).

Many types of delta have been recognized and they differ considerably in their sand-body geometry. Modern deltas can be subdivided on the strengths of the fluvial, wave and tidal input into three categories.

River-dominated deltas

The modern and ancestral Mississippi delta is the type example of a river-dominated delta. It is located within the Gulf of Mexico, which has a low tidal range and minimal wave action. There are two main subtypes of river-dominated delta, the *elongate* delta, exemplified by the modern Mississippi bird'sfoot delta and the *lobate* delta, typified by the ancestral Mississippi deltas.

Elongate deltas (Fig. 2.67) tend to have few distributaries; one major channel carries the sediment to the delta front where it then divides into several smaller channels. Bay sediments are well developed adjacent to the major channel and crevassing into the bay frequently takes place to produce small bay-fill deltas consisting of a thin coarsening-upward unit. Most sediment is deposited at the mouth bar/distal bar and as a result of seaward progradation, an elongate sand body is produced (a bar-finger or shoe-string sand). This will be several kilometres or more in length and relatively thick (70 m in the modern Mississippi). Cut into this will be the channel itself, with its coarse, fining-upward fill. Elongate deltas tend to form where river discharge is steady throughout the year and suspended-sediment load is high. They also tend to develop where the delta is prograding into relatively deep water.

Lobate deltas (Fig. 2.67) have many distributaries, some of which are inactive at any one time. Inter-distributary bays are absent or only small, and much swamp and marsh occurs between channels, also some lakes. Sand is deposited at the mouths of the small but numerous distributary channels around the delta front, to produce a thin but continuous sheet sand, passing offshore to the silty distal bar and muddy prodelta. Progradation of a lobate delta is relatively slow. Lobate deltas dominated the ancestral Mississippi, and produced delta-front sheet sands of 20–30 m thickness. After a period of lobate delta growth, the main river abandons this delta lobe and initiates a new one close by. By this process of avulsion and delta switching, many delta lobes were formed in the Mississippi delta area over the past 5000 years, until around AD 1500, when the modern elongate delta came into existence. After a constructional phase of delta growth, swamps and marshes develop over the delta in the abandonment phase, and then as compaction ensues, the lobe gradually subsides and is flooded by marine waters. The deltaic sediments may be reworked into beaches and barriers. A number of major coarsening-upward units can develop by successive lobate-delta growth and abandonment.

In river-dominated deltas, palaeocurrents are mostly directed offshore and are best measured from trough and planar cross-bedding of the delta mouth bar and channel sands. More dispersion can be expected from lobate-delta sands.

Wave-dominated deltas

Modern examples of this type are the Niger, Nile and Rhône deltas. Strong wave action at the delta front reworks mouth-bar sands to form coastal barriers and beach ridges (next section). A sand body is produced parallel to the delta front, although it will be cut through by distributary-channel deposits. A coarsening-upward unit is still produced through wave-dominated delta progradation but the sands of the upper part of the package should show low-angle cross-bedding and planar bedding through wave action on beaches, and some onshore-directed cross-bedding from dunes in the shoreface zone (Section 2.11.5). Palaeocurrents should be bimodal, onshore- and offshore-directed, and even polymodal or random, if there is much longshore drift. Ancient wave-dominated deltas may be difficult to distinguish from beach-barrier islands but an important difference will be in the back-barrier deposits; in a deltaic setting, these will be delta-plain facies whereas in a non-deltaic setting they will be lagoonal sediments.

Tide-dominated deltas

These occur where the tidal range is high and the reversing ebb-and-flood tidal currents are the principal mechanism of sediment dispersal at the delta front. The Ganges–Brahmaputra, Klang and Gulf of Papua are modern examples. Linear and digitate ridges are developed parallel to the direction of tidal currents, which may be normal or parallel to the delta front. The lower delta plain will have extensive tidal flats where mud is deposited. Tide-dominated deltas may be difficult to recognize in the geological record because their deposits will resemble tidal sand seas and tidal flats not associated with any major fluvial sediment input. The succession produced by progradation of a tide-dominated delta should be a thick coarsening-upwards package, muds passing up into sands showing tidal structures, cut through by major channels filled with sands, again with tidal bedding, and overlain by delta-plain facies, which may include tidal mud-flat facies.

Ancient deltas

Ancient deltaic successions are widespread in the geological record, although most described are from river-dominated deltaic systems. They are characterized by major coarsening-upward units (many tens of metres thick), from marine sediments, perhaps limestone, passing up into increasingly non-marine mudrocks and sandstones (Fig. 2.68). The latter are cut through by one or more channels, their sand fills overlain by seatearths and coals. Small-scale coarsening-upward cycles, a few metres thick, formed by the filling of bays and lakes, may terminate the whole deltaic package.

Deltaic sediments are particularly well developed in the Carboniferous of western Europe and eastern North America, and they contain economically important coal reserves. There has been much discussion over the causes of the repetition of the deltaic cycles in Carboniferous strata. Mechanisms put forward include (a) sedimentary processes of delta switching and compaction-induced subsidence, (b) tectonic processes involving periodic uplift or basin extension and (c) eustatic sea-level changes induced by orbital perturbations (Milankovitch rhythms), connected with the Permo-Carboniferous Gondwana glaciation. For descriptions and discussion see Heckel (1995), Hampson *et al.* (1996) and Miller & Eriksson (2000), and papers in Besly & Kelling (1988) and Whately & Pickering (1989).

Deltaic sandstones are important oil reservoirs in some basins. The Middle Jurassic Brent Sands of the North Sea are a case in point, and here the deltaic succession shows evidence of strong wave action in associated beach-barrier island facies. Thick deltaic sandstones in the Tertiary of the Gulf Coast of Texas and Louisiana, the Wilcox and Frio Groups, for example, are also major oil reservoirs (see Fig. 2.59 for their porosity–permeability patterns).

Reviews of deltas and their facies are provided by Bhattacharya & Walker (1992) and Reading & Collinson (1996). Papers concerned with deltas and their hydrocarbon and coal reserves are to be found in Whately & Pickering (1989) and Oti & Postma (1995), and papers on the Brent Delta of the North Sea can be found in Morton *et al.* (1992); also see Glennie (1998).

Fan deltas

Fan deltas, also called coarse-grained deltas, are cones of sediment being shed directly from a source area into the sea or into a lake (Fig. 2.69). They mostly occur near active fault zones where mountains are being uplifted and erosional debris is taken through canyons to alluvial fans that pass straight into the sea/lake as fan deltas. If the marine basin is deep enough, the fan delta may pass down into a submarine fan. Modern examples are prominent along the Gulfs of Aqaba and Suez in the Middle East, and in Baja California (e.g. Falk & Dorsey, 1998). Fan deltas mostly consist of gravels and sands, and material is transported on to the fan by debris flows, gravel slides, stream floods and sheet floods. Once on the fan delta, sediment can be reworked by waves and tidal currents or resedimented into deeper water by further debris flows and slumps. Internally, fan deltas consist mostly of steeply dipping beds (foresets) of coarse sand and gravel, that pass down into gently dipping to horizontal bottom-sets of finer sediment (Fig. 2.69). Modern and ancient fan deltas are described in Nemec & Steel (1988) and Collela & Prior (1990).

2.11.5 Marine shoreline environments and facies

Much siliciclastic sediment is deposited along marine shorelines in beaches, barrier islands, tidal flats, estuaries and the shoreface–shallow offshore. Sediment supply, tidal range, wave action, storm frequency, sea-level change, subsidence rates/tectonics and climate all affect the sedimentation. On the basis of tidal range, a distinction is made between microtidal (<2 m), mesotidal (2–4 m) and macrotidal (>4 m) shorelines (Fig. 2.70). Beaches and barriers are best developed in micro-/mesotidal areas of moderate to high wave action. Tidal flats and estuaries are best developed in macrotidal regions, as are the tidal ridges of open shelves (next section).

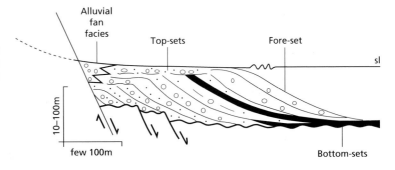

Fig. 2.69 Schematic facies model for a fan delta. Such deltas are characterized by coarse sediments and a single set of sandy or gravelly cross-beds (foresets) dipping at a high angle. Top-sets mostly consist of flat-bedded and smaller-scale, cross-bedded sand, and bottom-sets are finer sediments at the toe of the foresets. Fan deltas may prograde into a lake or the sea.

Two major depositional systems for sand along a shoreline are the beach-barrier island system, with associated lagoons and tidal inlets, and the beach-ridge/strandplain system.

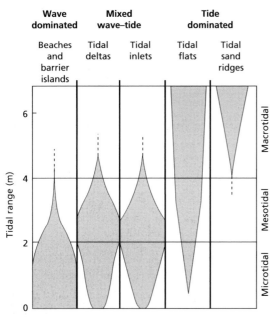

Fig. 2.70 Relationship between tidal range and shoreline environment.

Beach-barrier island and strandplain systems

These develop in microtidal to mesotidal areas, generally where tidal range is less than 3 m. A steady supply of sand and a moderately stable, low-gradient coastal plain favour barrier-island formation. A lagoon is located behind a barrier island and connected to the open sea via tidal inlets (Fig. 2.71). In microtidal areas, tidal inlets are widely spaced along the barrier, but in mesotidal locations tidal inlets are prominent and tidal deltas are usually developed at the ends of the inlets (Fig. 2.71). Mud flats and marshes are common around the lagoon, especially on the landward side. Where tidal inlets are sparse, then sediment is transported over the barrier during storms to be deposited in back-barrier washover fans. Well-documented barrier islands occur on the Dutch and German coasts of the North Sea and along the coast of Texas in the Gulf of Mexico, and Georgia and Virginia on the US eastern seaboard (Davis, 1994).

Where there is an abundant supply of sand, high wave energy and low tidal range, a strandplain of beach ridges forms. The ridges are separated by swales (depressions), with narrow lakes and marshes. Strandplains mostly develop at times of static sea-level or slow fall. They occur along the Nayarit coast of Mexico, along some parts of the southern US Gulf of

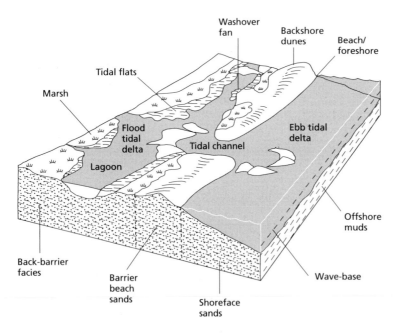

Fig. 2.71 Subenvironments for a beach-barrier island and lagoon shoreline system.

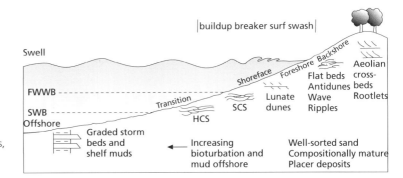

Fig. 2.72 Wave zones, subenvironments, sediments and sedimentary structures along a siliciclastic shoreline.

Mexico shoreline, in southern England, and western Australia.

Sediments along a siliciclastic shoreline show a general offshore to onshore grain-size increase and a succession of bedforms reflecting the changing wave conditions (Fig. 2.72). Offshore, sinusoidal waves (swell) are considerably modified on reaching shallow water. In the wave build-up and breaker zone of the shoreface, sand bars and troughs occur, with shoreward-directed lunate dunes and wave-formed ripples. These give rise to tabular and trough cross-bedding and cross-lamination. Swaley cross-stratification (SCS, Fig. 2.30) also may form here. In the surf and swash zones, shallow fast flows give planar surfaces with current lineation, leading to flat-bedded sands with a low-angle, offshore dip. Summer–winter changes in beach profile result in subtle, planar erosion surfaces between packages of flat-bedded sands (see Fig. 2.73). Rip currents in the nearshore zone generate shallow channels with offshore-directed bedforms. Shoreface and foreshore sands and gravels are reviewed in Hart & Plint (1995)

Offshore from the shoreface, sands with hummocky cross-stratification (HCS, Figs 2.30 & 2.31) generated by storm waves may occur between fairweather and storm wave-base (see Figs 2.72 & 2.74). Going into deeper water, these then pass into muddy sands with ripples and bioturbation, and graded sand layers ('tempestites') and shell lags produced by storm currents (see Figs 2.72 & 2.32). In the backshore, aeolian dunes have large-scale cross-beds orientated landwards. In microtidal regions, lagoons behind beach barriers are dominated by muds with wedges of flat-bedded and planar cross-bedded sand derived from storm washover. In mesotidal areas, tidal inlets rework the barrier sediments and migrate laterally if long-

Fig. 2.73 Foreshore (beach) stratification of flat beds in packets with a gentle dip (mostly in the seaward direction, to the right), separated by low-angle erosion/truncation surfaces. Pleistocene raised beach. Mallorca, Spain.

shore currents are strong. Inlet fills are fining-upward units of cross-bedded and flat-bedded sand upon a scoured surface (the channel floor). Flood-tidal deltas build into the lagoon and are covered by landward-directed dunes and ripples. Mud flats and marshes around the lagoon consist of fine-grained sediments, with rootlets and peat if the climate is humid, and evaporites if it is arid. Ebb-tidal deltas may occur on the oceanward side of the tidal inlets.

Sea-level changes, subsidence rates and sediment supply are important in determining the thickness of a beach-barrier island complex and also its migration direction. During a stillstand or slight sea-level fall, a thick sheet sand body can develop through seaward progradation of the beach-barrier system, if there is a continuous supply of sand. The result is that back-

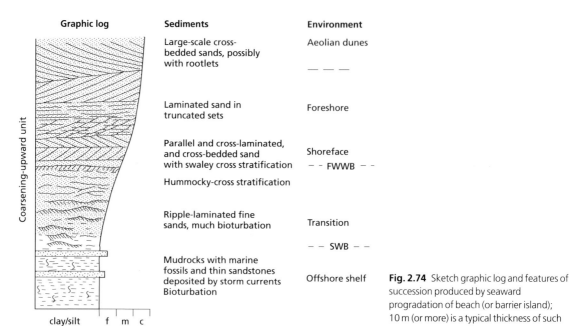

Graphic log	Sediments	Environment

Large-scale cross-bedded sands, possibly with rootlets — Aeolian dunes

— — —

Laminated sand in truncated sets — Foreshore

Parallel and cross-laminated, and cross-bedded sand with swaley cross stratification — Shoreface
– – FWWB – –

Hummocky-cross stratification

Ripple-laminated fine sands, much bioturbation — Transition

– – SWB – –

Mudrocks with marine fossils and thin sandstones deposited by storm currents Bioturbation — Offshore shelf

clay/silt   f   m   c
sand

Coarsening-upward unit

Fig. 2.74 Sketch graphic log and features of succession produced by seaward progradation of beach (or barrier island); 10 m (or more) is a typical thickness of such a coarsening-upward unit.

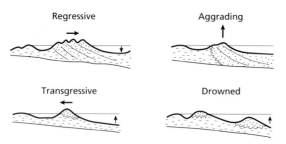

Regressive   Aggrading

Transgressive   Drowned

Fig. 2.75 Schematic stratigraphic models for barrier-island shorelines. The succession produced depends on rate of eustatic sea-level change, rate of subsidence and on sediment supply.

shore–foreshore–shoreface facies overlie offshore shallow-to-deep subtidal sediments (Figs 2.74 & 2.75). With some subsidence, lagoonal sediments may come to overlie the barrier-top sands. A slow sea-level rise can lead to the landward migration of the sand belt, and the generation of a transgressive succession (barrier sands overlying lagoonal sediments) (Fig. 2.75). In this instance the low-barrier sediments will consist of washover-fan sands, giving small coarsening-upward units (lagoonal muds to barrier sands) as the fans prograded into the lagoon.

In a balanced situation of relative sea-level rise and abundant sand supply, a barrier will aggrade to produce a beach-barrier stack (Fig. 2.75). Rapid transgression may (a) cause the barrier to migrate landwards so that the only record left is a disconformity (a ravinement) and a lag conglomerate, produced by surf-zone erosion, or (b) result in the barrier being drowned and abandoned on the shelf (Fig. 2.75). With strandplains, seaward progradation generates a coarsening-upward package capped by shoreface–beach–backshore sands (as in Fig. 2.74), but lacking any associated lagoonal muds.

Shoreline sands are typically quartz arenites (i.e. compositionally supermature), and they are usually texturally mature too, being well rounded and sorted. Heavy minerals may be concentrated in beach sands; some economically important sedimentary ores (placers) have formed in this way.

Ancient beach-barrier lagoon and strandplain successions are well represented in the Tertiary of the Gulf Coast, where stacking of barriers, aided by growth faulting, has produced successions up to 2.5 km thick. These shoreline sand bodies, which are major oil reservoirs with porosities up to 30%, are located between major delta systems that shed huge quantities of sediment into the ancestral Gulf of Mexico. At the present time, some Mississippi delta sediments are being reworked into transgressive barriers. Stacking of Cretaceous barrier complexes in New Mexico is described by Donselaar (1989), and descriptions have been given of stacked Devonian cycles of storm/tidal offshore sands to shoreline clastics in the Appalachians by Prave *et al.* (1996), and Cretaceous shoreface sands in the US Western Interior by Olsen *et al.* (1999). See Nummedal *et al.* (1987), Carter (1988) and Davis (1994) for more information. Tertiary shore platforms and beach conglomerates have been described by Gupta & Allen (1999) from the French Alps.

Tidal flats

Tidal flats reach several kilometres in width and occur around lagoons, behind barriers, and in estuaries and tide-dominated deltas. There is usually a decrease in sediment grain size across the flat from sand in the low intertidal zone to silt and clay in the higher part. The characteristic sedimentary structures of the mid–upper tidal flat are various types of ripple, usually showing interference patterns (see Fig. 2.29), and these give rise to flaser, wavy and lenticular bedding (see Fig. 2.27). Desiccation cracks form in the high intertidal/low supratidal zone, and, depending on the climate, marshes or sabkhas (Section 5.2) will occupy the supratidal area. Cross-bedded sands with some herringbone structure (Fig. 2.24) are deposited in the low tidal flat. Meandering tidal creeks dissect the tidal flats and their lateral migration produces sets of obliquely inclined mud and sand laminae and channel structures. Tidal rhythmites consisting of alternating sand and mud laminae, with thickening-upward and thinning-upward patterns produced by variations in tidal currents over the lunar cycles can be deposited on low tidal flats, in tidal channels, on offshore sand bars and also on distal mouth bars of deltas. Burrows and grazing trace fossils are common. Progradation of tidal-flat sediments typically forms a fining-upward unit capped by a soil or sabkha evaporite bed, with the thickness determined by the palaeotidal range. Modern and ancient tidal-flat deposits are discussed in Fleming & Bartholoma (1993) and Alexander *et al.* (1998).

Estuaries and incised valleys

Most modern estuaries are drowned river valleys resulting from the rapid, post-glacial, Holocene sea-level rise. Estuarine sand bodies are located within and adjacent to the main channel and consist of sediment brought down by the river and supplied from the adjacent marine shelf. Mud flats and swamps also occur in estuaries. Incised valleys are the result of strong falls in relative sea-level (forced regressions) and they are filled when sea-level rises subsequently. Incised-valley fills generally show a succession from fluvial-dominated sands in the lower part to tidal estuarine facies and then marine muds in the upper part (papers in Dalrymple *et al.*, 1994; Hunt & Gawthorpe, 2000; also Willis, 1997).

2.11.6 Shallow-marine shelves, epeiric seas and their facies

In the shallow-marine environment away from the coastline, where water depths range from 10 to 200 m, sands occur on continental shelves, such as on the present-day eastern shelf of North America, and in epeiric, epicontinental seas, such as the North Sea. However, many of the sediments in modern shallow seas are relict; they were deposited before or during the Holocene transgression in glacial, fluvial and shoreline environments. In many cases these sediments are being reworked by modern shelf processes. Sediments of continental shelves and epeiric seas are affected by tidal currents, wind-generated waves and storm-generated waves and currents. Although all processes operate on all shelves to a greater or lesser extent, tide-dominated and weather-dominated types are distinguished. Tide-dominated seas occur mostly on partially enclosed continental shelves bordering large oceans, and in seaways. Conditions vary considerably between weather-dominated shelves, depending on regional climatic factors. For example, meteorological input is high where an open shelf faces prevailing Westerlies (e.g. Washington–Oregon shelf), but low in the lee of such winds (e.g. Gulf of Mexico and eastern shelf of the USA). Seasonal variations are important on weather-dominated shelves, with fairweather processes distinct from foul.

The most significant and diverse deposits of tide-dominated seas and shelves are the tidal sand bodies, although muds are much more extensive, because large areas of these seas have relatively weak currents. Two broad types of tidal sand deposit are (a) sand sheets and (b) sand banks, both well known from studies in the North Sea. Offshore sand bodies also develop on weather-dominated shelves.

Tidal bedforms and sand sheets

Transverse bedforms developed in response to tidal currents are ripples, dunes, sand waves and sand patches. Ripples are ubiquitous where the sea bed is made of sand and tidal (and other) currents are moderate. Tidal sand waves, also referred to as giant ripples, megaripples or dunes, are mostly asymmetric bedforms. Large sand waves are 5–15 m in height and 150–500 m in wavelength, with a lee slope of 4–30° (mostly less than 20°). Many sand waves are sinuous crested and orientated normal to spring-peak tidal flow. In areas of limited sediment supply, sand waves are barchan shaped (lunate). Large sand waves are generated where tidal currents exceed 60 cm s⁻¹ and they usually are covered in dunes and ripples. Where tidal currents are relatively weak (less than 50 cm s⁻¹) and there is no continuous sand cover, sand patches occur a few metres thick. They appear to be controlled by storm-induced currents and waves, as well as tidal currents, and they do migrate laterally. Sand ribbons are longitudinal bedforms that form where peak tidal currents exceed 100 cm s⁻¹. Erosional scours and furrows occur in areas of fast tidal currents, also orientated parallel to the current.

Net sediment transport in shelf seas is generally in the direction of the stronger tidal current (ebb or flood), and zones of erosion (bedload parting) and deposition can be recognized. Non-tidal, especially storm-induced currents can considerably modify the tidal bedforms and direction of sediment movement. Where flow velocities decrease along sediment-transport paths, bedforms reflect this and erosional furrows pass into ribbons, sand waves and sand patches (Fig. 2.76). At the ends of tidal-current transport paths, mud predominates in the low-energy environment.

Sand sheets in the North Sea occupy up to 20 000 km² of sea floor and consist mostly of well-sorted, very fine to coarse sand. The dominant sedimentary structure is cross-bedding and this varies progressively along the sediment-transport path, along with grain size, as tidal currents weaken and bedforms change (Fig. 2.77). In coarser sands, major (master) bedding planes dipping less than 20° represent the lee slopes of large sand waves and internally, smaller-scale cross-bedding at steeper dips reflects the smaller parasitic sand waves/dunes. Most cross-bedding will be unidirectional, but some opposing sets can

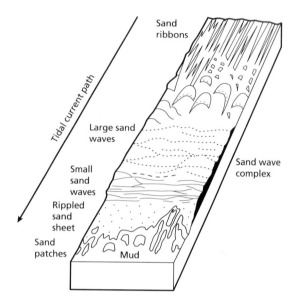

Fig. 2.76 Bedforms developed along the tidal current path of a shelf sea.

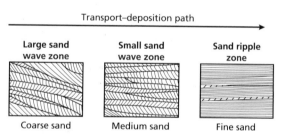

Fig. 2.77 Diagrammatic sketches of down-current changes in cross-bedding style in a tidal sand sheet: from large- to small-scale cross-bedding to cross-lamination as the tidal currents become weaker, reflecting the change from large sand waves (with parasitic dunes) to small sand waves to ripples.

be expected as well as reactivation surfaces. In medium to fine sands, along the transport path, high-angle cross-bedding from small sand waves will dominate, with mud drapes deposited from suspension at slack water. Tidal bundles may be present (Fig. 2.25). This subfacies will give way to flat and cross-laminated fine sands with ripples and then to lenticular muddy sands and muds. In time, a vertical succession of these subfacies may develop (to give a coarsening-upward package), as a result of the down-current migration of the subfacies belts.

Tidal sand banks

These structures, also called tidal current ridges, are the largest tidal bedforms, reaching 50 km in length, 6 km in width and 40 m in height in the North Sea. They appear to have formed at times of lower sea-level but are actively maintained in the modern tidal regime, notably where peak tidal currents exceed 50 cm s^{-1}. The tidal sand ridges are usually asymmetric in cross-section, with a steeper lee side up to 6° and a gentle stoss side (0.5°). Dunes are present on the bank surface. Internally they probably have a complex structure of gently dipping master beds and smaller-scale cross-beds (see Fig. 2.36), which may show current reversals.

Non-tidal offshore sand bodies

Much sediment can be transported by wind-induced waves and currents, especially during major storms. Particularly important are storm surges, which can transport much inshore sand (and skeletal material) many tens of kilometres out on to the shelf. Density currents generated by storms deposit graded beds with basal scours, flat bedding and ripples ('tempestites', Fig. 2.32). In areas frequently affected by storms, significant sand bodies can develop. Sand waves are formed, similar to those of tidal current origin, with internal cross-bedding. However, current reversals are likely to be rare. Hummocky cross-stratification (Fig. 2.30) and erosion surfaces may occur in the sands, produced by the passage of storm waves. Linear sand ridges, not unlike the tidal sand banks, occur on some shelves (especially the Atlantic shelf of the USA), and they are largely maintained by storm currents. Up to 10 m high and tens of kilometres long, they are dominated by unidirectional medium-scale cross-bedding, produced by sand waves upon the ridges.

Ancient shelf sandstones are common in the geological record and tend to be texturally and compositionally mature from the constant reworking. Many are thus quartz arenites. Fossils do occur, such as brachiopods and bivalves, and trace fossils, mainly burrows of suspension feeders in sands and sediment feeders in more muddy sediments. Ancient tidal sand-wave complexes typically consist of large-scale sets of cross-strata. Some current reversals may be recorded in 'herringbone' cross-bedding (Fig. 2.24), but commonly a sand-wave succession will consist of unidirec-tional cross-strata, with only reactivation surfaces or thin, rippled layers recording current reversals. Mud drapes may occur on the foresets and define tidal bundles (Fig. 2.25). Periodicities in the bundles reflect spring–neap tidal cycles. Examples of ancient sand waves include the Cretaceous Lower Greensand of southern England (Johnson & Levell, 1995). Papers on tidalites are contained in Flemming & Bartoloma (1995) and Alexander *et al.* (1998).

Ancient offshore sand-bar complexes are well known from the oil-productive Mesozoic of the Western Interior, North America, and include the Cardium Sandstone of Alberta, and the Shannon, Sundance and Sussex Sandstones of Wyoming. These shoreline-parallel elongate sandstone bodies generally consist of coarsening-upward units from 2 to 15 m thick. Some have been traced for distances of 20–50 km, and they all appear to have been located several tens of kilometres from the contemporary shoreline. Cross-bedding is a common feature of the sand-bank facies although much appears to be unidirectional. Like modern shelf sands, some of these Western Interior sandstones were first deposited at the shoreline and then, following a transgression, they were substantially modified by tidal and storm currents when they were placed in an offshore shelf environment.

For more information on shelf sands see Smith *et al.* (1991), Swift *et al.* (1991), Flemming & Bartoloma (1995), Batist & Jacobs (1996), Johnson & Baldwin (1996) and Bergman & Snedden (1999).

2.11.7 Continental margins and deep-water basins

Continental margins and deep-water basins are the depositional sites of sandstones and conglomerates derived from adjacent slopes and shelves. Transport downslope is through sliding and slumping, and through sediment gravity flows; in particular, turbidity currents, but also debris flows, and the less important grain flows and fluidized sediment flows. In addition, ocean-bottom currents produced through thermohaline density differences can transport and rework sediment. Apart from these resedimentation processes, deeper-water environments are primarily the sites of pelagic and hemipelagic deposition: cherts (Section 9.3), pelagic limestones (Section 4.10.7) and mudrocks (Section 3.7.2).

Slides and slumps involve small to large masses of sediment, with more internal deformation (folding and brecciation) occurring in slumps (see Section 2.3.3). Slumps may develop into sediment gravity flows. Slides and slumps are typical for slope environments and give rise to truncation surfaces and discontinuities in generally evenly bedded fine-grained sediments. Slope failure generating slumps and sediment gravity flows can be induced by earthquake shocks, which reduce the shear strength of the sediment. Other processes are rapid sedimentation, tectonic oversteepening, storm-wave loading, and sea-level change, especially a fall that reduces the pore-fluid pressure. Steep slopes *per se* are not necessary for failure. Off the Alaska continental shelf, slopes of less than 1.3° fail periodically. It also appears there that seismic shocks tend to generate slumps, whereas storm-wave loading, which may act upon the surficial sediment for several hours or more, so reducing the sediment shear strength, tends to produce sediment gravity flows.

Five types of sediment gravity flow are recognized, based on their rheology (liquid versus plastic behaviour) and particle-support mechanism (Table 2.13). In *turbidity currents*, the sediment is supported by the fluid turbulence and low- and high-density flows are distinguished. In *fluidized flows*, the sediment is supported by upward-moving pore fluid. With *liquefied flows*, the sediment is not fully supported; the grains settle through the fluid, which is displaced upwards. In *grain flows* the sediment is supported by the dispersive pressure arising from grain collisions. *Debris flows*, also called mudflows or cohesive flows, are ones where the sediment is supported by a cohesive matrix. Sediment is deposited from decelerating gravity flows by two different mechanisms. In fluidal flows, grains are deposited individually, either from the bedload (traction sedimentation) or from suspension, so that deposition takes place from the base of the bed upwards. With debris flows, the flow freezes as the applied shear stress falls below the yield strength of the moving material, so that deposition takes place *en masse* or from the outside inwards.

Turbidity currents are the most important of the sediment gravity flows, and they deposit sands with characteristic internal structures that change along the transport path. *High-density turbidity currents* can carry much gravel and coarse sand, mostly in the form of a traction carpet at the base of the flow and in suspension just above. Fluid turbulence, dispersive pressure from grain collisions, and finer sediment exerting a matrix buoyancy lift, keep the gravel and sand moving until the flow decelerates through decreasing slope or dilution. Sedimentation of the gravelly traction carpet takes place by freezing to give an inversely graded basal layer (designated R_2; see Fig. 2.78), which is overlain by a normally graded layer (R_3) deposited from suspension. In very proximal areas, crude cross- and flat bedding may develop in the gravels (R_1). With high-density sandy turbidity flows, deposition may also produce a basal cross- and flat-bedded layer (S_1), then inversely graded layers (S_2) from traction-carpet deposition. A massive or normally graded bed (S_3) is deposited out of suspension, and the rapid deposition

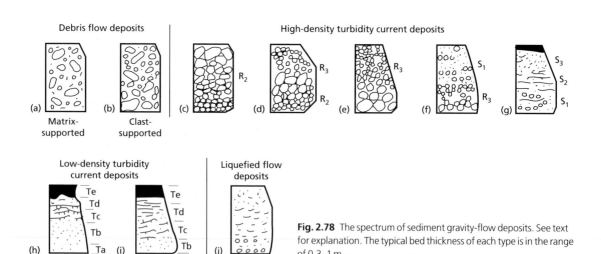

Fig. 2.78 The spectrum of sediment gravity-flow deposits. See text for explanation. The typical bed thickness of each type is in the range of 0.3–1 m.

Table 2.13 Classification of laminar sediment gravity flows based on flow rheology and particle-support mechanisms

Flow behaviour	Flow type	Sediment support mechanism
Fluid	turbidity current	fluid turbulence
	fluidized flow	escaping pore fluid
	liquefied flow	escaping pore fluid
Plastic	grain flow	dispersive pressure
	mudflow/cohesive debris flow	matrix density and strength

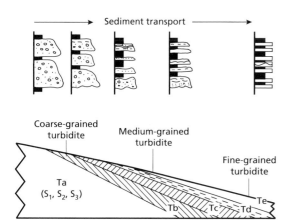

Fig. 2.79 Diagram illustrating the down-current changes (over many kilometres) in internal structure and thickness of a high- to low-density turbidity current deposit. See text and Fig. 2.78 for an explanation of letters. Typical thicknesses are from several tens of centimetres to a metre or more.

Fig. 2.80 A turbidite sandstone showing near-planar and sharp base, graded division (Ta), horizontally laminated division (Tb), cross-laminated division (Tc) and horizontally laminated division (Td). There is very little pelite division (Te) preserved, before the succeeding turbidite, which has a quite irregular and scoured base. Cambrian. North Wales.

may result in water-escape structures such as dish-and-pillar structures (Section 2.3.3). After deposition of the coarse sediment in a high-density turbidity current, any continuing flow may rework the sediment surface or deposit finer-grained sediments above the coarse bed or farther down-current (see Fig. 2.79).

Low-density turbidity currents transport sediment up to medium sand, kept in suspension by the fluid turbulence. As the flow decelerates sediment is moved as bedload in a traction carpet. The deposits of these currents are commonly 'classic' turbidites (e.g. Fig. 2.80), characterized by a definite succession of internal structures (a 'Bouma' sequence): a basal graded A division (Ta), passing up into a lower horizontally laminated B division (Tb), with parting lineation, overlain by a cross-laminated division (Tc), in some cases with stoss-side preservation (as in Fig. 2.26), an upper horizontally laminated division (Td) and a pelite division (Te). The sequence is the result of deposition from a waning flow: divisions A and B represent upper flow regime, C is lower flow regime and D and E are deposited from suspension (see Section 2.3.2). Turbidites deposited from low-density flows typically range in thickness from 0.05 to 0.5 m. However, commonly not all Bouma divisions are present, because their formation does depend on grain size and rate of deposition. Down-current, the lower divisions gradually disappear (see Fig. 2.79). Very distal turbidites are simply thin (a few centimetres), normally graded, fine sandstones and siltstones.

On the bases of turbidite beds there are usually sole structures (Section 2.3.1). On the bases of very coarse turbidites, they are gentle scours and groove casts (e.g. Fig. 2.12); medium-grained turbidite sandstones commonly possess flute marks (e.g. Fig. 2.11); and finer-grained beds tend to have more tool marks. Sole marks and internal structures (parting lineation and cross-lamination), as well as grain fabric, can all be used for palaeocurrent analysis (Section 2.4).

Turbidite sandstones are generally texturally and compositionally immature. Many are greywacke in composition, with a substantial matrix content, although much of the latter may well be diagenetic (see Plates 2b & 4a,b, and Section 2.7.4). Lithic grains are abundant in some turbidites, and in some cases, many are of volcanic origin. Limestone turbidites are common in deeper-water basins adjacent to carbonate platforms (Section 4.10.8).

Of the other sediment gravity-flow deposits, debris-flow deposits (debrites) are the most significant (Figs 2.78 & 2.9). The strength of the muddy matrix in some debris flows is sufficient to transport blocks many metres across. Debrites are typically mud-dominated with scattered clasts, in a matrix-support fabric. There is usually no sorting or grading of clasts. However, there are also clast-dominated debrites, with a pebble-support fabric. As little as 5% matrix is apparently sufficient to exert a buoyant lift and so reduce the effective weight of the large clasts. In these clast-rich debris flows, pebbles move by rolling and sliding, and so their deposits may show a preferred orientation of clast long axes. Slurry flows are watery, muddy sand flows transitional between the turbulent turbidity currents and viscous debris flows. Internally, they are massive and banded, and water escape structures are common (Lowe & Guy, 2000).

Liquefied and fluidized flows are less important in the submarine environment; they will mostly develop into turbidity currents. Their deposits are massive or poorly graded sandstones, but fluid-escape structures such as sand volcanoes and dish-and-pillar structures should occur (Section 2.3.3). True grain flows are not thought to be able to deposit beds thicker than a few centimetres, as dispersive pressure alone is considered unable to support a greater thickness of sediment against gravity. The avalanching of sand down subaqueous and subaerial dune lee-sides is a grain-flow process.

In addition to the sediment gravity flows, which generally move directly downslope to the basin floor, there are contour-following bottom currents that can transport and rework sediment. Their deposits, *contourites*, are usually thin (< 5 cm), laminated sandy siltstones with sharp, or gradational bases (see papers in Stow & Faugères, 1998). Lags of pelagic fossils are also formed by current reworking of the sea floor.

Deep-sea siliciclastic sediments have been the subject of several classification schemes to aid their description and interpretation. Pickering *et al.* (1986), for example, modifying earlier schemes, recognized seven facies classes (based on grain size), divided into 15 facies groups (based on internal organization of structures and textures), in turn divided into 41 individual facies.

Sediment gravity flows deposit their clastic material in the lower reaches of submarine canyons, on clastic ramps and slope aprons, on submarine fans located at the foot of slopes, in deep ocean trenches, and on basin plains. Submarine fans have received much attention and many ancient turbidite successions have been interpreted in terms of the fan model (Fig. 2.81). A fan occurs in a lower slope/basin-rise setting and usually is fed by one major channel, which divides into smaller and shallower channels on the fan itself. Channels have levees and may be active or abandoned, braided or meandering. Lobes of sediment are deposited at the ends of the channels, in the lower-mid-fan area. The outer part of the fan has a relatively smooth surface, passing into the basin plain. Sediments of the inner fan are mostly debrites and coarse turbidites, commonly channelized. Interbedded thin, fine-grained turbidites represent interchannel and levee facies. The mid-fan is characterized by packets (a few metres thick) of upward-thinning- and -fining turbidites, representing the fills of broad channels. In the lower part of the mid-fan, progradation of suprafan lobes gives rise to units of thickening-upward and coarsening-upward turbidites (Fig. 2.82). The outer-fan facies consists of more laterally persistent turbidites, interbedded with hemipelagic background sediment. Large-scale progradation of a submarine fan will generate a thick, overall coarsening-upward and bed-thickening-upward pile of sediments, with basinal mudrocks passing up into thin turbidites (outer fan), into packets of thicker turbidites (mid-fan), overlain by channelized coarse turbidites and debrites (inner fan). Such a package could be generated by a single fan in a small basin, which is then succeeded by shallow-water sedi-

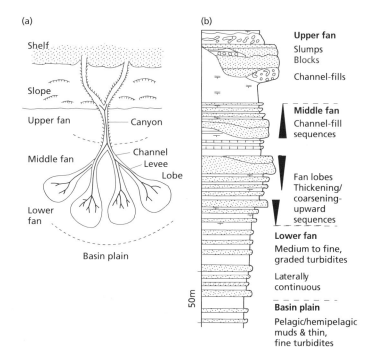

Fig. 2.81 Submarine-fan facies model. (a) Classic model of fan with suprafan lobes, some active, some abandoned. (b) Succession produced through fan growth and lobe progradation.

Fig. 2.82 Turbidite sandstones arranged into packages showing upward-increasing bed thickness (there is a corresponding upward increase in grain size too). These coarsening-upward, thickening-upward units are interpreted as submarine fan-lobe deposits. They are separated by hemipelagic mudrocks. Cretaceous. Great Valley, California.

ments, or it could be one subfan of a much larger fan, which is abandoned as a new subfan develops close by, in an analogous manner to delta switching. See Normark *et al.* (1998) for a description of a modern Californian fan.

Although many ancient siliciclastic turbidite forma-tions show some fan features, there has been much criticism of the model, in that many turbidite succes-sions do not have fan features, yet are forced into the model. There are deep-water sandstone formations that were deposited directly into the basin ('basin-fill turbidites'), rather than being concentrated in a basin-margin fan. In these formations, turbidites at one locality are commonly very similar, and large-scale proximal to distal trends can be observed. Such features are not shown by fans, where an interbedding of coarse and fine, thick and thin turbidites is characteristic of the mid-fan, from channel and levee deposition. There are also examples of turbidites and debrites being deposited in more-or-less continuous aprons at the toe of basinal slopes, rather than in a discrete fan (e.g. Lomas (1999) for a Cretaceous example from Antarctica). The point of interest here is whether the basin was fed by one major canyon (as in a fan), or by many channels along the shelf-margin/slope (as in an apron). Most resedimented limestones were deposited in slope aprons (see Section 4.10.8), because carbonate sediment is generated in abundance along shelf-margins. In the siliciclastic situation, sand aprons may form off narrow shelves or off fault-bounded shelves (Fig. 2.83). Turbidity currents may be derived directly from deltas, generating a clastic ramp, with channels and lobes (Fig. 2.83). Siliciclastic sediment gravity-flow successions, fans, aprons and ramps, tend to develop at times of relative sea-level lowstand, when rivers cross shelves and material is deposited near shelf margins. This contrasts with carbonate sediment gravity-flow successions, which are best developed during sea-level highstands when shallow-water carbonate productivity is at a maximum. Pickering *et al.* (1999) deduced a glacio-eustatic control on deep-marine fore-arc sedimentation from Japan, with turbidite sands preferentially deposited during glacial lowstands of sea-level.

Thick successions of turbidites occurring within mountain belts have been referred to as *flysch*, and in many cases flysch deposition took place during or just before the main orogenic deformational stage. Thick turbidite formations can develop in many different tectonic situations: within trenches at destructive plate margins, at passive margins, in back-arc basins and within failed rifts for example. The composition and texture of these deeper-water sandstones depend very much on the tectonic setting and on the amount of sediment supplied from the adjacent continent and/or volcanic sources (see Section 2.8).

For reviews and descriptions of deep-water siliciclastic sediments, consult Pickering *et al.* (1989) and Stow *et al.* (1996), and papers in Weimer & Link (1991), Weimer *et al.* (1994), Hartley & Prosser (1995), Winn & Armentrout (1995), Stoker *et al.* (1998), Stow & Faugères (1998) and Shiki *et al.*

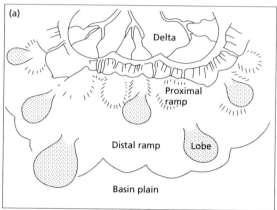

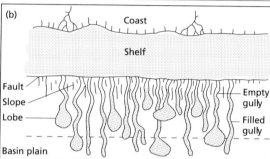

Fig. 2.83 Depositional models for siliciclastic sediment gravity-flow deposits. (a) Delta-fed submarine ramp. (b) Fault-controlled submarine ramp or apron.

(2000). Turbidity currents are reviewed in Kneller & Buckee (2000) and Stow & Mayall (2000), and the role of river floods in turbidite sedimentation is considered by E. Mutti (1999). Submarine channels are described in Clark & Pickering (1996). Deep-water sands and gravels have good reservoir potential, because they can be very porous and may be enveloped in impermeable mudrocks, which may also be organic-rich. In the North Sea, Upper Jurassic sandstones in the Brae Field and Palaeocene–Eocene sandstones in the Forties, Montrose and other fields are all broadly submarine fan in origin (see Glennie, 1998).

2.11.8 Glacial environments

Glacial environments include a wide range of depositional settings from continental to marine, subglacial to supraglacial and glaciofluvial to glaciolacustrine. Geomorphologically a number of different glacier types have been distinguished but a division on the thermal state into cold and temperate glaciers is more important in terms of glacial erosion and sedimentation. Cold glaciers are dry-based and have much rock debris incorporated into their basal parts; temperate glaciers are wet-based, contain less sediment, but move faster and so are more powerfully erosive.

Sediment deposited directly from a glacier, either subglacially while it is moving or when the glacier is stagnating and melting (ablating), is referred to as till or tillite if indurated. Subglacial tills and tillites are mostly massive (see Fig. 2.84) although some show a vague bedding. Typical features are: an extensive lateral development, usually on a regional scale; a thickness of several to tens of metres; a lack of stratification; much matrix which supports clasts, some of which may have striations and facets; clasts of both local and exotic origin; and a matrix that is largely comminuted rock fragments. Such sediments may be difficult to distinguish from debris-flow deposits (Sections 2.11.1 and 2.11.7), and indeed mass flow is involved in the deposition of some glacial deposits (flow till, for example, generated by wasting of a continental ice sheet and mass movement of supraglacial–englacial debris). Other evidence for glaciation is provided by cracked boulders, impressions of ice crystals, stone polygons produced by freezing and thawing, and striated pavements on bedrock.

Lenticular and stratified conglomerates and sandstones are commonly interbedded or associated with tillites. These are waterlain deposits resulting from glaciofluvial processes. They may be deposited subglacially, as in eskers, in supraglacial streams and lakes during ablation, and in proglacial outwash plains (sandar). Sediments of glacial lakes are sands and gravels where outwash streams construct deltas and, in deeper parts, rhythmically laminated mudrocks (varves, Section 3.7.2), probably with scattered clasts dropped from rafted ice (e.g. Fig. 2.85). Where ice sheets reach the sea, till is deposited on the sea floor, and it can be reworked by marine processes and trans-

Fig. 2.84 Tillite. A matrix-supported conglomerate with large and small clasts ('stones') in a sandy–muddy matrix. Late Precambrian. Death Valley, California, USA.

Fig. 2.85 Glacial-lake rhythmites ('varvites') with large dropstone from a melting iceberg depressing the lamination. Huronian, early Proterozoic. Ontario, Canada.

ported into deep water through slumping and turbidity currents. Glaciomarine tills may contain fossils. Icebergs calved from ice shelves release sediment as they melt and deposit dropstones over a wide area. These ice-margin deposits can be difficult to distinguish from true glacial deposits.

The description and interpretation of glacial deposits are more difficult than for some other facies because it is not easy to view the processes operating at the present time. As the words till/tillite imply a glacial origin, the term *diamictite* (also mixtite) is frequently used in a purely descriptive sense for pebbly mudstones and matrix-supported conglomerates, which may or may not have a glacial origin. A lithofacies code for describing glacial deposits has been devised by Eyles *et al.* (1983).

There have been five major periods of glaciation: an early Proterozoic glaciation mainly documented from North America; a late Precambrian glaciation, which affected all continents; a late Ordovician glaciation, evidence for which is particularly well displayed in North Africa; a Permo-Carboniferous glaciation, which affected Gondwanaland, that is South Africa, South America, India, Australia and Antarctica; and the late Cenozoic, chiefly Pleistocene glaciation which began in the Oligocene–Miocene, and resulted in the thick glacial and glacial-associated drift that now covers much of northern Europe, Asia and North America.

Deposits of the early Proterozoic glaciation in Ontario, eastern Canada, occur at several horizons in the Huronian Supergroup, but the Gowganda is the best known. In the north, a grounded continental ice sheet left tillites, some with cracked boulders, striated pavements and glacial-lake rhythmites with dropstones (Fig. 2.85). By contrast, to the south, diamictites were deposited from a floating ice shelf and much resedimentation took place through slumps, debris flows and high-density turbidity currents.

In the late Precambrian, four phases of glaciation are recorded in the period of about 900–600 Ma, and tillites of this period are found on all continents. One curious feature is that many of the tillites are closely associated with limestones and dolomites containing stromatolites, oolites and replaced evaporites, all features normally indicative of a warm climate. Evaporites are being precipitated today in some Antarctic glacial lakes, however. From palaeomagnetic studies, it appears that many of the tillites were deposited in low latitudes, another unusual feature. Rapid climatic changes probably are responsible, but more drastic mechanisms have been suggested: very rapid plate movements, equator to the poles and back; antigreenhouse effects; low-latitude but high-altitude glaciation; and an increase in the obliquity of the Earth's ecliptic (see review of Fairchild (1993) and the Hoffman (2000) 'snowball Earth' scenario). The late Ordovician and Permo-Carboniferous glaciations were both high-latitude events, occurring when continental plates were located at the South Pole.

Glacial environments and facies are reviewed in Dowdeswell & Scourse (1990), Anderson & Ashley (1991), Hambrey (1994), Menzies (1996), Miller (1996), Crowell (1999), Jones *et al.* (1999) and Mickelson & Attig (1999). Glacial sediments in a sequence stratigraphic context are explored by Brookfield & Martini (1999), and Pleistocene marine ice-margin facies are described by Plink-Bjorklund & Ronnert (1999) from Sweden. The glacial–deglacial Permo-Carboniferous strata of southern Africa and Australia have been described by Visser (1997) and Eyles *et al.* (1998) using a sequence stratigraphic/allostratigraphic approach.

2.11.9 Facies sequences and controls

Facies sequences and controls

Siliciclastic depositional environments give rise to characteristic sedimentary facies, as outlined above, with compositions, textures and sedimentary structures dependent mainly on sediment source and supply and depositional process(es). Tectonic context, position of sea-level and climate are major external factors affecting and controlling the pattern of sedimentation. The facies successions developed in an area depend on the depositional processes and changes in the external factors. Vertical facies sequences can be generated through depositional processes alone, particularly through the progradation and lateral migration of environments (as propounded by Johannes Walther, Section 1.2.2) and through vertical aggradation. Deltas building out into the sea (Figs 2.67–2.69), rivers meandering across a floodplain (Figs 2.65 & 2.66), beaches, barriers and tidal flats prograding seawards (Fig. 2.74), and the construction of alluvial (Fig. 2.62) and submarine fans (Fig. 2.81), all lead to lateral migrations of environments and subenvironments and the formation of characteristic depositional units and facies successions. Commonly, these are large-scale coarsening-upward, or fining-upward packages with associated changes in bed thickness and sedimentary structures. The filling of a sedimentary basin with turbidites, bringing the depositional surface into the shallow-marine environment, filling of a lake basin and sedimentation on a river floodplain are large-scale examples of vertical aggradation.

Many changes in siliciclastic facies successions, however, are brought about by the external factors. Tectonic movements can have a drastic and varied effect. Examples include:

1 the influx of coarse sediment through uplift in the source area—with rivers this could bring about a change from a meandering to braided character and in any siliciclastic succession it could be marked by a change in sediment composition and/or texture;

2 an increase in seismicity leading to more frequent sediment gravity flows, resedimentation and post-depositional sedimentary structures.

In the last few years there has been a great emphasis on the roles of tectonics in sedimentation. Facies and thickness patterns frequently have been explained as a result of contemporaneous fault movements, particularly in rift and extensional basins. Some depositional environments are closely tied to active faults, alluvial fans and fan deltas, for example. Sedimentary patterns also can be explained by much larger-scale plate movements, such as the formation of rifts, the closure of oceans and the uplift of mountain belts. The significance of plate-tectonic setting in sandstone composition has been discussed and the latter is very important to a sandstone's porosity evolution. Much effort has gone into understanding the formation of sedimentary basins themselves, their subsidence and thermal histories, and then their uplift and erosion (see the reviews of Miall, 2000; Einsele, 2000; Busby & Ingersoll, 1995; Allen & Allen, 2001). For information on sedimentary sequences in rifts and extensional basins see Frostick *et al.* (1986), Coward *et al.* (1987), Landon (1994) and Purser & Bosence (1998), in foreland basins see Allen & Homewood (1986) and Macqueen & Leckie (1992), in marginal basins see Leggett (1982), Kokelaar & Howells (1984) and Biddle (1991), and in strike-slip basins see Biddle & Christie-Blick (1985). See papers in Shanley & McCabe (1998) as well.

Changes in climate are very important in siliciclastic

deposition. With increased aridity, river flow becomes more erratic and braiding may result; red beds, calcretes, evaporites and aeolian sands are likely to develop, and subaerial weathering, which supplies much of the siliciclastic material for sedimentation, is more physical than chemical. With increased humidity, vegetation may be more lush, giving rise to coals; more clays may be produced through more intense chemical weathering in source areas and dissolution of grains may be more widespread during early diagenesis.

Sea-level changes have a profound effect on siliciclastic sedimentation. For example, with a rise in relative sea-level, deltaic cycles can be repeated, and clastic shorelines can be drowned and overlain by shelf or epeiric sea deposits; with a fall in sea-level, turbidity currents may be more frequent as fluvial sediments are deposited closer to shelf margins.

In recent years, the concepts of *sequence stratigraphy*, briefly introduced in Section 1.2.2, have been applied to more and more strata. Sedimentary packages (effectively sedimentary cycles), on both large and small scales, are identified, and their key surfaces, geometry, stacking patterns, etc., are studied with a view to deducing the controls on deposition.

Depositional sequences are packages of genetically related strata bound by unconformities (and their correlative conformities). It is thought that many sequences, but principally those of gently subsiding basin margins, are produced by second- to third-order (10^6–10^7 years) rises and falls of sea-level, against a background of steady subsidence (see Fig. 2.86). Within these sequences, which are of the order of hundreds of metres in thickness, there usually occurs a predictable series of sedimentary facies, laid down in specific depositional *systems tracts*, developed during particular intervals of the relative sea-level curve (see Fig. 2.86). The transgressive systems tract, established during the relative sea-level rise, is characterized by onlapping (retrogradational) strata of shallow-marine, shoreline and coastal-plain facies. The highstand systems tract is one of aggradation and progradation (offlap and downlap), again of shallow-marine to coastal-plain facies, with fluvial and deltaic facies becoming important later. During the transgressive and highstand systems tracts, outer shelves and basin floors are mostly starved of sediment and condensed horizons (with glauconite and phosphorite, for example) are formed. During the ensuing relative

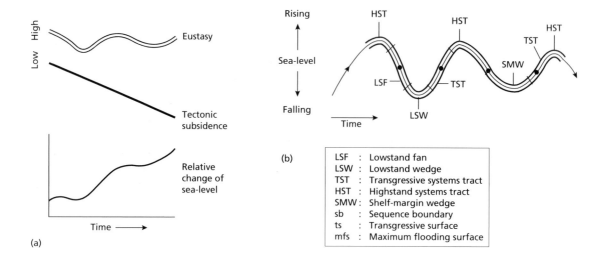

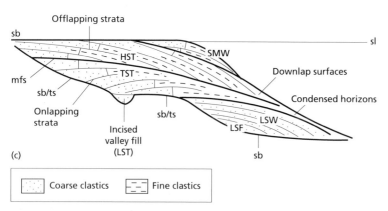

Fig. 2.86 Sequence stratigraphy, systems tracts and relative sea-level changes. (a) Simple graphs showing interaction of eustatic sea-level change (top) and uniform tectonic subsidence to give the relative change of sea-level. (b) Systems tracts and the relative sea-level-change curve. (c) Sequence stratigraphic depositional model for a siliciclastic shelf on a gently subsiding basin margin. A major sea-level fall, generating the lower sequence boundary (sb), drops sea-level below the shelf break. Exposure of the shelf causes fluvial incision there (incised valley) and deposition of a lowstand fan (LSF) at the toe of the slope. A lowstand wedge (LSW) of sediment gravity flow deposits accumulates as sea-level begins to rise. When the rate of sea-level rise increases, the incised valley is filled and the shoreline transgresses landwards across the shelf, depositing the transgressive systems tract (TST), as an onlapping package. The transgressive surface (ts) at the base of TST coincides with the sequence boundary over much of the shelf. The top of the TST is marked by the maximum flooding surface (mfs), succeeded by the offlapping and downlapping strata of the highstand systems tract (HST), as the sea-level rise reaches a maximum and begins to fall. The latter produces the next sequence boundary. If, as shown here, sea-level now does not fall below the shelf break, then a shelf-margin wedge (SMW) is generated. After Van Waggoner *et al.* (1988).

sea-level fall, which generates the sequence boundary, two types of systems tract are possible depending on the magnitude of the fall. If sea-level falls below the shelf-break, then a lowstand systems tract is estab- lished, with fluvial incision on the shelf, supplying sand to submarine fans and aprons developed along the toe of the shelf slope (lowstand fan (LSF) and low- stand wedge (LSW); Fig. 2.86). If the sea-level fall does not reach the shelf break, then a shelf-margin systems tract is established, with shoreline, coastal-plain and fluvial clastics on the shelf, which onlap as sea-level begins to rise again. These systems tracts may themselves be made up of small-scale cycles (parasequences), produced by relative sea-level changes on a 10^4–10^5 years scale.

One interpretation of sequences is that the second- to third-order sea-level rises and falls, operating on the scale of 1–10 million years, which produced them are global eustatic. If this is correct, it should be possible to correlate the sequences from one basin or passive margin to another. This is the rationale behind the Haq–Vail global sea-level curves. However, there are many that do not accept this assertion (see Miall (1997) for discussion), and advocate that the sequences are more the result of tectonic rifting and regional subsidence. Nevertheless, the larger-scale perspective given by sequence stratigraphy and the recognition of systems tracts within sequences, does help to give a better understanding of the overriding controls on sedimentation. The sequence approach can be taken further because many sequence boundaries represent extended periods of exposure of a continental shelf or craton, as a result of a major relative sea-level fall (however caused). At these times, and depending on climate, meteoric recharge can take place around the margins of the sedimentary basin, with important consequences for the diagenesis and porosity evolution of the clastic sediments.

Considering sedimentary successions in the context of relative sea-level changes reveals that certain depositional environments and their resulting facies models are developed preferentially, or even restricted, at particular times of the relative sea-level curve. Thus, deltas tend to be more common during the late highstand to early lowstand; incised valleys and turbidite fans and aprons are typical of lowstands; beaches, barrier islands and shelf sands are best developed during times of transgression and early highstand. Papers on clastic sequence stratigraphy can be found in Posamentier *et al.* (1993), Weimer & Posamentier (1993), Emery & Myers (1996) and Gradstein *et al.* (1998).

Further reading

Allen, P.A. (1997) *Earth Surface Processes*. Blackwell Science, Oxford, 404 pp.

Emery, D. & Myers, K. (1996) *Sequence Stratigraphy*. Blackwell Science, Oxford, 297 pp.

Galloway, W.E. & Hobday, D.K. (1983) *Terrigenous Depositional Systems*. Springer-Verlag, New York, 423 pp.

Leeder, M.R. (1999) *Sedimentology and Sedimentary Basins*. Blackwell Science, Oxford, 592 pp.

McDonald, D.A. & Surdam, R.C. (Eds) (1984) *Clastic Diagenesis*. Memoir 37, American Association of Petroleum Geologists, Tulsa, OK, 434 pp.

Miall, A.D. (1997) *Geology of Stratigraphic Sequences*. Springer-Verlag, Berlin, 433 pp.

Nichols, G. (1999) *Sedimentology and Stratigraphy*. Blackwell Science, Oxford, 355 pp.

Pettijohn, F.J., Potter, P.E. & Siever, R. (1987) *Sand and Sandstone*. Springer-Verlag, New York, 553 pp.

Pye, K. (Ed.) (1994) *Sediment Transport and Depositional Processes*. Blackwell Science, Oxford, 397 pp.

Reading, H.G. (Ed.) (1996) *Sedimentary Environments: Processes, Facies and Stratigraphy*. Blackwell Science, Oxford, 688 pp.

Scholle, P.A. (1979) *A Color-Illustrated Guide to Constituents, Textures, Cements and Porosities of Sandstones and Associated Rocks*. Memoir 28, American Association of Petroleum Geologists, Tulsa, Ok.

Scholle, P.A. & Spearing, D. (Eds) (1982) *Sandstone Depositional Environments*. Memoir 31, American Association of Petroleum Geologists, Tulsa, OK, 410 pp.

Walker, R.G. & James, N.P. (Eds) (1992) *Facies Models: Response to Sea-level Change*. Geological Association of Canada, 409 pp.

Weimer, P. & Posamentier, H.W. (1993) *Siliciclastic Sequence Stratigraphy*. Memoir 58, American Association of Petroleum Geologists, Tulsa, OK, 492 pp.

3 Siliciclastic sediments II: mudrocks

3.1 Introduction

Mudrocks are the most abundant of all lithologies, constituting some 45–55% of sedimentary rock successions. However, because mudrocks are easily weathered, they are frequently covered in vegetation and poorly exposed. In addition, as a result of their fine grain size, their study often requires detailed laboratory analyses. Mudrocks can be deposited in practically any environment although the major depositional sites are river floodplains and lakes, large deltas, the more distal areas of clastic shelves, basin slopes and deep-sea floors. The main constituents of mudrocks are clay minerals and silt-grade quartz. As these are largely detrital, the clay mineralogy to a greater or lesser extent reflects the climate and geology of the source area.

In terms of grain size, *clay* refers to particles less than 4 μm in diameter, whereas *silt* is between 4 and 62 μm (Table 2.3). Clay as applied to a mineral is a hydrous aluminosilicate with a specific sheet structure (Section 3.4.1); the typical size of clay minerals is less than 2 μm but they may reach 10 μm or more. The term *mud* (also lutite) loosely refers to a mixture of clay- and silt-grade material. *Mudstone*, the indurated equivalent of mud, is a blocky, non-fissile rock, whereas *shale* is usually laminated and fissile (fissility is the property of splitting into thin sheets). *Argillite* is used for a more indurated mudrock and *slate* for one that possesses a cleavage. A sedimentary rock dominated by clay-grade material is called a *claystone*, and one that contains more silt-grade particles than clay is called a *siltstone*. Calcareous mudrocks are *marls*. As a general term for all these rock types, mudrock is useful. Figure 3.1 gives a scheme for mixed sand–silt–clay deposits, and Fig. 2.1 gives terms for mud–sand–gravel mixtures.

To describe mudrocks, particularly in the field, the terms mudstone, shale, claystone and siltstone are best qualified by attributes referring to colour, degree of fissility, sedimentary structures, and mineral, organic and fossil content (Table 3.1). Detailed studies in the laboratory, especially the use of X-ray diffraction (XRD), may be required to determine the mineralogy.

3.2 Textures and structures of mudrocks

Fine terrigenous clastic sediments do not have the wide range of textures and structures so typical of coarser siliciclastic sediments (Chapter 2). This is mainly because of the finer grain size and cohesive properties of mud. The grain-size distribution of mudrocks is not easy to study, and there are problems of interpretation. The particle size of unconsolidated muds can be measured using a sedimentation chamber or settling tube (see McManus (1988) and Singer *et al.* (1988) for these techniques). The scanning electron microscope (SEM) is used for studying well-lithified mudrocks and the back-scattered mode in particular can give much useful information on textures (see Fig. 3.2 and Pye & Krinsley, 1984; Welton, 1984; White *et al.*, 1984; McHardy & Birnie, 1987). The interpretation of the grain-size data in terms of environmental conditions is complicated by the fact that clays are commonly deposited as floccules and aggregates, and that sediment-feeding organisms may generate pellets of mud. Bioturbation disrupts the depositional fabric of a mud but by examining mudrock microfabrics with the SEM and X-radiography, it is possible to determine whether the muddy sea floor was a soupground, softground or firmground (O'Brien, 1987).

A common texture is that of a *preferred orientation* of clay minerals and micas parallel to the bedding. This can be detected in thin-section by areas of common extinction. The texture is the result of deposition of clay flakes parallel to bedding and, perhaps to a lesser extent, the effect of compaction and dewatering. Modern marine muds do have preferred orientations of clays (see Kuehl *et al.* (1988) for a description of the microfabrics of muds from off the Amazon delta). Related to this is the property of *fissility*, possessed by shales (Fig. 3.3). Fissility is the ability of mudrocks to

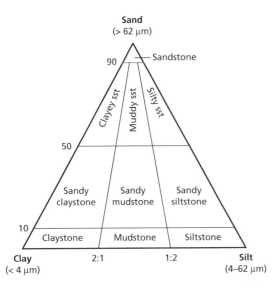

Fig. 3.1 Classification of siliciclastic sediments based on sand, silt and clay content.

Table 3.1 Features to note in the description of mudrocks

Attribute	Examples of descriptive adjectives
Colour	grey, green, red, brown, variegated, mottled
Degree of fissility	fissile, non-fissile, blocky, earthy, flaggy, papery
Sedimentary structure	bedded, laminated, slumped, bioturbated, massive
Mineral content	quartzose, illitic, kaolinitic, zeolitic, micaceous, calcareous dolomitic, gypsiferous
Organic content	organic-rich, bituminous, carbonaceous
Fossil content	fossiliferous, foraminiferal, ostracod, graptolitic

split along smooth planes parallel to the stratification. The origin of fissility is not fully resolved. A major factor is the compaction-induced alignment of clay minerals, but also important is the presence of lamination. Fissility may be poorly developed or absent, as it is in mudstones, because of bioturbation, the presence of much quartz silt or calcite, and the flocculation of clays during sedimentation producing a random fabric that is retained on compaction. The degree of fissility shown by mudrocks may be related to weathering at outcrop.

(a)

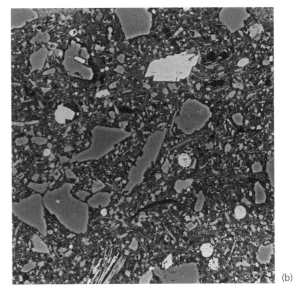

(b)

Fig. 3.2 Photomicrograph and back-scattered electron image of Upper Jurassic mudrock. Buckinghamshire, England. (a) Photomicrograph shows mudrock with quartz silt (white) and scattered pyrite (black grains). Field of view 1 × 1 mm. (b) Back-scattered scanning electron micrograph showing angular quartz silt grains (grey), coccoliths (round white grains) and pyrite (white crystals) in dark matrix of clay, quartz and amorphous organic matter. Field of view 200 × 200 μm. Courtesy of Joe Macquaker.

(a)   (b)

Fig. 3.3 Shale and mudstone. (a) Fissile black, organic-rich shale. Kimmeridge Clay, Jurassic. Yorkshire, England. (b) Massive. non-laminated mudstone with hackly, conchoidal fracture. Carboniferous. Glamorgan, Wales.

One common sedimentary structure of mudrocks is *lamination* (Fig. 3.4). Lamination is the result mainly of variations in grain size and/or changes in composition. Size-graded laminae may be deposited from low-density turbidity and suspension currents or from decelerating storm currents, in relatively short periods of time (hours or days). Other laminae may develop over much longer periods of time (months or years) if there is a seasonal or annual fluctuation in sediment supply and/or biological productivity. Organic laminae in mudrocks, for example, may be produced by seasonal microbial blooms, and the varved couplets of glacial lakes (Section 3.7.2) are taken to reflect the annual spring melting.

Siltstones may show the same sedimentary structures as finer grades of sandstone (Section 2.3.2). Small-scale current ripples occur and give rise to cross-lamination. Planar beds with parting lineation (Fig. 2.17) can develop at higher stream powers. Symmetrical wave-formed ripples also can form. Where mud and sand are deposited alternately, through fluctuating current regimes and/or sediment supply, as in tidal-flat, delta-front and other environments, then flaser and lenticular bedding are produced (Fig. 2.27). Different types of lamination, small-scale scour-and-fill structures, and micro-cross-lamination, have been described from modern marine muds by Rine & Ginsburg (1985) and Kuehl *et al.* (1988).

Some mudrocks do not possess sedimentary structures but are *massive*. This massive nature may be a result of the depositional mechanism. It is a common feature where sediment–water flows have a high viscosity at the time of deposition, as in mudflows and debris flows (Sections 2.11.1 and 2.11.7). In other cases mudrocks have become massive subsequent to initial deposition through bioturbation, mass sediment movement (slumping), dewatering, soil processes (pedogenesis) and root growth. The structures in muds and mudrocks can be studied using X-radiography; this may reveal lamination or bioturbation in an otherwise massive-looking rock (see Bouma (1969) for details of the technique).

Other sedimentary structures that occur in mudrocks, and have been described in Chapter 2, include: the erosional structures cut into muds and preserved on the soles of sandstones (grooves and flutes); slump structures; desiccation cracks formed through subaerial exposure; synersis cracks formed through subaqueous sediment shrinkage; biogenic structures and rain-spot prints. Although in the field, few sedimentary structures are seen in mudrocks, detailed examination of thin-sections and polished blocks can reveal different types (e.g. Schieber, 1989). One particular feature of Precambrian shales is that they are undisturbed by bioturbation, so that primary depositional structures are well developed and preserved. Carbonaceous shales are also common in the Precambrian, and Schieber (1986) presented evidence to suggest that they formed beneath benthic microbial mats (in an analogous manner to stromatolites in limestones). Macquaker & Gawthorpe (1993) documented a variety of microfacies in the apparently

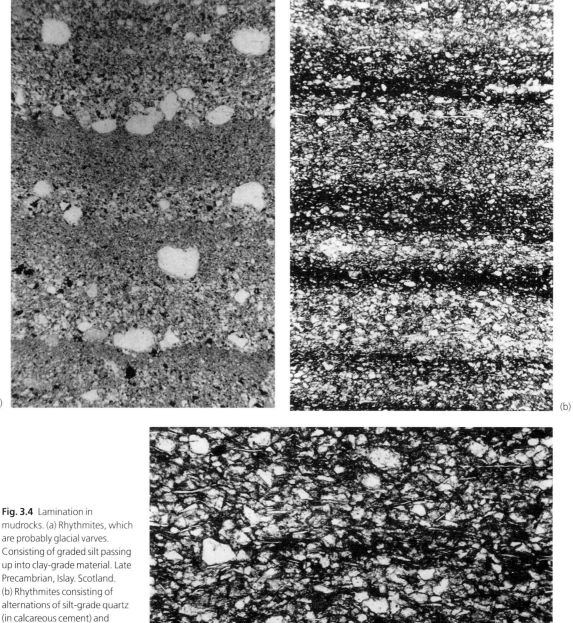

Fig. 3.4 Lamination in mudrocks. (a) Rhythmites, which are probably glacial varves. Consisting of graded silt passing up into clay-grade material. Late Precambrian, Islay. Scotland. (b) Rhythmites consisting of alternations of silt-grade quartz (in calcareous cement) and clay–organic matter; the result of seasonal deposition in a non-glacial lake. Caithness Flagstones, Devonian, Scotland (Fig. 2.16). In plane-polarized light.

(a)

(b)

(c)

homogeneous Kimmeridge Clay of the Wessex Basin, southern England, and Schieber (1999) documented the mudstone facies in the Devonian of New York. See papers in Schieber *et al.* (1998) and Aplin *et al.* (1999).

3.2.1 Nodules and concretions

Many mudrocks contain *nodules*, also called *concretions* (Fig. 3.5). These are regular to irregular, spherical, ellipsoidal to flattened bodies, commonly composed of calcite, siderite (Section 6.4.2), pyrite (Section 6.4.3), chert (Section 9.4) or calcium phosphate (Section 7.4), together with some original sediment. Nodules grow within the sediment during diagenesis and this may take place just below the sediment–water interface or much deeper in the sediment column.

Early diagenetic nodules form in sediments that were still soft and uncompacted. These can be recognized from the presence of uncrushed fossils within the nodules and from the folding of laminae in the mudrock around the nodule, showing that compaction took place after growth. Some early formed nodules are reworked into conglomerates, and they may be bored and encrusted. Nodules also may form after compaction of the host sediment during burial diagenesis, and with these, laminae in the sediment pass unaffected through the nodule. Nodules containing an internal network of cracks, particularly ones that widen towards the interior, are referred to as *septarian nodules*. The cracks form through shrinkage and excess pore pressure, and usually they are filled with

coarse crystals, commonly calcite (see Hounslow, 1997).

The growth of nodules arises from the localized precipitation of cement from pore waters within the sediment. The composition, Eh and pH of these pore waters are important in controlling nodule mineralogy and growth rates. In some cases nodules formed around a nucleus, a fossil, for example, as a result of the local chemical conditions. More commonly nodules are without a nucleus, and form along definite horizons or within particular beds, reflecting a level at which supersaturation of pore waters was achieved. Elongate nodules may show a preferred orientation, reflecting the direction of pore-water movement (e.g. McBride *et al.*, 1994). Descriptions and discussions of the origin of nodules include Scotchman (1991) and papers in Cope & Curtis (2000).

A poorly understood structure occurring in mudrocks is *cone-in-cone structure*. It consists of nested cones of fibrous calcite, rarely ankerite or siderite, orientated normal to the bedding, the whole constituting a nodule or impersistent layer. There is often evidence that the structure has formed by displacement. Cone-in-cone may be analogous to fibrous gypsum (Section 5.2.5), being formed through hydraulic jacking-up of the sediment, with the fibrous nature of the calcite reflecting growth under stress and high pore-fluid pressures.

3.3 The colour of mudrocks

The colour of a mudrock, and indeed of any rock, is a function of its mineralogy and geochemistry. Colour can be very useful in field mapping to distinguish between various mudrock units. The main controls on colour are the organic matter and pyrite content and the oxidation state of the iron. With an increasing amount of organic matter and pyrite, mudrocks take on a darker grey colour and eventually become black. Many marine and deltaic mudrocks are various shades of grey or even black as a result of finely disseminated organic matter and pyrite.

Red and purple colours result from the presence of ferric oxide, hematite, occurring chiefly as grain coatings and intergrowths with clay particles. It generally is accepted that the red colour develops after deposition, through an ageing process of a hydrated iron oxide precursor, as is the case with red sandstones (Section 2.9.6). The origin of the precursor, however —

Fig. 3.5 Calcareous nodules in red mudrock. Relatively deepwater, marine shale. Devonian. Harz Mountains, Germany.

whether it has formed through *in situ* dissolution of metastable mafic grains or is detrital — is still a matter of discussion and both mechanisms will doubtless apply in different cases. The impermeable nature of many red mudrocks might suggest a detrital source. Where iron oxide coatings on grains are patchily developed or less intense, then brown colours may result. Many non-marine mudrock successions of arid-zone playa lakes and floodplains are red in colour, reflecting the dominantly oxidizing nature of the depositional and early diagenetic environment.

Green mudrocks contain no hematite, organic matter or iron sulphides, but the colour comes from ferrous iron within the lattices of illite and chlorite. The green colour may be original but in many instances it develops in mudrocks that originally were red, through the reduction of hematite by migrating groundwaters. The green colour may thus occur in more porous silty bands, in mudrocks close to porous sandstones, or adjacent to faults and joints. Green spots and patches in some red mudrocks are sites of iron reduction from local occurrence of organic matter.

Other colours in mudrocks result from a mixing of the colour-producing components. Olive and yellow mudrocks, for example, may owe their colour to a mixing of green minerals and organic matter. Some mudrocks have a colour mottling. In different shades of grey, it simply may be the result of bioturbation, but in yellows/reds/browns it can be the result of pedogenic processes, of water moving through a soil causing an irregular distribution of iron oxide/hydroxide and/or carbonate, and the effect of roots. The term *marmorization* has been applied to this process. Colour mottling is common in lacustrine and floodplain muds and marls, especially those of palustrine facies (see Section 4.10.1). This feature is well developed in Eocene palaeosoils developed in fluvial mudrocks of Wyoming, where a range of soil textures and profiles is seen, depending on the length of time pedogenesis was operating (Bown & Kraus, 1987).

3.4 Mineral constituents of mudrocks

3.4.1 Clay minerals

Clay minerals are hydrous aluminosilicates with a sheet or layered structure; they are phyllosilicates, like the micas. The sheets of a clay mineral are of two basic types. One is a layer of silicon–oxygen tetrahedra with three of the oxygen atoms in each tetrahedron shared with adjacent tetrahedra and linked together to form a hexagonal network (Fig. 3.6). The basic unit is Si_2O_5 but within these silica layers aluminium may replace up to half the silicon atoms. The second type of layer consists of aluminium in octahedral coordination with O^{2-} and OH^- ions so that in effect the Al^{3+} ions are located between two sheets of O/OH ions (Fig. 3.6). In this type of layer, not all the Al (octahedral) positions may be occupied, or Mg^{2+}, Fe and other ions may substitute for the Al^{3+}. Layers of Al–O/OH in a clay mineral are referred to as *gibbsite layers* because the mineral gibbsite $(Al(OH)_3)$ consists entirely of such layers. Similarly, layers of Mg–O/OH are referred to as *brucite layers* after the mineral brucite $(Mg(OH)_2)$ composed solely of this structural unit. Clay minerals, then, consist of sheets of silica tetrahedra and aluminium or magnesium octahedra linked together by oxygen atoms common to both. The stacking arrangement of the sheets determines the clay-mineral type, as does the replacement of Si and Al ions by other elements. Structurally, the two basic groups of clay minerals are the kandite group and smectite group.

Members of the kandite group have a two-layered structure consisting of a silica tetrahedral sheet linked to an alumina octahedral (gibbsite) sheet by common O/OH ions (Fig. 3.6). Replacement of Al and Si does not occur so that the structural formula is $(OH)_4Al_2Si_2O_5$. Members of the kandite group are *kaolinite*, by far the most important, the rare dickite and nacrite, which have a different lattice structure, and halloysite, which consists of kaolinite layers separated by sheets of water. Related structurally to kaolinite are the alumino-ferrous silicates berthierine and chamosite and the ferrous silicate greenalite (see Section 6.4.4). Kaolinite has a basal spacing, i.e. distance between one silica layer and the next, of 7 Å.

Members of the smectite group have a three-layered structure in which an alumina octahedral layer is sandwiched between two layers of silica tetrahedra (Fig. 3.6). The typical basal spacing is 14 Å but smectites have the ability to adsorb water molecules and this changes the basal spacing; it may vary from 9.6 Å (with no adsorbed water) to 21.4 Å. This feature of smectites, as a result of which they are often called 'expandable clays', is utilized in their X-ray identification. The common smectite is *montmorillonite*; it approximates

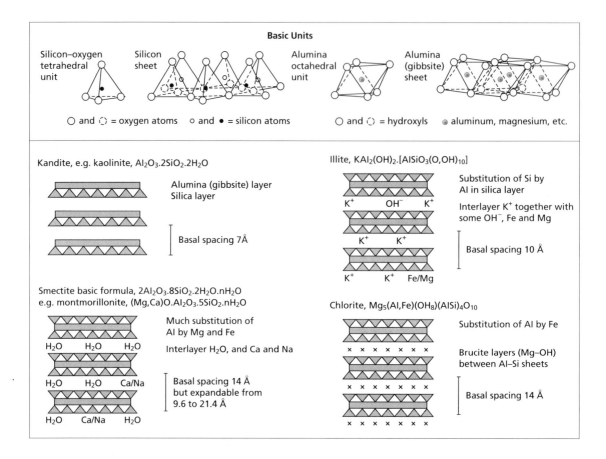

Fig. 3.6 Diagrams illustrating the structures of clay minerals.

to $Al_4(Si_4O_{10})_2(OH)_4.nH_2O$ but substitution of the Al^{3+} by Fe^{2+}, Mg^{2+} and Zn^{2+} can take place. A net negative charge resulting from such substitutions is balanced by other cations, especially Ca^{2+} and Na^+, which are contained in interlayer positions. Nontronite, saponite and stevensite are other smectites found in sediments; for nontronite Fe^{3+} replaces Al^{3+} in the octahedral layers, and in saponite and stevensite, Mg^{2+} substitution of Al^{3+} has taken place. Vermiculite has a structure similar to smectite, although it is less expandable, with all the octahedral positions occupied by Mg^{2+} and Fe^{2+} and much substitution of Si^{4+} by Al^{3+}.

Illite, the most common of the clay minerals in sediments, is related to the mica muscovite. It has a three-layered structure, like the smectites, but Al^{3+} substitution for Si^{4+} in the tetrahedral layer results in a deficit of charge, which is balanced by potassium ions in interlayer positions (Fig. 3.6). Some hydroxyl OH^-, Fe^{2+} and Mg^{2+} ions also occur in illite. The basal spacing is about 10 Å.

Other clay minerals are chlorite, glauconite, sepiolite and palygorskite. Chlorite, like the smectites and illite, has a three-layered structure, but with a brucite (Mg–OH) layer between (Fig. 3.6). Substitution by Fe^{2+} occurs in chlorite (imparting the green colour) and the basal spacing is 14 Å. Glauconite (Section 6.4.4) is related to illite and the micas, but contains Fe^{3+} substituting for Al^{3+}. Sepiolite and palygorskite are magnesium-rich aluminosilicates.

In addition to the four common clay minerals— illite, kaolinite, montmorillonite and chlorite— *mixed-layer clays* also are common. These consist of an interleaving of sheets of the common clays, in particular illite–montmorillonite and chlorite–

montmorillonite. Specific names have been applied where there is a regular mixed-layering: *corrensite* for a chlorite–montmorillonite mixed-layer clay, for example.

During weathering and diagenesis, interlayer cations can be leached out of the clay minerals by percolating waters. Such non-stoichiometric clays are termed *degraded* and in fact many illites and smectites in modern sediments are degraded, as well as some chlorites, biotites and muscovites.

In view of their fine crystal size and the presence of unsatisfied bonds, clay minerals are important in the process of ion exchange. Ions in aqueous solutions can be adsorbed onto and desorbed from clays, with the water chemistry controlling the exchange process. Some elements, such as iron and manganese (Section 6.2) can be transported by adsorption on clays.

The identification of clay minerals in a mudrock (or any other rock) normally is undertaken through X-ray diffraction of the less than 2 µm fraction of the sediment. See Hardy & Tucker (1988) for details. The basal spacings of the clay minerals are deduced from the X-ray reflections for the untreated sample, and then the sample treated (a) with glycol, which causes expansion of the lattice of any smectites present and (b) with heat, which reduces the lattice spacing of smectite. The clay minerals in mudrocks can rarely be identified with the petrological microscope because of their fine crystal size, only where authigenic in sandstones (Section 2.9.5; Plates 2a & 5c). A clearer picture of clay-mineral crystal shape can be obtained from the scanning electron microscope, which is now increasingly used for clay mineral identifications (see Figs 2.54 & 2.55) and mudrock textural studies, especially in the back-scattered mode (see Fig. 3.2 and Welton, 1984; Trewin, 1988).

For further information on clay minerals and their structure see Brindley & Brown (1984), Chamley (1989) and Weaver (1989).

3.4.2 Quartz

Quartz in mudrocks is chiefly of silt-grade, although coarser, sand-size grains do occur, especially where the mudrocks grade laterally or vertically into sandstones. Quartz silt is invariably angular in comparison with typically more rounded quartz sand. Silt-grade quartz is derived from grain collisions in aqueous and aeolian media and from glacial grinding. Some quartz in mudrocks is diagenetic rather than detrital. The regional variation in grain size and percentage of quartz in a mudrock formation can be used in palaeogeographical studies (e.g. Macquaker & Gawthorpe, 1993).

3.4.3 Other constituents

Feldspars are generally present in low concentrations in mudrocks in view of their lower mechanical and chemical stability relative to quartz (Section 2.5.3). However, because mudrocks are less permeable than sandstones, feldspars and other less stable grains may be preserved preferentially in muddy sediments, so that the non-clay fraction can be used for provenance information.

Muscovite is common but biotite much less so. Calcite (or aragonite) may occur in the form of skeletal debris. Diagenetic calcite, as well as dolomite and siderite, occur as microscopic crystals, evenly disseminated or concentrated into nodules. Mudrocks may grade into calcareous mudrocks (marls). Pyrite occurs as cubes, framboids and nodules in dark, organic-rich mudrocks (Section 6.4.3). Other minerals present locally are glauconite, berthierine, hematite, gypsum–anhydrite and halite. Organic matter is common in mudrocks, particularly black shales (Section 3.7.2) and if present in sufficient concentrations it may give rise to oil shales (Section 8.8). Finely comminuted rock debris (rock 'flour') is a major constituent of muddy sediments in glacial environments.

3.5 The formation and distribution of clay minerals in modern sediments

Clay minerals in a sediment or sedimentary rock have three origins: (i) inheritance, (ii) neoformation and (iii) transformation. In the first, the clays are detrital and have been formed in another area, perhaps at a much earlier time, but they are stable in their present location. In the second, the clays have formed *in situ*, and they have either been precipitated from solution or formed from amorphous silicate material. With transformation, inherited clays are modified by ion exchange or cation rearrangement. In the study of mudrocks, it is clearly important to be certain of the origin of the clays if meaningful interpretations are to be made. Inherited clays will give information on the provenance of the deposit and probably the climate there, whereas neoformed clays reflect the pore-fluid chemistry, degree of leaching and temperature that existed within the sample at some stage. Transformed clays will carry a memory of inherited characteristics from the source area, together with information on the chemical environment to which the sample was later subjected.

There are three major locations where clay-mineral formation takes place:

1 in the weathering and soil environment;
2 in the depositional environment;
3 during diagenesis and into low-grade metamorphism.

The major site of clay-mineral formation is in the weathered mantles and soil profiles developed upon solid bedrock and unconsolidated sediments. Soils develop through physico-chemical and biological processes (*pedogenesis*) and usually possess distinct horizons, A, B and C (see Fig. 3.7), which often can be subdivided further. The upper or A horizon of a soil is chiefly a zone from which material has been transferred down to the B horizon. This process of *eluviation* transfers clay minerals, colloidal organic matter and ions in solution. The A horizon may contain much organic matter decomposing into humus. The B horizon is the zone of *illuviation* (i.e. accumulation) and so contains a higher clay content, precipitated hydroxides of iron and manganese, and carbonates. The C horizon consists of partly altered bedrock or sediment, passing down into fractured and then fresh parent material. The thickness, development of horizons, and mineralogy of soils vary considerably, being dependent largely on climate, nature of the source material, topography, vegetation and time. The common modern soil types are given in Table 3.2.

Practically all clay-mineral types can develop within soils through pedogenesis, and within weathering mantles. The clays form by the alteration and replacement of other silicate minerals such as feldspars and micas, the transformation of detrital clay minerals, and direct precipitation (neoformation/authigenesis). The degree of leaching and the pH–Eh of the soil water, both determined largely by the climate, are the two main factors controlling clay-mineral formation and stability. The nature and composition of the host rock or sediment is also important.

Where the degree of leaching is limited, as with many soils in temperate areas, then illite is the typical clay mineral formed. Chlorite also forms during intermediate stages of leaching in temperate soils but it is more easily oxidized and so occurs preferentially in acid soils. It also forms in soils of arid regions, both high and low latitude, where chemical processes are minimal. Montmorillonite is a product of inter-

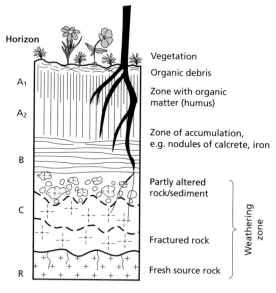

Fig. 3.7 Schematic diagram of a soil profile; there are many variations depending on sediment grain size, mineralogy, vegetation, climate and topography.

Table 3.2 Common types of modern soil

Soil type	Features and conditions of formation
Oxisols	extensive alteration of silicate minerals under wet equatorial climates, with the formation of clays such as kaolinite and iron oxides
Calcisols	precipitation of calcium carbonate as nodules and layers under hot, semi-arid climate
Gleysols	waterlogged soils with poor drainage and reducing conditions in temperate and polar regions
Argillisols	layers and grain coatings of clay forming in wet mid-latitude areas
Vertisols	poorly developed horizons but strong vertical fabric and fractures owing to repeated desiccation of expandable clays in relatively dry climates
Histisols	high contents of organic material forming peat layers in moist temperate and some humid equatorial areas
Protosols	poorly developed and incipient soils, especially in cold or highland areas

mediate leaching and moderate weathering conditions, being common in temperate soils with good drainage and neutral pH, in gley (poorly drained) soils and in arid-zone soils that are highly alkaline. Mixed-layer clays form mostly through the leaching of pre-existing illite and mica. Kaolinite and halloysite are characteristic of acid tropical soils where leaching is intensive. Further leaching of kaolinitic soils and the removal of silica gives rise to gibbsite and other aluminium hydroxides that form *bauxite*. Iron-rich soils of the humid tropics, the *laterites* or ferricretes, also are formed through extreme weathering. They are composed of hydrated iron oxides and kaolinite. Some less common clay minerals are developed in particular soil types, such as palygorskite and sepiolite in calcretes, Mg-rich soils and silcretes.

After neoformation and transformation in the provenance region, clay minerals are available for erosion, transportation and deposition. Many, perhaps most, of the clay minerals in modern and ancient clayey sediments are detrital. There are other processes, however, whereby clays are formed. Clay minerals can be precipitated directly from water or pore waters in surficial sediments. Sepiolite, palygorskite and corrensite, for example, are formed in this manner in Mg-rich, alkaline lakes. Clay minerals form readily in the alteration of volcanic material (Sections 3.7.3 and 10.7). Volcanic glass is metastable and in time it devitrifies and is replaced by smectites, chlorite and illite, as well as zeolites, and these can be reworked into younger sediments. Clay minerals can be precipitated within coarser siliciclastic sediments during diagenesis (Section 2.9.5), and the clays of mudrocks are themselves altered and replaced by other clays during diagenesis (Section 3.6). Clays formed through these non-pedogenic processes can amount to more than the detrital component in a muddy sediment.

The distribution of clay minerals in modern sediments is largely a reflection of the climate and weathering pattern of the source area. This is well illustrated by the clay mineralogy of the world's ocean-floor sediments, which also reflects contemporaneous volcanic activity. Kaolinite is dominant in low-latitude areas, particularly off major rivers draining regions of tropical weathering, and illite is more common in ocean-floor muds of higher latitudes. The distribution of smectites, derived largely from alteration of volcanic material, is related to the active mid-ocean ridge systems and volcanic oceanic islands.

Two less important factors in the distribution of clay minerals in the oceans are grain size and clay-mineral transformations. There are differences in the size of clay-mineral flakes, with kaolinite being the largest (up to 5 μm), illite typically 0.1–0.3 μm and montmorillonite even finer grained. Montmorillonite commonly occurs as floccules or aggregates up to several microns in diameter. Clay-mineral flocculation arises from changes in water chemistry, with Eh–pH and salinity the major factors. In relatively nearshore areas and on continental shelves, a zonation of clay on the basis of grain size frequently can be detected, with a more kaolinite-rich zone inshore of montmorillonite and illite.

It has been suggested that clay minerals from freshwater environments are modified and altered while suspended in sea water or soon after deposition on the sea floor (a process referred to as *halmyrolysis*). Postulated transformations include: montmorillonite to illite, chlorite or palygorskite, and kaolinite to illite, chlorite or smectite. Although it is difficult to assess the importance of these clay-mineral changes, the good correlation between clay mineralogy and climate/weathering of adjacent land masses and ocean-floor volcanicity for modern ocean-floor sediments tends to suggest that such mineralogical transformations are not on a significant scale. They may be more important in areas of very slow sedimentation, however. Berthierine and glauconite do form in sediment-starved locations. Glauconite may form from alteration of degraded micaceous clays by adsorption of K^+ and Fe^{2+} or from neoformation within pre-existing particles such as carbonate grains, clay minerals or faecal pellets (see Section 6.4.4). There is, however, much evidence for chemical alteration of clays on the sea floor. Na^+, Mg^{2+} and K^+ may all be adsorbed, commonly exchanged for Ca^{2+}.

As clay minerals can be related to source areas and climate, they can be used to monitor sediment dispersal patterns in estuaries and along shelves, and to give information on the palaeoclimate. In a vertical sense and allowing for clay diagenesis (see below), variations in the clay mineralogy of marine muds can be related to climatic changes on adjacent continents. As examples, Ruffel & Batten (1990) related depletions in kaolinite, and other features in Cretaceous strata of western Europe, to arid climatic phases linked to lowering of sea-level, and Gibson et al. (2000) related an enrichment in kaolinite in Palaeocene–

Eocene boundary strata in northeastern USA to intensified weathering owing to higher temperature and precipitation.

3.6 Diagenesis of clay minerals and mudrocks

Clay minerals can be modified and altered during early and late diagenesis, and into metamorphism. The main physical post-depositional process affecting the mudrocks as a whole is *compaction*. Compaction in mudrocks expels water and reduces the thickness of the deposited sediment by a factor of up to 10. When muds are deposited they contain in the region of 70–90% water by volume. Compaction through overburden pressure soon removes much of the water so that at depths of 1000 m or so, the mudrocks contain around 30% water (Fig. 3.8). Much of this is not free pore water but is contained in the lattice of the clay minerals and adsorbed onto the clays. Further compaction through water loss requires temperatures ap-

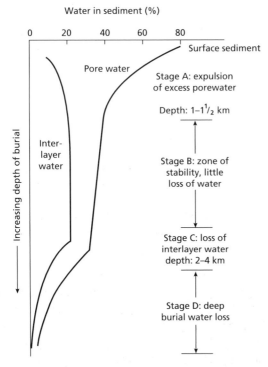

Fig. 3.8 Diagram illustrating the stages of water loss from muddy sediments with increasing depth of burial.

proaching 100°C and these are attained through burial at depths in the region of 2–4 km (Fig. 3.8). Dehydration of clays then takes place, accompanied by some changes in the clay mineralogy (see below). Final compaction to give a mudrock with only a few per cent water requires a much longer period of overburden pressure with elevated temperatures. Evidence of compaction in mudrocks is provided by the fracture of shells, flattening of burrows, and the bending of laminae around shells and early diagenetic nodules (see Plate 5d). The sandy fills of mudcracks and sandstone dykes may be ptygmatically folded through compaction of host mudrocks.

Changes in clay mineralogy during diagenesis take place principally through the rise in temperature accompanying increased depth of burial. Studies of long cores through thick mudrock successions show that the main change is an alteration of smectites to illite via mixed-layer clays of smectite–illite (Fig. 3.9). This alteration involves the incorporation of K^+ ions into the smectite structure and loss of interlayer water. The process is largely temperature dependent and the temperature at which smectite begins to disappear is of the order of 70–95°C; that is, at depths of 2–3 km in areas of average geothermal gradient (30°C km^{-1}). At slightly higher temperatures and greater depths, kaolinite is replaced by illite and chlorite.

There is a change in the clay mineralogy of mudrocks through geological time (Fig. 3.10). Mudrocks back to the Lower Palaeozoic contain a variety of clay minerals, but Lower Palaeozoic and Precambrian mudrocks are dominated by illite and chlorite. The more stable composition of the older sediments probably arises from diagenetic alteration through burial, as noted above, with their greater age allowing more time for diagenetic reactions. A further factor that may be significant is the evolution of land plants during the Late Devonian and into the Carboniferous, and the effects these plants must have had on soil-forming and weathering processes (such as enhanced leaching through organic decomposition and creation of acidic pore waters). The somewhat more irregular pattern of smectite abundance (Fig. 3.10) may relate to orogenic periods. At these times, volcanism is generally more widespread, leading to the formation of much smectite.

Passing into the realm of *incipient metamorphism* (catagenesis or anchimetamorphism) clay minerals are further altered and replaced (Fig. 3.9). The phyllosili-

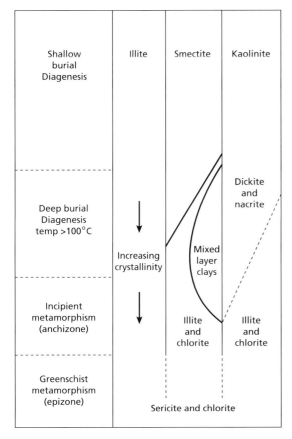

Fig. 3.9 Diagram illustrating the changes of clay minerals with increasing depth of burial and into metamorphism.

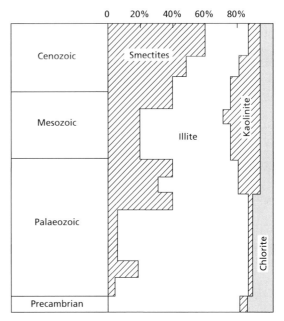

Fig. 3.10 The distribution of clay minerals through time.

cate pyrophyllite (related to talc) and laumontite (a zeolite) may develop at the expense of the clay minerals. Although smectites, mixed-layer clays and kaolinite do not survive into metamorphism, illite and chlorite do. With increasing degree of incipient and low-grade metamorphism, there is an increase in the order or crystallinity of the illite lattice (measured from XRD by a sharpness ratio of the peak, see Hardy & Tucker, 1988). There also are changes in the chemical composition (an increase in the $Al/(Fe + Mg)$ ratio). Illite is then replaced by sericite, a finely crystalline variety of muscovite. The percentage of smectite–illite mixed layering also decreases with increasing burial. Studies of clay mineralogy can thus indicate the late diagenetic to early metamorphic grade, and if combined with measurements of the rank of associated coal and vitrinite reflectance (Section 8.6.2), they can give an in-

dication of the temperatures to which the formation as a whole has been subjected. Clay minerals were studied by Moss (1994) as an indicator of burial diagenesis and the effects of thrust loading in the French Alps foreland basin.

3.7 Mudrocks and their depositional environments

Three major groups of mudrock in the geological record are:
1 those formed *in situ* through contemporaneous processes of weathering and soil formation upon preexisting rocks and sediments;
2 those formed through the normal sedimentary processes of erosion, transportation and deposition;
3 those formed through *in situ* weathering and/or later alteration of volcaniclastic deposits (the latter are discussed at length in Section 10.7).

The second group is by far the most important. See Schieber *et al.* (1998) for papers on shales and mudstones.

3.7.1 Residual mudrocks and soils

Ancient soils are relatively common in the geological

record but weathering mantles developed on bedrock are quite rare. They should be preserved at unconformities but in the majority of cases they have been eroded before deposition of the overlying beds. Weathering profiles on igneous and metamorphic rocks are preserved locally and noteworthy ancient examples include a 2300-Ma Huronian palaeosoil on Archaean greenstone in Ontario, the weathered Lewisian gneiss beneath the Torridonian Sandstones in northwest Scotland and Proterozoic soils on Archaean basalts in southern Africa. Weathering profiles are reviewed by Nesbitt & Young (1989) and papers in Thiry & Simon-Coinçon (1999).

Soils developed upon/within sediments have a long geological record, back to the early Precambrian, but with the advent of land plants in the Siluro-Devonian and their widespread development in the Carboniferous, many new types of soil formed from this mid-Palaeozoic time onward. Of these soils forming within sediments, two familiar types are calcrete (or caliche) and seatearth. *Calcretes* vary from scattered to densely packed nodules of $CaCO_3$ (Fig. 4.39) and are typical of semi-arid climatic areas where evaporation exceeds precipitation (Section 4.10.1). They occur in many river floodplain sediments and clay minerals formed in these soils include sepiolite, palygorskite and smectite. They extend back to the early Proterozoic, although many of the early ones are dolomitic. The calcretes of the Old Red Sandstone (Devonian) in Britain, occurring in floodplain mudrocks (see Fig. 4.31) are particularly well-documented (e.g. Allen, 1986).

In deltaic successions, such as occur in the Upper Carboniferous of Europe and the Mississippian of eastern North America, clayey soils informally known as *seatearths* or underclays occur beneath coal seams. These are commonly *vertisols*, formed under a humid tropical climate with seasonal shrink–swell processes. They are typically massive with rootlets and siderite nodules, but they commonly have polygonal and vertical crack systems. If subjected to excessive leaching they may be rich in kaolinite. Soils that form within more sandy sediments may be leached intensely, to form a *ganister*, a rock of nearly pure quartz (Sections 2.7.1 and 8.7). Some kaolinite-rich beds occurring within or above coal seams and referred to as *tonsteins*, have formed from the alteration of volcanic ash (Section 10.7).

Soils are common throughout the Mesozoic and Cenozoic, and where formed in muddy sediments,
they show colour mottling, blocky textures, rootlets, rhizocretions and nodules of various sorts. The clay mineralogy of soils can be used to give an indication of the prevailing climate. In Cretaceous-to-Tertiary palaeosoils of northwest USA, decreasing kaolinite and increasing illite and chlorite up through the succession is attributed to a change in climate from more humid to more arid (Retallack, 1986).

The literature on soils and palaeosoils is vast, and soil scientists have invented much jargon for their description. A review of soils through the geological record is provided by Retallack (1990), and many papers on palaeosoils are to be found in Wright (1986) and Reinhardt & Sigleo (1988). Also see the colour atlas of palaeosoils by Retallack (1997).

3.7.2 Detrital mudrocks

The majority of clay minerals and silt-grade quartz in mudrocks are derived from the erosion of contemporary land masses. These fine terrigenous clastic particles are transported largely in suspension by water, with deposition taking place in quiet, low-energy environments. Rivers in particular transport vast quantities of silt and clay in suspension for deposition in floodplains, lakes, deltaic environments and nearshore and offshore marine environments. Mud in the form of aggregates is also transported as bedload by rivers (e.g. Maroulis & Nanson, 1996). Wind is another major agent of transportation, carrying dust up to thousands of kilometres from source areas, typically deserts and glacial-outwash plains. The dust may be deposited on land as *loess* (aeolian silt, see later this section) or carried into the oceans where it contributes towards hemipelagic sedimentation. On continental shelves and slopes mud can be resuspended by storms to form clay- and silt-rich clouds referred to as *nepheloid layers*, which are thought to be an important mechanism for transporting fine sediment to ocean floors and abyssal plains. Slumping of slope and outer-shelf muds and the generation of clay-laden, low-density turbidity currents are also important in supplying ocean basins with fine sediment. On alluvial fans, in association with volcanoes (Section 10.3), in glacial–proglacial regimes and on submarine fans, mud can be transported in viscous, sediment-rich, water-poor flows known as mudflows. Glaciers themselves can transport fine terrigenous clastic material for deposition in tills.

Apart from colour and sedimentary structures, the differentiation of one detrital mudrock from another may depend on detailed mineralogical and geochemical studies. Where there are significant variations in the clay-mineral assemblage through a mudrock succession, these usually will indicate some major event in the source area, such as climatic, tectonic or volcanic origin. One avenue of mudrock geochemistry that has been explored is the use of the boron content of illite as a palaeosalinity indicator (higher values in marine than freshwater muds). Although the technique has been much criticized, it can give meaningful results. For many mudrocks, interpretations of depositional environment may rely heavily on the fauna within the mudrocks and on the characteristics of associated and interbedded sediments. Determination of rare-earth elements (REEs) and platinum group elements (PGEs) can help in mudrock provenance studies.

Non-marine mudrocks

The mudrocks of river floodplains are best identified by their association with fluvial-channel sandstones. In many cases the mudrocks are overbank deposits and constitute the upper part of fining-upward units generated through lateral migration of the river channel or by channel abandonment (Section 2.11.1). Floodplain silts and clays deposited under a semi-arid climate invariably are red and many contain calcareous nodules of pedogenic origin (calcretes; Sections 3.7.1 and 4.10.1).

Mudrocks deposited in lakes vary considerably depending on the chemistry of the lake waters, organic productivity and climate. In the majority of cases the clay minerals are detrital, but clay minerals and the related zeolites can be precipitated on the lake floor or within the surficial sediments. Clays that may be neoformed in this way include sepiolite, palygorskite and corrensite, chiefly occurring in alkaline and hypersaline lacustrine sediments. Lacustrine formations with such clays are the Eocene Green River Formation of western USA and the Triassic Mercia Mudstone Group (Keuper Marl) of western Europe. In the latter case, a detrital (inherited) illite and chlorite assemblage occurring throughout the succession can be distinguished from an assemblage of neoformed Mg-rich clays (sepiolite, palygorskite, chlorite, smectite and corrensite) that occurs at particular horizons. The neoformed clays are the result of changes in water chemistry within the basin, probably arising from the influx of marine waters.

One of the characteristic features of lacustrine sediments is a millimetre-scale rhythmic lamination (Figs 2.16 & 3.4). In many non-glacial lakes, it results from a seasonal clastic influx, coupled with phytoplankton growth. Some oil shales possess a lamination of this type (Section 8.8). Glacial-lake sediments are typically *varved*; that is, they consist of a rhythmic alternation of coarse and fine laminae (Fig. 3.4a). The coarse laminae, of silt to fine-sand grade, are deposited from low-density suspension currents during spring melting, and the fine laminae, of clay-grade material, are deposited from suspension during summer and winter. Thermal stratification of the lake is of critical importance in the formation of varves. Organic productivity in lakes can supply much sediment. Photosynthesis by planktonic algae and microbial processes can cause precipitation of $CaCO_3$, giving rise to marl, or if there is a paucity of clay, then limestone. The accumulation of the organic matter itself can lead to the formation of bituminous mudrocks, black shales and oil shales (see below and Section 8.8). Diatoms inhabit many freshwater lakes and the accumulation of their tests produces siliceous muds (diatomaceous earths) and diatomites (Section 9.5). Papers describing laminated sediments, including glacial rhythmites, can be found in Kemp (1996).

Marine mudrocks

In the marine environment, mudrocks are deposited in five main locations: muddy coastlines; nearshore and mid-shelf mud belts; open-shelf mud blankets; basinal slopes; and basin floors.

Muddy coastlines are mostly close to major river systems debouching large quantities of suspended sediment into the sea. Tidal mudflats, salt marshes and mangrove swamps are typical environments here. Muds are also deposited in lagoons behind barrier islands. Off major deltas, marine shelves may be covered completely in mud to give a *shelf mud blanket*. The structures, fabrics and facies of modern muds from off the Amazon delta have been described by Kuehl *et al.* (1988) and from the Surinam coastal mud belt by Rine & Ginsburg (1985). Ancient mudrocks of shoreline environments can be identified by their fauna and flora, sedimentary structures (rippled lenses and desiccation cracks, for example) and associated coarser clastic deposits, such as channel, beach or barrier sandstones. Restricted fossil assemblages may suggest brackish water or hypersaline conditions, and the presence of rootlets may indicate emergence. These mudrocks are generally a dark grey colour, a function of the high organic content.

The *nearshore mud belt* occurs in water depths of 5–20 m, beyond the foreshore–shoreface sand belt of many coastlines (Section 2.11.5). Mud is deposited out of suspension below fairweather wave-base. Along higher-energy coasts, the mud belt is displaced farther offshore into the deeper-water mid-shelf area. Large parts of modern continental shelves are covered in relict sands and gravels deposited in coastal, fluvial and glacial environments from when sea-level was lower or the shelf exposed during the Pleistocene glaciation. The low suspended mud concentration on many mid- to outer shelves together with periodic storms mean that these coarse sediments have not been covered in mud. Many modern siliciclastic shelves are thus not in equilibrium with their sea-floor sediment; if they were, they would be covered in a thinning mud blanket from the nearshore sand belt to the shelf-margin/slope, with the rate of deposition of mud decreasing markedly offshore, as the amount of suspended sediment decreased.

The typical mudrocks of the *open-marine shelf*, deposited mainly below wave-base, are various shades of grey and in some cases rich in fossils. The fossils are both epifaunal (living on the sediment surface) and infaunal (living within the sediment) with some pelagic forms (free-swimming and free-floating species). Bioturbation is common. Thin, sharp-based and graded beds of sandstone and limestone within the mudrocks may represent storm deposits (Section 2.3.2). Fine laminae of silt and mud may be distal storm deposits. Mudrocks of the marine shelf are a major component of all shelf and epeiric-sea successions of the geological record and usually they pass laterally (shorewards) or vertically into limestones or sandstones of the shallow-water, nearshore zone. A good example of a muddy shelf deposit is the Cretaceous Mowry Shale of Wyoming, deposited in the Western Interior shelf sea. It contains graded siltstone and mudstone units a few millimetres thick, deposited from distal storm currents, as well as bioturbated horizons (Davis *et al.*, 1989). In the centre of this basin, anaerobic bottom waters permitted organic-rich sediments to accumulate. The apparently homogeneous Jurassic Oxford

Clay of England has a millimetre-scale lamination from weak, waning-flow currents and metre-scale coarsening-upward units capped by shelly layers, which are upward-shoaling parasequences (Macquaker & Howell, 1999).

Muds and mudrocks deposited in deeper water largely from suspension are termed *hemipelagic*. Such muds (hemipelagites) cover the sea floor on the deep, outer parts of continental shelves, on continental slopes and over vast areas of the ocean basins. The deep-ocean floors are usually well oxygenated. The reason for this is that cold, dense oxygen-rich waters are produced in the polar regions, and these descend and flow to lower latitudes, thus keeping the ocean floor ventilated. These deep-sea currents are locally responsible for erosion of the sea floor, so that skeletal lags and orientated fossils (Fig. 2.44) may form in these areas, and Fe–Mn oxides may be precipitated there. Many hemipelagic mudrocks are thus grey in colour, although red, brown, green and black varieties also occur. They are characterized by a fauna that is dominantly pelagic, such as diatoms, planktonic foraminifers and Coccolithophoridae from the Mesozoic to the present day; radiolaria from the Palaeozoic; cephalopods in the later Palaeozoic and Mesozoic; and graptolites in the early Palaeozoic. Hemipelagic mudrocks commonly are interbedded with siliciclastic and carbonate turbidites (Sections 2.11.7 and 4.10.8; Figs 2.82 & 4.50). The muds themselves also may be resedimented and transported by fine-grained turbidity currents, which deposit massive, laminated and graded beds (effectively Tde beds, see Section 2.11.7 and Stow & Shanmugan, 1980). Contour currents also rework muds and lead to lamination and cross-lamination (see Section 2.11.7 and Stow & Faugères, 1998).

Hemipelagites also may grade laterally or vertically into pelagic limestones (Section 4.10.7) or radiolarites (Section 9.3), which form in areas or at times of minimal clay and silt sedimentation. Modern hemipelagic sediments accumulating below the carbonate compensation depth (Section 4.10.7) are red and brown clays. These cover the abyssal plains of the central Pacific and occur in the Atlantic and Indian Oceans. They consist of detrital clay and silt, clay minerals and zeolites derived from alteration of volcanic ash (Section 10.7), and radiolarians, diatoms and sponge spicules. Examples of hemipelagic mudrocks are to be found in all basinal successions throughout the geological record.

The graptolitic shales of the Lower Palaeozoic of western Europe (e.g. Dimberline *et al.*, 1990), and the Cretaceous–Tertiary foraminiferal marls of the Alpine–Tethyan region are good examples. Ancient hemipelagic slope sediments commonly are poorly preserved as a result of tectonic deformation, which is often more intense at basin margins. Slides, slump folds, chaotic bedding, channels and debrites are typical features. Eocene abyssal-plain hemipelagites, associated with radiolarites, are part of an accretionary prism exposed on Barbados (Gortner & Larue, 1986). Mud turbidites also occur in the Barbados succession (Kasper *et al.*, 1987).

Organic-rich mudrocks and black shales

One particularly important group of mudrocks comprises those rich in organic matter; these are black shales and carbonaceous and bituminous mudrocks, which typically contain 3–10% organic carbon. With an increasing organic content, organic-rich mudrocks pass into oil shales, which yield a significant amount of oil on heating (Section 8.8). In many depositional environments organic matter is decomposed and destroyed at the sediment surface but if the rate of organic productivity is high, then organic matter can be preserved. Much of this organic matter is sapropelic, supplied by phytoplankton (Section 8.2). The accumulation of organic matter is favoured if the circulation of water is restricted to some extent so that insufficient oxygen reaches the bottom sediments to decompose the organic material (Fig. 3.11). Locations where this commonly takes place are lakes, fjords, silled basins (such as the Black Sea), sediment-starved basins (such as the Gulf of California) and deep-ocean trenches (such as the Cariaco Trench). As a result of the poor circulation and restriction, the water body becomes stratified and the sea or lake floor may become oxygen deficient (dysaerobic) or totally anoxic. Dysaerobic conditions also occur where the sea floor is within the oxygen-minimum zone, generally in the depth range 100–1000 m. This zone of low O_2 results from the bacterial decomposition of organic matter sinking from fertile, oxic, surface waters. Irrespective of the degree of anoxia, however, the major control on organic-carbon accumulation does appear to be the primary production rate.

Where there is an oxygen deficiency on the sea floor, organic matter will be preserved, but the surface sedi-

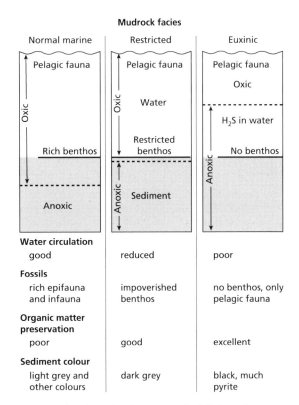

Fig. 3.11 The relationships between mudrock facies and oxicity–anoxicity, fauna and organic matter content.

ments could still support a benthic epifauna, although of low diversity. Where there are anoxic conditions on the sea floor, there usually is much H_2S in the water and benthic organisms are absent (Fig. 3.11). This is the case with the Black Sea and Cariaco Trench at the present time. Mudrocks deposited in an anoxic environment would contain only pelagic fossils. Pyrite is common in marine organic-rich mudrocks and siderite in non-marine ones (Section 6.3).

There are many organic-rich mudrocks in the geological record; their importance lies in their potential as source rocks for petroleum, if buried to suitable depths and subjected to appropriate temperatures (Section 8.10). Examples include Jurassic and Cretaceous bituminous shales, with organic matter derived from radiolarians and foraminifers, which have supplied much of the petroleum for the Middle East oilfields, and the organic-rich Kimmeridge Clay, the chief source rock for North Sea oil. Many descriptions of hydrocarbon source rocks are contained in Brooks & Fleet (1987) and Fleet *et al.* (1988), Katz & Pratt (1993) and Huc (1995).

Thin horizons of Cretaceous black shales and organic-rich mudrocks have been cored from the Atlantic, Pacific and Indian Oceans, and equivalent horizons can be found in sections on land. These organic-rich layers, which signify short-lived periods of poor circulation and oxygen deficiency in the world's oceans, are thought to result from increased organic productivity associated with transgressive events that produced widespread shallow seas (e.g. Bellanca *et al.,* 1999). These *oceanic anoxic events* (OAEs) tend to occur at times of global equable climate, when there is little dense, cold water produced at the poles for ventilation of deep, lower-latitude ocean basins. Oxygen-minimum zones may expand and become anoxic at these times. Black mud sapropels also occur in the Miocene–Pleistocene of the Mediterranean; the Pleistocene stagnations were caused by stratification of the water at times of glacial expansions.

One feature of organic-rich sediments is that they contain high concentrations of certain trace elements, in particular Cu, Pb, Zn, Mo, V, U and As. The trace elements are adsorbed onto the organic matter and also onto the clay minerals. It is likely that the source of these elements is sea water and that they are scavenged by the organisms and organic matter. An example is the organic-rich Permian Marl Slate of northeast England and the North Sea and its equivalent in Germany and Poland, the Kupferschiefer, which has a high metal-sulphide content. It was deposited in a thermally stratified sea with anoxic bottom waters.

Black shales are reviewed by Arthur & Sageman (1994) and Wignall (1994), and organic matter in sediments has been covered by Tyson (1995).

Loess and loessite

Loess is a yellow- to buff-coloured clastic deposit composed principally of silt-sized quartz grains, generally in the size range of 20–50 μm. A distinctive feature is the well-sorted nature of the silt, together with a dominantly angular shape to the grains. Loess is usually unstratified and unconsolidated, but it may contain shells of land snails and concretions formed around roots. Loess deposited during the late Pleistocene occurs over vast areas of central Europe, the Mississippi valley region of the USA, eastern South America and China.

There are few unequivocal accounts of pre-Pleistocene wind-blown silt, but two well-documented loessites are from the Pennsylvanian/Permian of Colorado (Johnson, 1989) and the Triassic of Utah (Chan, 1999).

Loess is regarded primarily as an aeolian deposit, but once deposited it can be considerably modified by fluvial reworking and pedogenesis. Two types of loess are distinguished by many authors: loess of cold, periglacial regimes derived from deflation of glacial-outwash plains (this accounts for most of the late Pleistocene occurrences) and loess derived from hot, arid, desert areas. Glacial grinding is considered the most effective mechanism for producing vast quantities of quartz silt. Loess from desert weathering and deflation is probably of minor geological importance. For further information on loess and wind-blown dust see papers in Frostick & Reid (1987) and Pye (1995).

3.7.3 Mudrocks of volcaniclastic origin

Mudrocks formed from the alteration of volcaniclastic material are known as bentonites (also fuller's earth) if montmorillonite is the main clay mineral present and tonsteins if kaolinite is dominant. Zeolites usually are present too. The volcaniclastic deposit may be subaerial or subaqueous, but because of the metastable nature of volcanic glass, devitrification soon takes place and clay minerals and zeolites form. The identification of these mudrocks is based on the presence of glass-shard pseudomorphs, euhedral crystals and geochemistry (Section 10.7).

Further reading

Aplin, A.C., Fleet, A.J. & Macquaker, J.H.S. (Eds) (1999) *Muds and Mudstones: Physical and Fluid-flow Properties.* Special Publication 158, Geological Society of London, Bath, 200 pp.

Brindley, G.W. & Brown, G. (1984) *Crystal Structures of Clay Minerals and their X-ray Identification.* Mineralogical Society, London, 495 pp.

Chamley, H. (1989) *Clay Sedimentology.* Springer-Verlag, Berlin, 623 pp.

Potter, P.E., Maynard, J.B. & Pryor, W.A. (1980) *Sedimentology of Shale.* Springer-Verlag, New York, 270 pp.

Schieber, J., Zimmerle, W. & Sethi, P. (Eds) (1998)

Shales and Mudstones. Schweizerbart'sche Verlagsbuchhandlung, Stuttgart.

Singer, A. & Muller, G. (1983) Diagenesis in argillaceous sediments. In: *Diagenesis in Sediments and Sedimentary Rocks 2* (Ed. by G. Larsen & G.V. Chilingar), pp. 115–212. Elsevier, Amsterdam.

Velde, B. (1985) *Clay Minerals*. Elsevier, Amsterdam, 427 pp.

Weaver, C.E. (1989) *Clays, Muds and Shales*. Elsevier, Amsterdam, 890 pp.

Wignall, P.B. (1994) *Black Shales*. Clarendon Press, Oxford, 127 pp.

Also see recent papers in the journals *Clay and Clay Minerals* and *Clay Minerals Bulletin*.

4 Limestones

4.1 Introduction

Biological and biochemical processes are dominant in the formation of carbonate sediments, although inorganic precipitation of $CaCO_3$ from seawater also takes place. Once deposited, the chemical and physical processes of diagenesis can considerably modify the carbonate sediment. Limestones occur throughout the world in every geological period from the Cambrian onwards and reflect the changing fortunes, through evolution and extinction, of invertebrates with carbonate skeletons. In the Precambrian, carbonates are also abundant, but they are commonly dolomite and many contain stromatolites, produced largely by microbes, especially the cyanobacteria ('blue–green algae').

The economic importance of limestones today lies chiefly in their reservoir properties, as about half of the world's major petroleum reserves are contained within carbonate rocks. Limestones also are hosts to epigenetic lead and zinc sulphide deposits of the Mississippi Valley type and they have a wide variety of chemical and industrial uses, including the manufacture of cement.

As a result of recent geological events, notably the Pleistocene glaciation and a global sea-level lowstand, shallow-marine carbonate sediments are not widely developed at the present time. In the past, shallow epeiric seas periodically covered vast continental areas so that limestones were deposited over many thousands of square kilometres. On a broad scale, extensive carbonate deposition correlates with global sea-level highstands. Organisms with carbonate skeletons occur throughout the world's seas and oceans so that carbonate sediments can develop anywhere. However, there are several factors, of which the most important are temperature, salinity, water depth and siliciclastic input, that control carbonate deposition. Many carbonate skeletal organisms, such as the reef-building corals and many calcareous green algae, require warm waters in which to flourish. The majority of carbonate sediments therefore occur in the tropical–subtropical belt, some 30° north and south of the Equator, and most limestones of the Phanerozoic formed in low latitudes. Biogenic carbonate productivity is highest in sea water of normal salinity in the shallow (less than 10 m), agitated part of the photic zone (the depth down to which light penetrates, of the order of 100–200 m). Skeletal carbonate sands do occur in higher latitudes, such as along the western coast of Ireland and Norway where calcareous red algae (especially *Lithothamnion*) and Mollusca dominate the sediments, and also off southern Australia, where bryozoans are especially important. However, there are few ancient examples of these so-called *cool-water carbonates* (see papers in James & Clarke (1997) and Lukasik *et al.* (2000) for a Miocene example from South Australia).

Non-skeletal grains, such as ooids and lime mud, are precipitated only in the warm shallow waters of the tropics. In the deeper-water pelagic environment, calcareous oozes are developed extensively and they are composed principally of the skeletons of pelagic organisms, Foraminifera and coccoliths, which live in the photic zone. High rates of carbonate dissolution at depths of several thousand metres result in little carbonate deposition below this, the *carbonate compensation depth*. Limestones also form in lakes and soils. One overriding control on carbonate deposition is the input of siliciclastic material. Many carbonate-producing organisms cannot tolerate the influx of large quantities of terrigenous mud.

For the petrographic study of limestones thin-sections are examined routinely, but acetate peels are useful too and quick to make. The surfaces of a limestone are polished and then etched with dilute acid (5% HCl or acetic acid for 30–60 s). The surface is then covered in acetone and a piece of acetate sheet is rolled on, with care taken not to trap air bubbles. After at least 10 min, the acetate is peeled off, and placed between glass. It is now ready for the microscope. All textural details are faithfully replicated, but of course polarizers cannot be crossed. See Miller (1988a) for further details.

Much petrographic information is obtained from thin-sections viewed in transmitted light but further detail can be revealed through the technique of cathodoluminescence (CL): bombarding a polished thin-section (no cover slip) with electrons in a special small vacuum chamber mounted on a microscope stage (see Marshall (1988) and Miller (1988b) for details). An example of CL is given in Plate 13c. Examining a thin-section under UV light and observing the fluorescence can also reveal 'hidden' textures (see Dravis & Yurewicz, 1985). It is common practice to stain a thin-section of a limestone, or the polished surface of a hand specimen before taking a peel, to show the mineralogy (calcite or dolomite), and the iron content (ferroan or non-ferroan). Alizarin Red S and potassium ferricyanide are used (for the recipe see Miller, 1988a), and calcite (non-ferroan) stains pink and ferroan calcite blue to mauve; dolomite (non-ferroan) does not take up the stain whereas ferroan dolomite is turquoise–blue. Table 4.1 summarizes the features to look for in a carbonate thin-section and Table 4.2 gives a simple table for notes when describing a slide.

4.2 Mineralogy of carbonate sediments

In Recent sediments, two calcium carbonate minerals dominate: aragonite (orthorhombic) and calcite (trigonal). Two types of calcite are recognized depending on the magnesium content: low-magnesium calcite with less than 4 mol.% $MgCO_3$ and high-magnesium calcite with greater than 4 mol.%, but typically ranging between 11 and 19 mol.% $MgCO_3$. By comparison, aragonite normally has a very low Mg content (less than 5000 p.p.m.) but it may contain up to 10 000 p.p.m. (1%) strontium, substituting for calcium. The mineralogy of a modern carbonate sediment depends largely on the skeletal and non-skeletal grains present. Carbonate skeletons of organisms have a specific mineralogy or mixture of mineralogies (Table 4.3), although the actual magnesium content of the calcites may vary, being partly dependent on ambient water temperature.

Aragonite is unstable at surface temperatures and pressures and in time high-Mg calcite loses its Mg. Thus all carbonate sediments with their original mixed mineralogy are converted to low-Mg calcite during diagenesis. Grains and cement composed originally of low-Mg calcite generally are perfectly preserved in limestones; those originally of high-Mg calcite are mostly well preserved, like low-Mg calcite, but they may show some microstructural alteration and minor dissolution. Grains of aragonite are either replaced by calcite with some retention of original structure (the process of *calcitization*), or dissolved out completely to leave a mould, which later may be filled with calcite (a cement). A limestone also may be *dolomitized*, whereby dolomite, $CaMg(CO_3)_2$, replaces the $CaCO_3$ minerals and is precipitated as a cement (dolomitization is discussed in Section 4.8). Non-carbonate minerals in limestones include terrigenous quartz and clay, and pyrite, hematite, chert and phosphate of diagenetic origin. Evaporite minerals, in particular gypsum–anhydrite, may be closely associated with limestones (see Chapter 5).

4.3 Components of limestones

Limestones are very varied in composition but broadly the components can be divided into four groups: (i) non-skeletal grains, (ii) skeletal grains, (iii) micrite and (iv) cement. The common cement, sparite, and others are discussed in the section on diagenesis (Section 4.7).

4.3.1 Non-skeletal grains

Ooids and pisoids

Modern ooids are spherical–subspherical grains, consisting of one or more regular concentric lamellae around a nucleus, usually a carbonate particle or quartz grain (Figs 4.1 & 4.2). Sediment composed of ooids is referred to as an *oolite*. The term ooid (formerly oolith) has been restricted to grains less than 2 mm in diameter and the term pisoid (formerly pisolith) is used for similar grains of a larger diameter. If only one lamella is developed around a nucleus, then the term superficial ooid is applied (Fig. 4.1). Composite ooids consist of several small ooids enveloped by concentric lamellae. *Coated grain* is a general term frequently used for ooids and pisoids, and includes *oncoids*, grains with a microbial coating (see Section 4.3.3).

The majority of modern ooids range from 0.2 to 0.5 mm in diameter. They typically form in agitated

Table 4.1 Scheme for petrographic description for carbonate rocks

Hand specimen

Note the colour; type(s) of grain(s); grain size; grain shape if significant; evidence of fossil breakage, disarticulation, abrasion, packing, etc.; presence of micrite and/or sparry calcite if observable; physical and biogenic sedimentary structures; stylolites, etc.

Any evidence of diagenetic alteration, e.g. dolomitization, silicification, dissolution, compaction

Thin-section

Check macroscopic features of thin-section by holding up to light and noting any lamination or large fossils or grains

Grains: type(s) — bioclasts, ooids, peloids, intraclasts/aggregates, etc.; size, sorting, shape, packing of grains; mineralogy/composition of grains (in stained sections — ferroan/non-ferroan calcite/dolomite). Identify bioclasts from shape, internal structure and preservation. Ooids — determine original composition (aragonite or calcite). Peloids — micritized grains or faecal pellets?

Micrite (lime mud): grains mostly less than 4 µm and normally almost opaque and brownish in colour, but may be peloidal. Any aggrading neomorphism?

Cements: Usually coarser than 10 µm; identify types of cement: fibrous calcite, bladed calcite, syntaxial overgrowths, drusy calcite spar, poikilotopic calcite spar; note geometry: meniscus, isopachous, pore-filling; spar may be non-ferroan, ferroan, or zoned (use staining). Determine original mineralogy (aragonite, calcite, high-Mg calcite). Timing of cement precipitation, early versus late, pre- versus post-compaction. Where cemented — marine, meteoric, burial?

Replacement, recrystallization and neomorphism: minerals such as dolomite, silica (chert) or phosphate may replace calcite grains, micrite or cement. Neomorphism of grains, e.g. aragonitic shells, ooids or cements replaced by calcite (calcitization) and micrite replaced by microspar — aggrading neomorphism. Timing of replacement relative to compaction?

Dolomitization: scattered rhombs or pervasively dolomitized? Any fabric control: grains or micrite preferentially dolomitized? Determine dolomite texture and crystal shape/size/zonation. Is dolomite early or late, pre- or post-compaction; timing relative to calcite diagenesis. Any dedolomitization?

Compaction: look for evidence of mechanical compaction (broken bioclasts, broken micrite envelopes, spalled oolitic coatings or early cements). Check for chemical compaction: sutured contacts between grains, pre-burial spar cementation, and stylolites: through-going pressure dissolution seams, post-spar cementation

Porosity: identify type: primary (intergranular, intragranular, cavity, growth, etc.) and secondary (fracture, dissolutional, intercrystalline — as through dolomitization)

Classify using the Dunham scheme (or Folk) after assessing the proportion of grains, cement and micrite, and dominant grain types. Where diagenetic alteration is significant, then qualify name, e.g. partly dolomitized bioclastic grainstone, silicified oosparite, recrystallized lime mudstone

Interpretation

Depositional environment: use grain types and texture, e.g. most grainstones represent moderate to high energy shallow subtidal, many lime mudstones–wackestones represent lower-energy, lagoonal or outer-shelf/ramp environments. Check bioclasts; these can indicate open-marine, restricted, deep-water, shallow, non-marine, etc.

Diagenesis: identify early (near-surface) and late (burial) processes, determine timing of cementation. Assess degree of compaction; any dolomitization? Try to interpret in terms of pore-water chemistry (e.g. marine or freshwater from cement fabric, and Eh/pH from ferroan/non-ferroan nature of calcite/dolomite). Deduce succession of diagenetic events and porosity evolution

---

waters where they are frequently moved as sandwaves, dunes and ripples by tidal and storm currents, and wave action. On the Bahama platform, the ooids form shoals close to the edge of the platform. In the Arabian Gulf the ooids form in tidal deltas at the mouth of tidal inlets between barrier islands along the Trucial Coast. Along the Yucatan shoreline (northeast Mexico), ooids are being precipitated in the shoreface and foreshore zones. Depths of water where ooids precipitate usually are less than 5 m, but they may reach 10–15 m.

Practically all marine ooids forming today, such as in the Bahamas and Arabian Gulf, are composed of aragonite and they have a high surface polish (Fig.

4.2). Bimineralic high-Mg calcite–aragonite ooids have been recorded from Baffin Bay, Texas. Relict early Holocene ooids of high-Mg calcite occur off the Great Barrier Reef of Queensland and on the Amazon Shelf.

The characteristic microstructure of modern aragonitic marine ooids is a tangential orientation of acicular crystals or needles, 2 µm in length. Lamellae of microcrystalline aragonite and of randomly oriented aragonite needles also occur. The sub-Recent high-Mg calcite ooids have a radial fabric. Ooids contain organic matter, located chiefly between lamellae and in the microcrystalline layers.

Table 4.2 Scheme for describing carbonate rocks in thin-section

Features	Thin-section 1	Thin-section 2
Grains present and percentage Bioclasts Ooids Pellids Intraclasts Micrite: uniform, peloidal, microsparitic		
Texture Roundness, sorting, fabric, binding, framework		
Cavity structures Umbrella, geopetal, fenestrae — birdseye, laminar, burrow, rootlet		
Cements Fibrous, bladed, drusy calcite, sparite, geometry		
Replacements Calcitization, neomorphism, microspar, micritization, dissolution		
Dolomite Crystal shape, size, distribution, texture, fabric preservation, timing		
Evidence of compaction Broken grains and micrite envelopes, sutured contacts, stylolites		
Porosity Intergranular, intragranular, framework, mouldic, intercrystalline, fracture, etc.		
Name/microfacies		
Depositional environment		
Diagenetic events	1: 2: 3:	

Ooids and pisoids can also form in quieter-water marine locations, such as in lagoons (e.g. Baffin Bay, Texas) and on tidal flats (e.g. the Trucial Coast). Ooids–pisoids also are formed in association with tepee structures on intertidal–supratidal flats (see Section 4.6.1), being precipitated in local pools and beneath cemented crusts. These low-energy coated grains commonly have a strong radial fabric, so that they do break relatively easily (e.g. Fig. 4.3). Those forming in a vadose situation, such as in association with tepees, are commonly asymmetric; they may also show downward thickening laminae and a fitted fabric from *in situ* growth. Such *vadose pisoids* are a feature of the back-reef facies in the Capitan Reef Complex of Texas–New Mexico.

Ooids can form in high-energy locations in lakes, as in the Great Salt Lake, Utah and Pyramid Lake, Nevada. These lacustrine ooids are commonly dull, and they may have a cerebroid (bumpy) surface. The Great Salt Lake ooids are composed of aragonite and many have a strong radial fabric. The Pyramid Lake ooids are bimineralic (low-Mg calcite–aragonite).

Structures resembling ooids–pisoids do form in calcareous soils. They are usually composed of fine-grained calcite and have a poorly developed concentric lamination, which may be asymmetric. They form

Organism	Mineralogy			
	Aragonite	Low-Mg calcite	High-Mg calcite	Aragonite + calcite
Mollusca				
Bivalves	X	X		X
Gastropods	X			X
Pteropods	X			
Cephalopods	X		(X)	
Brachiopods		X	(X)	
Corals				
Scleractinian	X			
Rugose + Tabulate		X	X	
Sponges	X	X	X	
Bryozoans	X		X	X
Echinoderms			X	
Ostracods		X	X	
Foraminifera				
Benthic	(X)		X	
Pelagic		X		
Algae				
Coccolithophoridae		X		
Rhodophyta	X		X	
Chlorophyta	X			
Charophyta		X		

Table 4.3 The mineralogy of carbonate skeletons (X, dominant mineralogy; (X), less common). During diagenesis, these mineralogies may be altered or replaced; in particular, aragonite is metastable and is invariably replaced by calcite, and high-Mg calcite loses its Mg

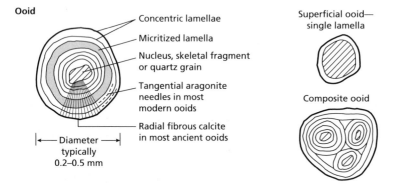

Ooid
- Concentric lamellae
- Micritized lamella
- Nucleus, skeletal fragment or quartz grain
- Tangential aragonite needles in most modern ooids
- Radial fibrous calcite in most ancient ooids

Diameter typically 0.2–0.5 mm

Superficial ooid— single lamella

Composite ooid

Aggregate

e.g.

A collection of grains cemented together

Peloid – composed of micrite

(a) A pellet, typically 0.1–0.5 mm diameter

(b) Amorphous grain, many are micritized skeletal grains

Fig. 4.1 The principal non-skeletal grains in limestones: ooids, peloids and aggregates.

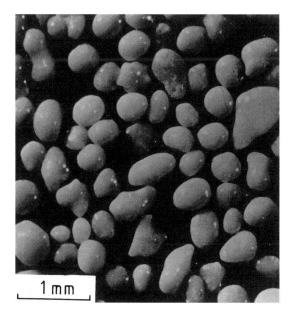

Fig. 4.2 Modern aragonitic ooids from the Bahamas with polished surface.

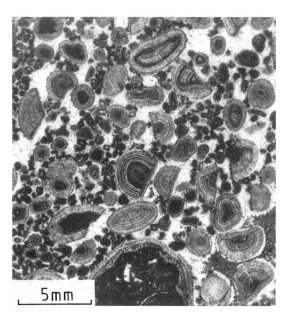

Fig. 4.3 Pisoids, probably of vadose origin (vadoids) showing strong radial structure and broken and recoated grains. Smaller grains are peloids. Rock is a grainstone, but it is 100% dolomite. The sediment was dolomitized with perfect preservation of original texture. The clear spar is a dolomitic cement. Dürrenstein Formation, Triassic. The Dolomites, Italy.

largely by microbial processes and calcification of fungal and bacterial filaments. These *calcrete pisoids* are commonly associated with laminated crusts, formed by the calcification of root mats (see Section 4.6.1).

Ancient marine ooids

Ooids in the rock record are composed of calcite (low Mg), unless dolomitized or silicified. However, although there has been much discussion over the matter, it is clear that some were originally calcite, whereas others were originally aragonite. Former bimineralic ooids also have been reported. Primary calcite ooids, whether in high-energy or low-energy facies, typically have a radial texture of wedge-shaped, fibrous crystals (see Plate 6c). Under crossed polars, an extinction cross is seen. The cortex of larger, originally calcitic ooids may have an inner radial part and an outer radial–concentric part. It is not easy to determine whether calcitic ooids originally had a low or high Mg content. The presence of small dolomite crystals (microdolomites), evidence of minor dissolution of the cortex or a moderate to high iron content may indicate an original high-Mg calcite composition (see Section 4.7.1).

Ancient ooids originally of aragonite will have been altered during diagenesis to a greater or lesser extent (see Plate 6d). They may be replaced by calcite with some retention of the original tangential structure (if high energy) or radial structure (if quiet water) through the presence of minute inclusions of organic matter and/or relics of the aragonite. This replacement process is termed *calcitization* (also see Section 4.7.4). Alternatively, the aragonite of the ooids may be dissolved out completely, to leave oomoulds. These holes may be left empty to give the limestone an oomouldic porosity, or filled with calcite cement (see Plate 6d). Some ancient ooids have a fine-grained micritic texture. As with modern ooids, this may be the result of micritization by endolithic microbial organisms (see Section 4.3.3) or it may result from diagenetic alteration (*neomorphism*, see Section 4.7.3).

Origin of ooids

There has been much discussion on the origin of ooids; current ideas invoke biochemical or inorganic processes, a direct microbial origin being largely discarded now. Although a precise mechanism of inorganic precipitation has not been demonstrated, seawater in shallow tropical areas is supersaturated with respect to $CaCO_3$, so that this, together with water agitation, CO_2 degassing and elevated temperature, might be sufficient to bring about carbonate precipitation on nuclei. A biochemical origin hinges on the organic mucilage that coats and permeates the ooids. One view is that bacterial activity within the organic matter creates a microenvironment conducive to carbonate precipitation. Some ooids have a proteinaceous matrix, also suggesting a biochemical process, because in organisms it is amino acids that induce calcification. A biological origin is supported by SEM examination of modern marine ooids, which shows that the aragonite rods and nanograins that form the cortex are identical to those associated with endolithic and epilithic bacteria and mucilaginous films occurring on and within the ooids. Laboratory synthesis of ooids has suggested that organic compounds in the water are instrumental in the formation of quiet-water ooids with their radial fabric, but that ooids formed in agitated conditions are precipitated inorganically.

The factors determining the primary mineralogy of ooids are water chemistry, especially P_{CO_2}, Mg/Ca ratio and carbonate saturation, and possibly the degree of water agitation. It is believed that aragonite and high-Mg calcite ooids are precipitated when P_{CO_2} is low and Mg/Ca ratio high, and that low-Mg calcite ooids form when P_{CO_2} is high and Mg/Ca ratio low. It is unclear what controls the precipitation of aragonite ooids as opposed to high-Mg calcite ooids but carbonate supply rate has been implicated. High carbonate supply, as would occur in high-energy locations, is thought to favour aragonite precipitation.

Surveys of the original mineralogy of ooids through the Phanerozoic have shown that there is a secular variation, with aragonitic ooids, which may be associated with calcitic ooids (presumably high-Mg calcite originally) in the late Precambrian/early Cambrian, mid-Carboniferous through Triassic and Tertiary to Recent, and calcitic ooids (presumed to have had a low-to-moderate Mg content) dominant in the mid-Palaeozoic and Jurassic–Cretaceous (Fig. 4.4). This pattern suggests that there have been subtle fluctuations in seawater chemistry through time, in P_{CO_2} and/or the Mg/Ca ratio. See Sandberg (1983) and the review in Stanley & Hardie (1998). Some anomalies

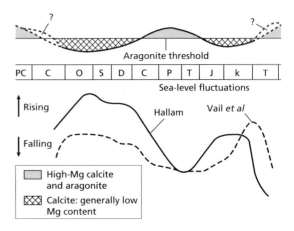

Fig. 4.4 Mineralogy of marine, abiogenic, carbonate precipitates through the Phanerozoic, compared with the first-order global sea-level curve.

do occur in the broad trend, notably in Upper Jurassic strata where aragonitic ooids are recorded, as in the Smackover Formation of the US Gulf Coast subsurface (see Plate 6d; Heydari & Moore, 1994).

The general trend shown in Fig. 4.4 appears to tie in with the first-order, global sea-level curve, suggesting that a geotectonic mechanism(s) is causing the subtle variations in seawater chemistry, giving rise to the secular variation in ooid mineralogy. High sea-level stands, correlating with calcite seas, are times of high rates of sea-floor spreading, when P_{CO_2} can be expected to be relatively high from increased metamorphism at subduction zones and the Mg/Ca ratio relatively low from increased extraction of Mg^{2+} at mid-ocean ridges as sea water is pumped through (see Stanley & Hardie, 1998). Temporal exceptions to the broad trend, as in the Late Jurassic, could be the result of local conditions, the most likely being a raised Mg/Ca ratio from evaporite deposition, leading to aragonite precipitation.

Peloids

Peloids are spherical, ellipsoidal or angular grains, composed of microcrystalline carbonate, but with no internal structure (Fig. 4.1 and Plates 6a, 7a,c & 8c). The size of peloids may reach several millimetres but the majority are in the range of 0.1–0.5 mm in diameter. Most peloids are of faecal origin and so can be referred to as pellets. Organisms such as gastropods, crustaceans and polychaetes produce pellets in vast quantities. Faecal pellets have a regular shape and they are rich in organic matter. They are most common in the sediments of protected environments such as lagoons and tidal flats. Pellets are very common in limestones and many micritic limestones, seemingly without sand-sized grains, may actually be pelleted. The definition of pellets is commonly lost as a result of diagenetic processes, and the limestones may show a flocculent or clotted texture.

The term peloid includes micritized bioclastic grains formed by alteration of skeletal fragments by microboring microbes and recrystallization (Section 4.3.3). They are more irregular in shape and are an important component of modern Bahamian carbonate sediments.

Aggregates and intraclasts

Aggregates consist of several carbonate particles cemented together by a microcrystalline cement or bound by organic matter. Such grains in the Bahamas are known as grapestones and form in relatively protected shallow subtidal areas, usually beneath a thin, surficial microbial mat.

Intraclasts are fragments of lithified or partly lithified sediment. A common type of intraclast in carbonate sediments is a micritic flake or chip, derived from desiccation of tidal-flat muds or disruption by storms of partially lithified or cemented subtidal lime muds (Fig. 4.5). The latter are particularly common in the Precambrian and Cambrian. An abundance of these flakes produces flat-pebble or edgewise conglomerates, also called flakestones. They may show an imbrication of clasts (see Section 2.2.4).

One distinctive intraclast is a *black pebble*. These are carbon-impregnated pebbles, commonly with soil fabrics, which are associated with palaeosoils, laminated crusts and palaeokarsts, and may be reworked into intraformational conglomerates. They form in soil horizons and may be the result of forest fires, or more likely organic-matter impregnation in water-logged, reducing conditions (Shinn & Lidz, 1988).

4.3.2 Skeletal components (excluding algae)

The skeletal components of a limestone are a reflection of the distribution of carbonate-secreting invertebrates through time and space (Fig. 4.6). Environmental factors, such as depth, temperature, salinity, substrate and turbulence, control the distribution and development of the organisms in the various carbonate environments. Throughout the Phanerozoic, various groups expanded and evolved to occupy the niches left by others that were declining or becoming extinct. The mineralogy of carbonate skeletons also varies through the Phanerozoic, like the inorganic precipitates (Section 4.3.1), and this probably is also a reflection of tectonically forced shifts in seawater chemistry (see Stanley & Hardie, 1998).

The main skeletal contributors to limestones are discussed in the following sections, with comments on their recognition. Detailed accounts of skeletal struc-

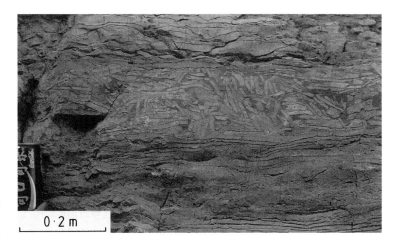

Fig. 4.5 Intraclasts. Storm bed composed of rip-up clasts of micritic limestone. Cambrian. Qinhuandao, northeast China.

0·2 m

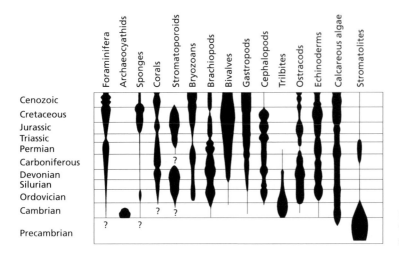

Fig. 4.6 Age range and generalized taxonomic diversity of principal carbonate-secreting organisms.

ture and thin-section appearance are given in Bathurst (1975), Scholle (1978), Flügel (1982), Adams *et al.* (1984) and Adams & Mackenzie (1998). For the identification of skeletal particles in thin-section the points to note are:

1 shape (and size), bearing in mind that under the microscope a two-dimensional view only is given; look for other sections of the same fossil to determine the three-dimensional shape;

2 internal microstructure, which may be modified or obliterated by diagenesis;

3 mineralogy—although in a limestone everything will be calcite, unless dolomitized or silicified, fabric evidence can be used to decide if a skeletal particle was originally aragonitic (staining a thin-section for ferroan calcite and dolomite using Alizarin red S and potassium ferricyanide (see Section 4.1) may provide additional information—for example, skeletal components originally of high-Mg calcite may be replaced preferentially by ferroan calcite);

4 other features likely to be diagnostic, such as presence of spines or pores.

Mollusca

Bivalves, gastropods and cephalopods occur in limestones from the Early Palaeozoic onwards. The bivalves are a very large group with species occupying most marine, brackish and freshwater environments. Bivalves have been important contributors to marine carbonate sediments, particularly since the Tertiary following the decline of the brachiopods. The modes

of life are very varied, too, including infaunal (living within the sediment), epifaunal (attached to a hard substrate), vagile (crawlers), nektonic (free-swimming) and planktonic (free-floating). Certain bivalves, such as oysters, may form reef-like structures. During the Cretaceous, masses of aberrant, coral-like bivalves called rudists formed reefs in Mexico, southern USA, the Mediterranean region and the Middle East, for example. Fresh- and brackish-water limestones may be composed largely of bivalves; examples occur in the Upper Carboniferous, Upper Triassic (Rhaetic) and Upper Jurassic (Purbeck) of western Europe.

The majority of bivalve shells are composed of aragonite; some are of mixed mineralogy (the rudists, for example, Plate 7e); others, such as the oysters and scallops, are calcitic. Bivalve shells consist of several layers of specific internal microstructure, composed of micron-sized crystallites (see Plate 7b,c). One common shell structure is of an inner nacreous layer consisting of sheets of aragonite tablets, and an outer prismatic layer of aragonite (or calcite). If composed originally of aragonite, the internal structure of a fossil bivalve shell is likely to be poorly preserved or not preserved at all (see Fig. 4.7). The aragonite may be dissolved out completely to leave a mould, which subsequently may be filled by calcite (a cement). This is the most common mode of preservation so that most bivalve fragments in limestones are composed of clear, coarse drusy sparite (see Plate 7d; also Plates 6c & 9e). Alternatively, the aragonite of the shell may be replaced by calcite (calcitized) so that faint relics of the

Plate 1

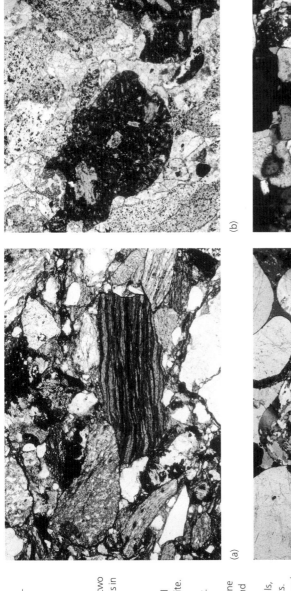

(a)

(b)

(c)

(d)

(a) Lithic grains of sedimentary origin — laminated and silty shale, also angular quartz grains. Plane-polarized light. Carboniferous fluvial sandstone. Cantabrians, Spain. Field of view 3 × 2 mm.

(b) Lithic grains of volcanic origin. The two dark grains consist of mafic phenocrysts in a very altered, originally glassy, dark groundmass. The other grains consist of minute feldspar laths in a glassy groundmass, and have numerous small dark iron-rich spots. The cement is calcite. Plane-polarized light. Triassic shallow-marine sandstone. The Dolomites, Italy. Field of view 3 × 2 mm.

(c) Quartz grains: several monocrystalline quartz grains showing both uniform and unit extinction; polycrystalline quartz grains with several and many subcrystals, some of the latter with sutured contacts. The cement is poikilotopic calcite. Crossed polars. Permian aeolian sandstone. Durham, England. Field of view 3 × 2 mm.

(d) Feldspar grains: microcline on left with grid-iron twinning and orthoclase on right with bright specks of alteration material (sericite). Quartz grains also present (mostly monocrystalline with unit extinction), the one in lower centre with overgrowth. Muscovite flake extreme right showing blue colour. Grains are coated with hematite, giving red/brown rim. Crossed polars. Precambrian fluvial sandstone. Torridon, Scotland. Field of view 1.2 × 0.8 mm.

Plate 2

(a) Muscovite and kaolinite in quartz arenite. Muscovite shows orange colour. Kaolinite forms the small dark crystals between the quartz grains, which are mostly monocrystalline with unit extinction; the one on the left has minute fluid and mineral inclusions. Crossed polars. Carboniferous fluvial sandstone. Northumberland, England. Field of view 1.2×0.8 mm.

(b) Biotite mica (brown) showing effects of compaction, and light brown matrix consisting of minute, unresolvable clay minerals, iron minerals (dark brown/black), and silt-grade quartz. Also present are many angular quartz grains (white), and feldspar and lithic grains. Plane-polarized light. Silurian turbidite greywacke. Scotland. Field of view 1.2 × 0.8 mm.

(c, d) Quartz arenite with well-developed overgrowths on quartz grains, which are mostly monocrystalline with unit extinction. Dusty looking quartz grain on left is of hydrothermal origin and is full of fluid inclusions. Note clear overgrowth. Red hematite coating around grains. Two feldspar grains show effects of dissolution. Rock impregnated with blue resin to show porosity (reduced intergranular and dissolutional intragranular). (c) Plane-polarized light; (d) crossed polars. Permian aeolian sandstone. Cumbria, England. Field of view 3 × 2 mm.

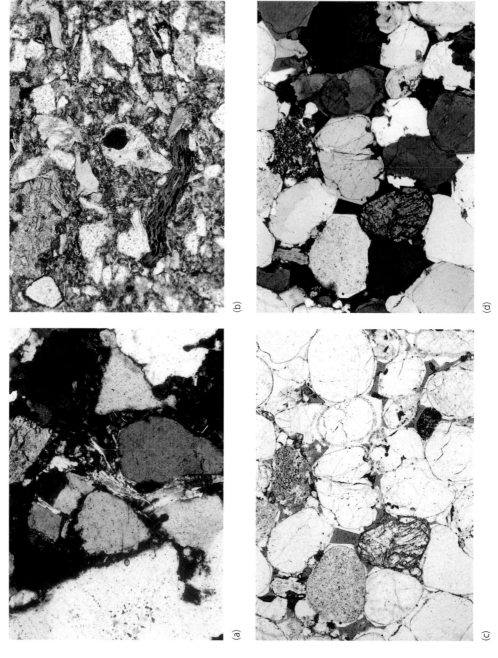

Plate 3

(a) Quartz arenite with angular looking quartz grains from overgrowth cement. Dust-line around grains visible in some cases. Supermature sandstone consisting of monocrystalline, unit-extinguishing quartz grains. Zircon, a heavy mineral, upper left (red). Crossed polars. Carboniferous marine sandstone. Durham, England. Field of view 1.2 × 0.8 mm.

(b) Litharenite with sedimentary rock fragments of fine sandstone and mudrock (some showing lamination). Quartz grains also present (clear). Note the effects of compaction: a tight-fitting arrangement of grains, and some interpenetration and squashing of grains too. Plane-polarized light. Carboniferous fluvial sandstone. Cantabrians, Spain. Field of view 6 × 4 mm.

(c) Arkose with many feldspar grains (dusty/dirty looking compared with clearer quartz grains), and hematite coatings around sand grains; note that hematite is absent where grains are in contact. Plane-polarized light. Precambrian fluvial sandstone. Torridon, Scotland. Field of view 1.2 × 0.8 mm.

(d) Arkose with feldspar grains (orthoclase, microcline) showing incipient replacement by sericite (minute bright crystals). Crossed polars. Precambrian fluvial sandstone. Torridon, Scotland. Field of view 1.2 × 0.8 mm.

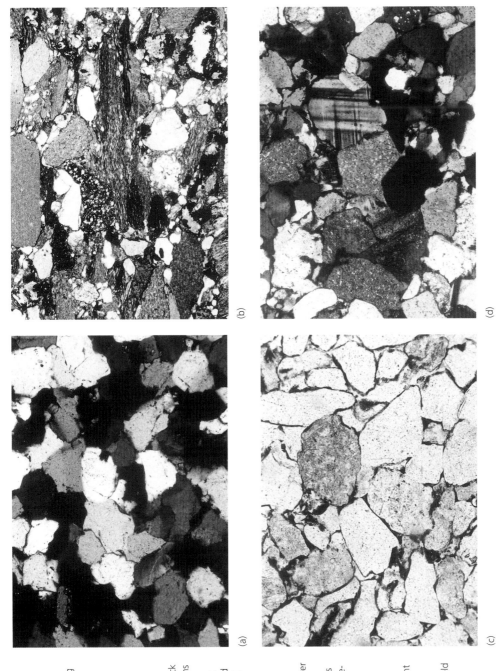

(a)

(b)

(c)

(d)

Plate 4

(a, b) Greywacke. Quartz, feldspar and lithic grains are contained in a fine-grained matrix of chlorite and silt-grade quartz: **(a)** plane-polarized light; **(b)** crossed polars. Silurian turbidite greywacke. Southern Uplands, Scotland. Field of view 3 × 2 mm.

(c, d) Quartz overgrowths on quartz grains. The grain surface is shown by the red hematite coating. Better overgrowths on monocrystalline quartz than polycrystalline quartz grains. Feldspar grain upper right has no overgrowth and shows some effects of dissolution. Rock impregnated with blue resin to show porosity (reduced intergranular): **(c)** plane-polarized light; **(d)** crossed polars. Permian aeolian sandstone. Cumbria, England. Field of view 1.2 × 0.8 mm.

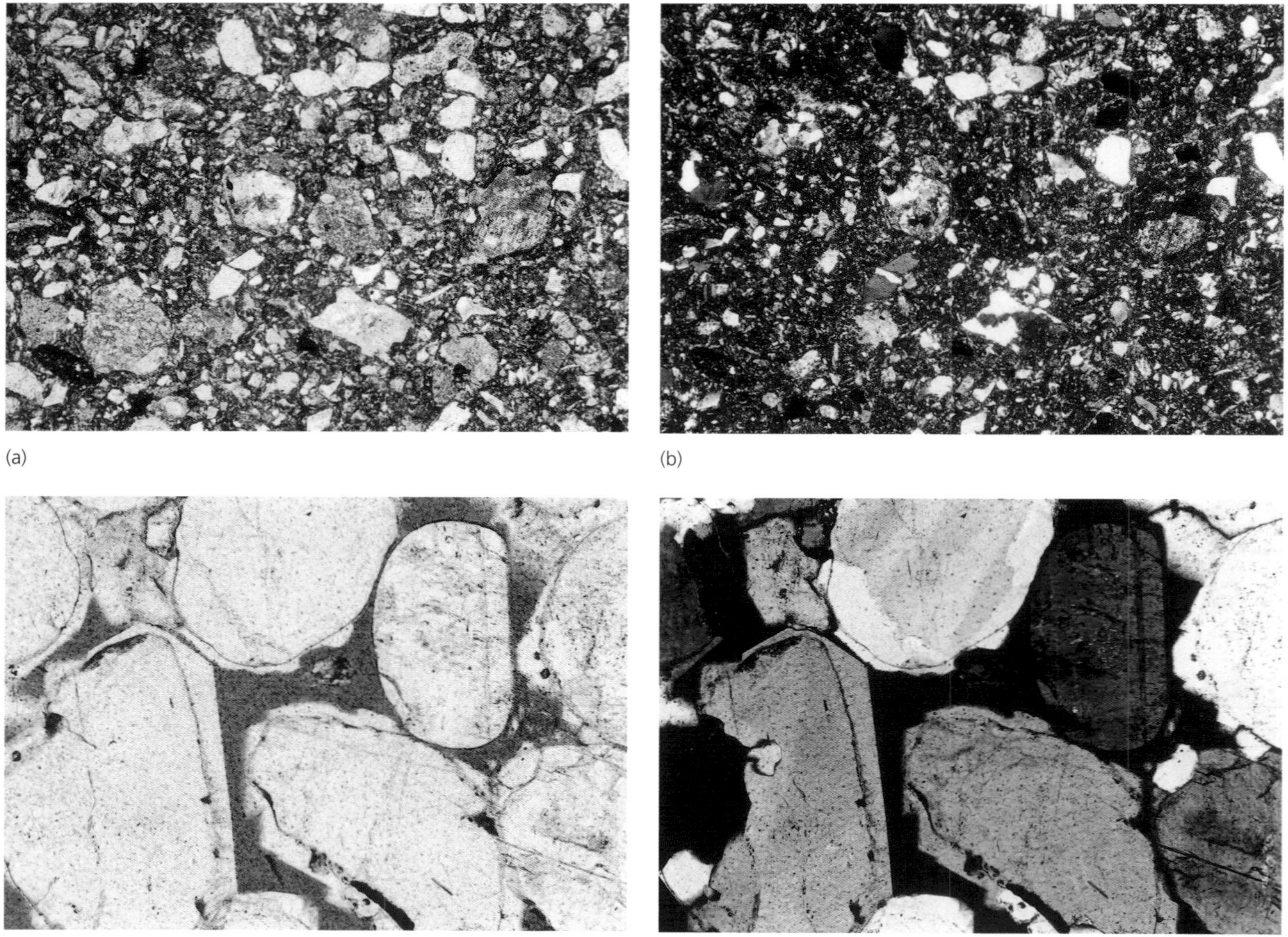

Plate 5

(a, b) Calcite cement in quartz arenite. Large poikilotopic calcite crystals enclosing several grains. Quartz grains are well-rounded monocrystalline, with unit and undulose extinction, and polycrystalline. Feldspar grain showing splitting by calcite crystal; detail in (b). Crossed polars. Permian aeolian sandstone. Durham, England: **(a)** field of view 3 × 2 mm; **(b)** field of view 1.2 × 0.8 mm.

(c) Kaolinite (small black and white crystallites) between quartz grains (mono- and polycrystalline), probably replacing a feldspar grain. Distorted muscovite flake (showing blue colour) between quartz grains. Crossed polars. Carboniferous fluvial sandstone. Northumberland, England. Field of view 1.2 × 0.8 mm.

(d) Mudrock showing effects of compaction with fracture of shells (thin brachiopods), folding of laminae around shells and flattening of burrows (upper left). Minute silt-grade quartz and shell debris disseminated throughout mud; round white grain (centre left) is crinoid ossicle. Carboniferous marine mudrock. Northumberland, England. Field of view 6 × 4 mm.

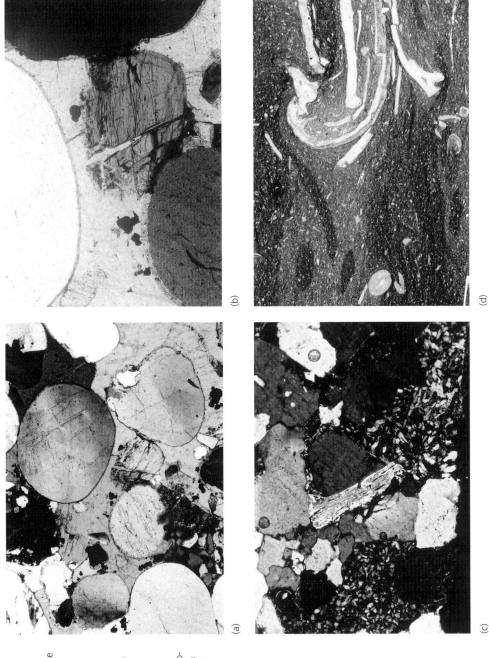

Plate 6

(a, b) Holocene ooids composed of aragonite showing concentric structure and a nucleus, several of which are peloids, and one in the centre of the biggest ooid is a bioclast. Structureless oval grain at lower left is a peloid. Also present is a meniscus cement of calcite precipitated in the meteoric vadose zone. The white areas between grains in (a) and the black areas in (b) are pore space:
(a) plane-polarized light; **(b)** crossed polars. Joulters Cay, Bahamas. Field of view 1.2 × 0.8 mm.

(c) Primary, calcitic ooids with strong radial-concentric structure and nuclei mostly of peloids. Bivalve fragment, now composed of clear calcite spar crystals, is also coated. Notice contact between grains; a little interpenetration indicating some burial compaction before cementation. The ooids are contained in a very large poikilotopic calcite cement (here appearing white). Oolitic grainstone, Jurassic. Lincolnshire, England. Field of view 3 × 2 mm.

(d) Formerly aragonitic ooids, now composed of calcite with poor preservation of original concentric structure and oomoulds (filled with blue resin). Some compaction of oomoulds indicating that the drusy calcite spar cement is a burial precipitate. Oolitic grainstone, Smackover Formation, Jurassic. Subsurface Arkansas, USA. Field of view 3 × 2 mm.

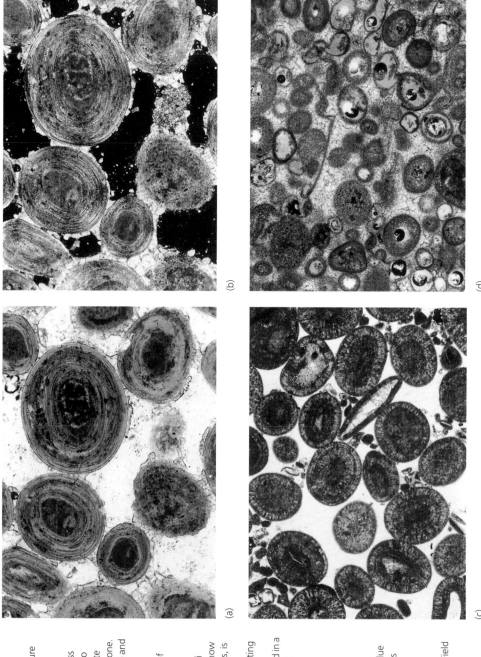

(a)

(b)

(c)

(d)

Plate 7

(a) Peloids; many are micritized bioclasts and ooids, some are faecal pellets. Micrite envelope defines a bivalve shell that has dissolved away; bivalve shell within coated grain is replaced by coarse calcite crystals. The sparse cement consists of small, stubby calcite crystals of probable meteoric phreatic origin. Small dolomite rhombs also present. Blue resin shows porosity. Jurassic. Dorset, England. Field of view 0.8 × 0.8 mm.

(b,c) Modern bivalve fragment with micrite envelope. Shell, composed of aragonite, consists of minute crystallites giving a sweeping extinction under crossed polars (c). Abu Dhabi, UAE. Field of view 0.8 × 1.0 mm.

(d) Bivalve fragment (the elongate grain) in centre with a prominent micrite envelope and shell now composed of drusy calcite spar, a cement. The micrite envelope has fractured (in the centre) as a result of compaction. Also present are numerous peloids, most of which are micritized bioclasts, and some crinoid fragments, in a sparite cement. Urgonian, Cretaceous. Vercors, France. Field of view 3 × 2 mm.

(e) Hippuritid rudist bivalve (the conical, attached valve) showing brown, well-preserved fibrous calcite outer wall, and thinner inner wall with tabulae, which originally were aragonite, but are now composed of calcite spar. Round areas of sediment (micrite) are the fills of sponge borings. Cretaceous. Provence, France. Field of view 6 × 4 mm.

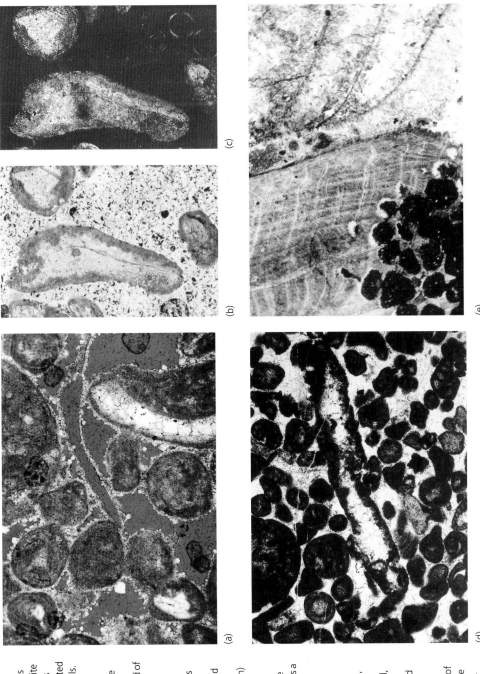

Plate 8

(a, b) Calcitized bivalve shells. Shells originally composed of aragonite but replaced by calcite with some retention of original shell structure. Calcite crystals cross-cut the shell structure and are pseudopleochroic (different shades of brown on slide rotation): **(a)** plane-polarized light and **(b)** crossed polars. Jurassic. Dorset, England. Field of view 0.8 × 0.8 mm.

(c) Gastropods, in long- and cross-section, defined by thin micrite envelopes (black). The shells, originally aragonite, dissolved out and voids were filled by marine fibrous calcite cement (pale brown). A later clear calcite cement filled remaining pores (white). Jurassic. Sicily, Italy. Field of view 4 × 4 mm.

(d) Brachiopod shell with puncti, some filled with lime mud sediment, showing preservation of internal structure consisting of obliquely arranged fibres. Other grains are micritized bioclasts (peloids). Bioclastic grainstone. Cretaceous. Vercors, France. Field of view 3 × 2 mm.

(e) Brachiopod shells and spines with well-preserved shell structure, but limestone has suffered compaction and many shells are broken. Thin-section is stained with Alizarin Red S and potassium ferricyanide; bioclasts are pink (calcite) and cement is blue (ferroan calcite). Cement occurs within cracks showing that it is a burial precipitate. Crinoids and clay (brown) also present. Bioclastic packstone. Carboniferous. Northumberland, England. Field of view 6 × 4 mm.

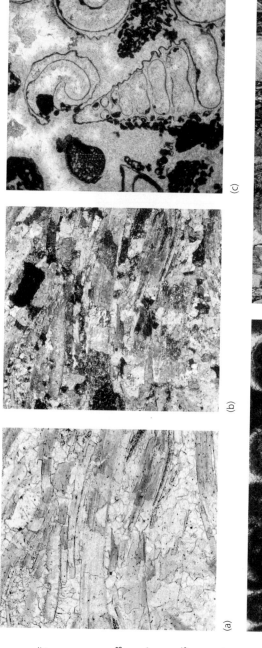

Plate 9

(a) Rugose coral (*Lithostrotion* sp.) showing internal plates (septa, tabulae and dissepiments). The pores within the coral are partly filled with an initial isopachous fibrous marine cement and then by internal sediment. Carboniferous. Durham, England. Field of view 6×4 mm.

(b) Scleractinian coral from a Jurassic patch reef showing variable preservation of structure as a result of replacement of the original aragonite by calcite. Also present, on the right, is a boring made by a lithophagid bivalve (shells present); the boring shows a geopetal structure, with sediment below (dark) and drusy calcite spar above (white). Yorkshire, England. Field of view 6×4 mm.

(c, d, e) Foraminifera. **(c)** Endothyracid foram, crinoid to left, cut by a calcite vein. Carboniferous. Clwyd, Wales. Field of view 4×2 mm. **(d)** *Nummulites*. Bioclastic grainstone, Eocene. Tunisia. Field of view 6×4 mm. **(e)** Miliolids, also bivalve fragments. Cretaceous. Vercors, France. Field of view 4×2 mm.

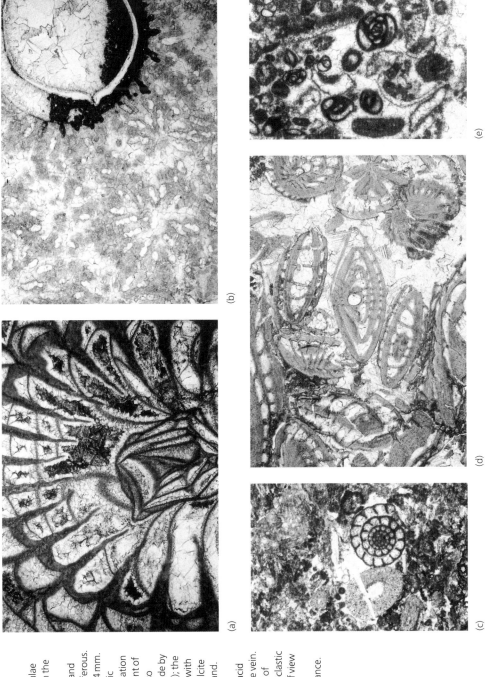

(a)

(b)

(c)

(d)

(e)

Plate 10

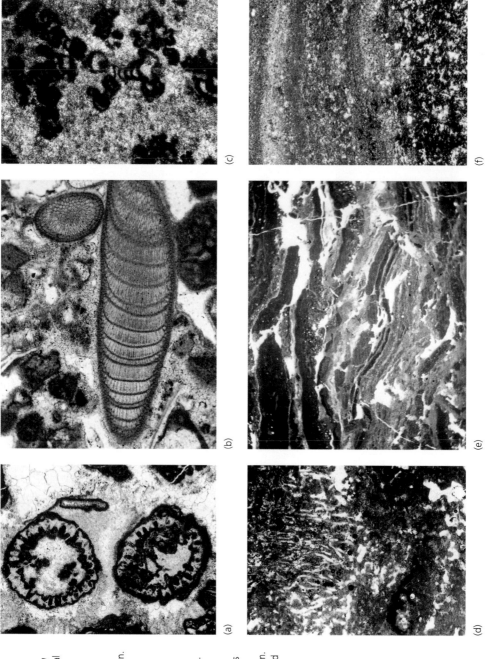

(a) Dasyclad algae. Capitan, Permian. Texas, USA. Field of view 4 × 2 mm.

(b) Calcareous red algae, *Lithothamnion* in longitudinal section (showing seasonal growth zones) and cross-section. The grains here are cemented by isopachous high-Mg calcite marine cement. Recent. Belize. Field of view 1.0 × 0.8 mm.

(c) Calcified microbes, *Renalcis*. Devonian. Guilin, China. Field of view 4 × 2 mm.

(d) Modern microbial mat composed of dolomite. The filaments of the cyanobacteria are clearly visible, but the dolomite crystals are submicroscopic. Recent. Bahamas. Field of view 4 × 2 mm.

(e) Stromatolite (microbial mat) composed of micrite laminae and laminoid fenestrae, with some intraclasts from desiccation. Carboniferous. Glamorgan, Wales. Field of view 5 × 4 mm.

(f) Stromatolite composed of micritic and grainy laminae, and small spar-filled fenestrae. Precambrian. Flinders, Australia. Field of view 4 × 2 mm.

Plate 11

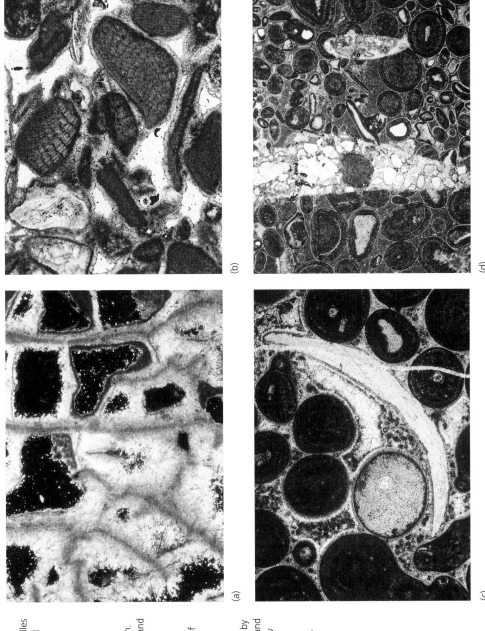

(a) Coral with aragonite cement — needles and a botryoid within the corallites, and lime mud internal sediment (black) and peloids in other pores. Crossed polars. Recent. Belize. Field of view 6 × 4 mm.

(b) Isopachous high-Mg calcite cement around skeletal grains (including calcareous red algae). Recent fore-reef debris. Belize. Field of view 1.2 × 0.8 mm.

(c) Oolitic grainstone with brachiopod and crinoid fragments (with thin micrite envelope — black) cemented by early isopachous fibrous marine cement (calcite), then some internal sediment of peloids, followed by drusy calcite spar cement. Carboniferous. Glamorgan, Wales. Field of view 6 × 4 mm.

(d) Hardground with ooids surrounded by thin isopachous marine cement fringe and then pore space filled by lime mud, now micrite. Cemented rock then cut by annelid borings, which later filled with quartz grains. Jurassic. Gloucestershire, England. Field of view 6 × 4 mm.

Plate 12

(a,b) Fusulinid foraminifer cemented by radiaxial fibrous calcite. The crystals are columnar and cloudy with inclusions, and have undulose extinction under crossed polars. Small area of clear calcite spar:
(a) plane-polarized light, field of view 6×4 mm; (b) crossed polars, field of view 3×2 mm. Capitan, Permian. Texas, USA.
(c,d) Syntaxial calcite overgrowth cement on crinoid grain. Early part of overgrowth calcite cloudy with inclusions is probably a marine precipitate; clear later overgrowth a burial precipitate. Grains mainly peloids, micritized bioclasts and faecal pellets, showing concavo-convex/interpenetrative contacts, indicating some compaction. Very thin isopachous calcite cement fringes around grains seen in (d) are probably marine precipitates. Rock also cut by thin calcite veins. (c) Plane-polarized light; (d) Crossed polars. Cretaceous. Alps, France. Field of view 3×2 mm.

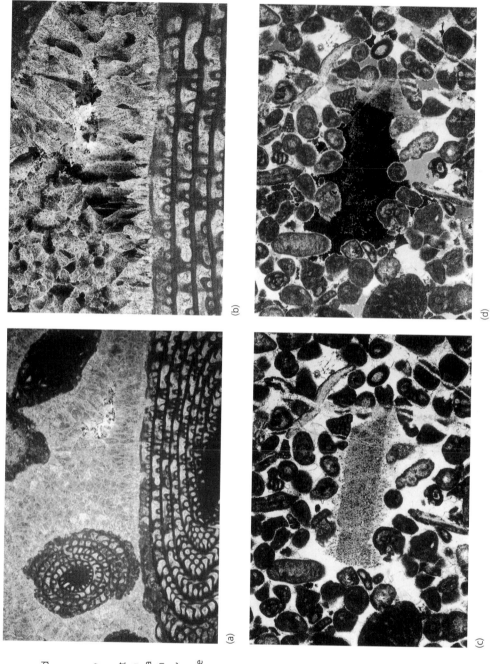

(a)

(b)

(c)

(d)

Plate 13

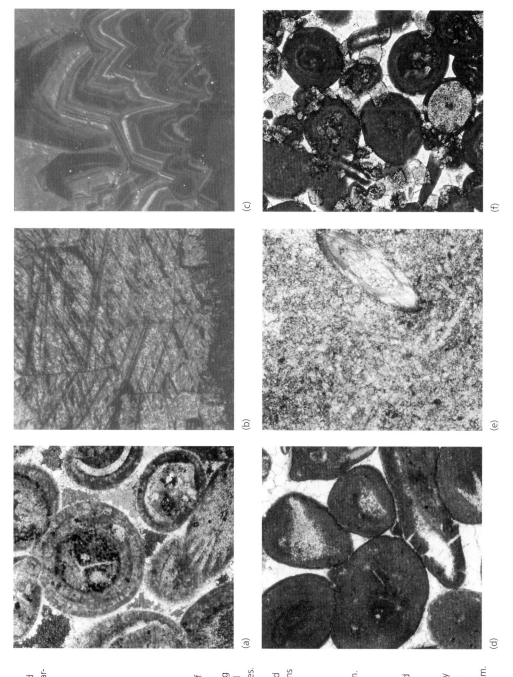

(a) Calcite cement at grain contacts and irregularly around grains, indicating near-surface, meteoric vadose environment. Later precipitation of large poikilotopic calcite (black, in extinction) took place during burial after some compaction. Crossed polars. Carboniferous. Glamorgan, Wales. Field of view 0.8 × 0.8 mm.

(b, c) Calcite spar.

(b) Calcite spar under plane-polarized light showing drusy fabric (crystal-size increase away from the substrate) and prominent twin planes. (c) Same field of view under cathodoluminescence showing delicate growth zones resulting from subtle variation in manganese and iron contents. Triassic. Glamorgan, Wales. Field of view 2 × 2 mm.

(d) Sutured (microstylolitic) contacts and concavo-convex contacts between grains (micritized ooids and bioclasts) and mechanical fracture of lower grain. Calcite spar between grains is a post-compaction burial cement. Jurassic. Burgundy, France. Field of view 2 × 2 mm.

(e) Microspar–pseudospar formed through aggrading neomorphism with fossil relics (brachiopod spine on right). Carboniferous. Yorkshire, England. Field of view 2 × 2 mm.

(f) Oolitic grainstone with scattered dolomite rhombs precipitated after early compaction (see grain contacts), before calcite spar cement. Carboniferous. Glamorgan, Wales. Field of view 2 × 2 mm.

Plate 14

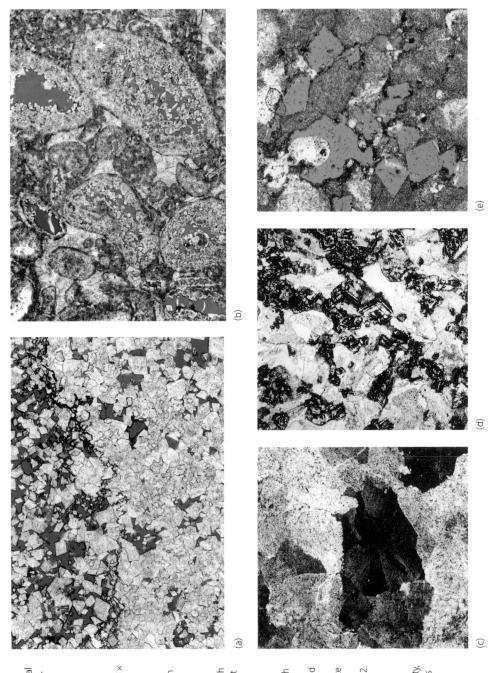

(a) Dolomitized oolite (no relics of original grains) with stylolite, highlighted by iron-rich clay. Xenotopic dolomite (anhedral crystals) below stylolite and idiotopic dolomite (euhedral crystals) above. Intercrystalline porosity shown through impregnation with blue resin. Arab Formation. Offshore UAE. Field of view 3 × 2 mm.

(b) Dolomitized grainstone with moderate preservation of original ooids. Intercrystalline porosity is present, shown up by impregnation with blue resin. Cretaceous. Offshore Angola. Field of view 6 × 4 mm.

(c) Baroque dolomite: coarse crystals with undulose extinction. Rock is an oolite but there are no relics. Crossed polars. Carboniferous. Glamorgan, Wales. Field of view 2 × 2 mm.

(d) Dedolomite: crinoidal grainstone with overgrowths containing scattered dolomite rhombs that have been replaced by calcite. The dark material is iron oxide/hydroxide, suggesting the dolomite was originally ferroan. Carboniferous. Northumberland, England. Field of view 2 × 2 mm.

(e) Dolomite moulds: grainstone with scattered dolomite rhombs that have been dissolved out to give a good porosity, as shown by blue resin. Stylolitic contacts between grains. Jurassic. Burgundy, France. Field of view 0.8 × 0.8 mm.

Plate 15

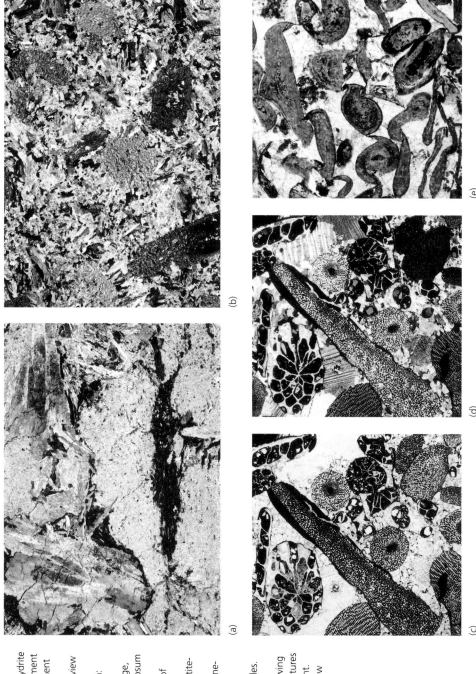

(a) Anhydrite: minute crystals of anhydrite forming nodules with some clay sediment between (brown) and large replacement anhydrite crystals. Crossed polars. Permian. Cumbria, England. Field of view 6×4 mm.

(b) Anhydrite and secondary gypsum: small crystals and laths of anhydrite (bright colours) being replaced by large, low birefringence, porphyrotopic gypsum crystals. Crossed polars. Permian (Zechstein). Teesside, England. Field of view 6×4 mm.

(c, d) Hematitic ironstone with hematite-impregnated crinoid and bryozoan fragments in a calcite cement: (c) plane-polarized light; (d) crossed polars. Bioclastic grainstone, Carboniferous Rhiwbina Ironstone. Glamorgan, Wales. Field of view 2×2 mm.

(e) Berthierine–chamosite ooids showing distorted shapes and elephantine features in calcite cement. Plane-polarized light. Jurassic. Raasay, Scotland. Field of view 2×2 mm.

(a)

(b)

(c)

(d)

(e)

Plate 16

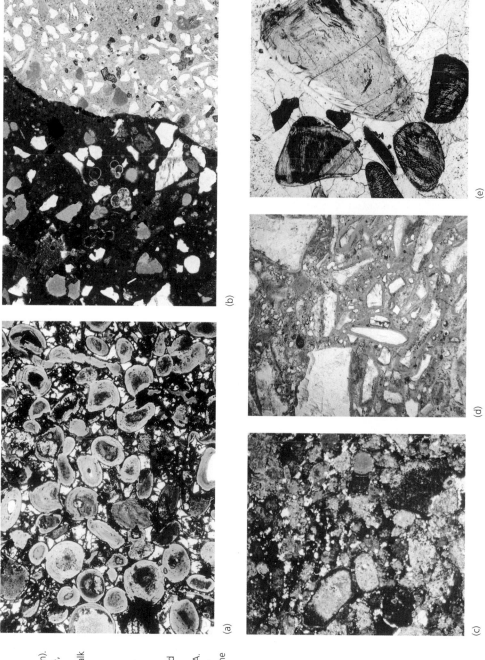

(a) Berthierine–chamosite ooids, with some shape distortion, in a siderite cement partly altered to goethite (brown). Plane-polarized light. Jurassic. Yorkshire, England. Field of view 3 × 2 mm.

(b) Glauconite grains (green) in dark chalk with planktonic foraminifers and in a pebble (on right) of phosphate (pale brown), with angular quartz grains in both. Plane-polarized light. Cretaceous. Bornholm, Denmark. Field of view 3 × 2 mm.

(c, d, e) Phosphorites: all plane-polarized light; all fields of view 2 × 2 mm. (c) Phosphorite pellets. Permian. Idaho, USA. (d) Fish scales and bones in phosphorite mudstone. Permian. Idaho, USA. (e) Bone fragments. Triassic. Gloucestershire, England.

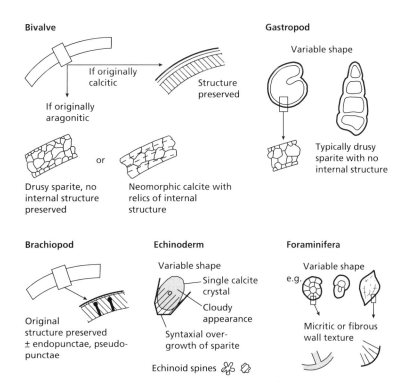

Fig. 4.7 Typical thin-section appearance of bivalve, gastropod, brachiopod, echinoderm and foraminiferal skeletal grains in limestone.

internal structure (growth lines) are preserved (see Plate 8a), and there are minute inclusions of aragonite left in the calcite (see Section 4.7). Calcitic bivalves normally will retain their original structure and the most common types are foliaceous (thin parallel sheets) and prismatic. Bivalve fragments in thin-section will be seen as elongate, rectangular to curved grains, typically disarticulated.

Gastropods are ubiquitous throughout shallow-marine environments. They also occur in vast numbers, but low species diversity, in hypersaline and brackish waters, such as on tidal flats and in estuaries, because certain species are able to tolerate fluctuations and extremes of salinity. Most gastropods are benthic, vagile creatures. The encrusting vermetiform gastropods, often confused with serpulids, form reef-like structures in the tropics and Carboniferous. The small, conical pteropods are important in Cenozoic pelagic sediments.

The majority of gastropods have shells of aragonite with similar internal microstructures to bivalves. The internal microstructure of fossil gastropods also is rarely seen because the original aragonite is mostly dissolved out and the void filled by calcite cement. Gas-tropod fragments can be recognized easily under the microscope by their shape, although this is very much dependent on the plane of section (Fig. 4.7 and Plate 8b). Gastropods may resemble certain foraminifers but the latter are usually much smaller and composed of dark micritic calcite. (See Plate 9c,d,e.)

Of the *cephalopods*, nautiloids and ammonoids are relatively common in limestones of the Palaeozoic and Mesozoic and belemnites occur in Mesozoic lime-stones. They were wholly marine animals with a domi-nantly nektonic or nekto-planktonic mode of life, as with the modern *Nautilus*, octopus and cuttlefish. The cephalopods are more common in pelagic, relatively deep-water deposits. Examples include the Ordovi-cian–Silurian *Orthoceras* limestones of Sweden, the Devonian Cephalopodenkalk and Griotte of western Europe and the Jurassic Ammonitico Rosso of the European Alpine region. Nautiloid and ammonoid shells were originally aragonitic and so in limestones they are typically composed of calcite spar with little internal structure. The shape, normally large size and presence of septa are the features to note. Belemnite guards were made of calcite and have a strong radial-fibrous fabric in cross-section.

Brachiopods

Brachiopods are particularly common in Palaeozoic and Mesozoic limestones of shallow-marine origin. These were largely benthic, sessile organisms; a few species were infaunal. Only in rare cases, such as in the Permian of west Texas, did the brachiopods contribute to reef development. At the present time brachiopods are an insignificant group of marine invertebrates.

Although in section brachiopod shells are similar to those of bivalves in shape and size, most articulate brachiopods were composed of low-Mg calcite, so that the internal structure is invariably well preserved. The common structure is a very thin, outer layer of calcite fibres orientated normal to the shell surface, and a much thicker, inner layer of oblique fibres (Fig. 4.7). It can be difficult to distinguish brachiopod shells from bivalves with foliaceous calcite. However, certain brachiopods have modifications to the shell, with punctae and pseudopunctae. In the punctate brachiopods, such as the terebratulids, fine tubes (endopunctae) perpendicular to the shell surface perforate the inner layer and are filled with sparite or micrite. Pseudopunctae, as occur in the strophomenid group, are prominent rod-like prisms within the shell. Inarticulate brachiopods, composed mostly of chitin or chitinophosphate, are rare in limestones. (See Plates 8c,d & 11c.)

Cnidaria (especially corals)

The Cnidaria include the Anthozoa (corals), of which two ecological groups exist today: hermatypic corals that contain symbiotic dinoflagellate algae (zooxanthellae) in their polyps and ahermatypic corals without such algae. Because of the algae, hermatypic corals require shallow, warm and clear seawater. They are the reef-forming corals at the present time, being mainly responsible for the reef framework, which is reinforced by red algae. Ahermatypic corals can occur at much greater depths and tolerate colder waters. They locally form build-ups. The rugose and tabulate corals were important in Silurian and Devonian reefs, and many Triassic reefs contain scleractinian corals. Some of the latter may well have been ahermatypic. Corals, both solitary and colonial, and coral debris occur in many non-reefal limestones.

The Palaeozoic rugose and tabulate corals were composed of calcite, most probably high-Mg calcite, so that preservation generally is very good (see Plate 9a). Scleractinian corals (Triassic to Recent), on the other hand, have aragonitic skeletons and so normally are poorly preserved in limestones (see Plate 9b). Identification of coral is based on such internal features as septa and, where present, other internal plates in the Rugosa and Scleractinia, and tabulae in the Tabulata. Corallite form and colonial organization are also important. The microstructure of Palaeozoic calcitic and later aragonitic corals is very similar, chiefly consisting of fibres in spherulitic or parallel arrangements, which form linear structures called trabeculae, or sheets.

Echinodermata

Echinoderms are wholly marine organisms that include the echinoids (sea-urchins) and crinoids (sea-lilies). In modern seas, echinoids inhabit reef and associated environments, locally in great numbers, but crinoids are restricted to deeper waters and are insignificant as producers of carbonate sediment. In the Palaeozoic and Mesozoic, fragments of echinoderms, especially the crinoids, are a major constituent of bioclastic limestones. Many deep-water limestone turbidites are composed of crinoidal debris, derived from shallow platforms.

Echinoid and crinoid skeletons are calcitic; modern forms generally have a high Mg content. Echinoderm fragments are easily identified because they are composed of large, single calcite crystals, individual grains thus showing unit extinction. In many cases, a sparite cement crystal has grown syntaxially around the echinoderm fragment (Fig. 4.7). Echinoderm grains have a dusty appearance, especially relative to a sparite cement overgrowth, and they may show a porous structure filled with micrite or sparite. (See Plates 9c, 12a,b & 15c,d.)

Bryozoa

Although these small, colonial marine organisms are significant suppliers of carbonate sediment only locally at the present time (notably to cool-water carbonates, as off southern Australia), they have in the past contributed to the formation of reef and other limestones, particularly in the Palaeozoic. Examples include the Mississippian mud-mounds of southwestern USA and Europe, the Permian reefs of Texas and western Europe, and the Danian Chalk of Denmark.

Modern bryozoan skeletons are composed of either aragonite or calcite (commonly high-Mg calcite) or a mixture of both. There are many types of bryozoans but the fenestrate variety, including the fenestellids, are seen most frequently in sections of Palaeozoic limestones. The skeleton consists of foliaceous calcite with round holes (the zooecia, where the individuals of the colony used to live) filled with sparite or sediment (see Plate 15c).

Foraminifera

Foraminifera are dominantly marine Protozoa, mostly of microscopic size. Planktonic foraminifers dominate some pelagic deposits, such as the *Globigerina* oozes of ocean floors and some Cretaceous and Tertiary chalks and marls. Benthic foraminifers are common in warm, shallow seas, living within and on the sediment, and encrusting hard substrates.

Foraminifera are composed of low- or high-Mg calcite, rarely aragonite. Foraminifera are very diverse in shape but in section many common forms are circular to subcircular with chambers. The test wall is dark and microgranular in many thin-walled foraminifers such as the endothyracids and miliolids, but light-coloured and fibrous in larger, thicker species, such as the rotaliids, nummulitids and orbitolinids. (See Plate 9c,d,e.)

Other carbonate-forming organisms

There are many other organisms that have calcareous skeletons but contributed in only a minor way to limestone formation, or were important for only short periods of geological time.

Sponges (Porifera). Spicules of sponges, which may be composed of silica or calcite, occur sporadically in sediments from the Cambrian onwards. The importance of spicules is as a source of silica for the formation of chert nodules and silicification of limestones (Section 4.8.4). At times, sponges provided the framework for reefs and mounds; examples include lithistid sponges in the Ordovician, calcisponges in the Permian of Texas and Triassic of the Alps, and silicisponges (now calcitized) in the Jurassic of southern Germany. Sclerosponges are the dominant reef-forming organ-

ism in some modern Caribbean reefs. *Stromatoporoids*, once considered hydrozoans, are now classified as a subphylum of the Porifera. Stromatoporoids were marine colonial organisms that had a wide variety of growth forms, ranging from spherical to laminar, depending on species and environmental factors. Stromatoporoids were a major reef organism in the Silurian and Devonian, commonly in association with rugose and tabulate corals, and grew up to a metre or more in size. *Archaeocyathids*, also probably sponges, formed reefs in the lower Cambrian of North America, Morocco, Siberia and South Australia.

Arthropods. Of this group, the *ostracods* (Cambrian to Recent) are locally significant in Tertiary limestones and some others. They live at shallow depths in marine, brackish or freshwater environments. Ostracods have small (around 1 mm in length), thin bivalved shells, smooth or ornamented, composed of calcite with a radial-fibrous structure. *Trilobites* (Cambrian to Permian), with a similar skeletal structure, occur locally in Palaeozoic shelf limestones, but never in rock-forming quantities.

Calcispheres. These are simple spherical objects, up to 0.5 mm in diameter, composed of calcite (usually sparite), in some cases with a micritic wall. They are probably some form of alga, although an affinity with Foraminifera has been suggested. They occur in many Palaeozoic limestones, particularly fine-grained micrites of back-reef or lagoonal origin.

4.3.3 The contribution of algae and microbes to limestones

Algae and microbes make a major contribution to limestones by providing skeletal carbonate particles, trapping grains to form laminated sediments and attacking particles and substrates through their boring activities. Many of the Precambrian limestones were at least in part produced by microbes and algal–microbial limestones are widely distributed throughout the Phanerozoic. Four groups of algae are important: red algae (Rhodophyta), green algae (Chlorophyta), yellow–green algae (Chrysophyta) and cyanobacteria (formerly blue–green algae). Relevant texts are Walter (1976), Flügel (1977), Wray (1977), Monty (1981), Toomey & Nitechi (1985), Riding (1990) and Riding & Awramik (2000).

Rhodophyta (red algae)

Calcareous algae of the Rhodophyta, such as the Corallinaceae (Carboniferous to Recent) and Solenoporaceae (Cambrian to Miocene) have skeletons composed of cryptocrystalline calcite, which is precipitated within and between cell walls. In section, a regular cellular structure is present (see Plates 10b & 11b). Modern coralline algae have a high Mg content in the calcite, which is related to water temperature (higher values for a given species in warmer waters). Many of the coralline algae encrust substrates and if this is a pebble or shell then nodules, referred to as *rhodoliths*, develop. Encrustations may be massive and rounded, or delicately branched, depending on ecological factors. One of the most important roles of these red algae is in coating, binding and cementing the substrate, particularly in modern reefs. In temperate and arctic carbonate sands, the red alga

Fig. 4.8 Calcareous green algae. (a) *Penicillus*, which on death and disintegration produce micron-sized aragonite needles. (b) *Halimeda*, which gives rise to sand-sized grains. Recent. Florida Keys, USA.

Lithothamnion is a major contributor. In the Palaeozoic and Mesozoic, red algae of the Solenoporaceae are locally abundant and in some cases participate in reef formation.

Chlorophyta (green algae)

Three algal groups are important: the Codiaceae, Dasycladaceae (both Cambrian to Recent) and Characeae (Silurian to Recent). The Characeae are incompletely calcified (low-Mg calcite), so that only stalks and reproductive capsules are found in limestones. Modern (and fossil) forms are restricted to fresh or brackish water.

With the dasyclad algae, calcification is also incomplete and involves the precipitation of an aragonitic crust around the stem and branches of the plant. Dasyclads are marine algae that tend to occur in shallow, protected lagoonal areas of the tropics. Under the microscope, circular, ovoid or elongate shapes are seen, representing sections through the stems or branches (see Plate 10a).

The codiacean algae include *Halimeda* and *Penicillus* (Fig. 4.8), two common genera of Caribbean and

(a)

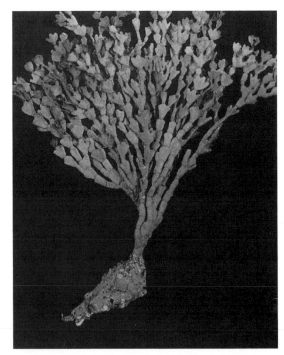

(b)

Pacific reefs and lagoons. *Halimeda* is a segmented plant that on death and disintegration generates coarse, sand-sized particles. In thin-section, they look like Swiss cheese. *Penicillus*, the shaving-brush alga, is less rigid, consisting of a bundle of filaments coated in needles of aragonite. Death of this and other algae provides fine-grained carbonate sediment (lime mud or micrite, Section 4.3.4).

Phylloid algae are a group of late Palaeozoic algae that have a leaf or potato-crisp shape. Some belong to the codiaceans, whereas others have more affinities with red algae. Phylloid algae are one of the main components of shelf-margin mud-mounds in the Upper Carboniferous to Lower Permian of southwest USA.

Chrysophyta (coccoliths)

Coccolithophorids (Jurassic to Recent) are planktonic algae that have a low-Mg calcite skeleton consisting of a spherical coccosphere (10–100 μm diameter) composed of numerous calcareous plates, called coccoliths. In view of their size, these algae are studied with the scanning electron microscope. The coccoliths are chiefly disc-shaped, commonly with a radial arrangement of crystals (Fig. 4.9). Coccoliths are a significant

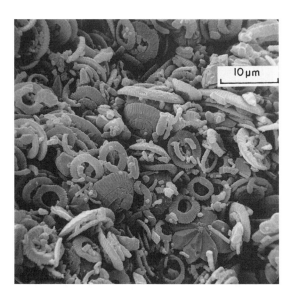

Fig. 4.9 Scanning electron micrograph of coccoliths from pelagic ooze, Shatsky Rise, northwest Pacific. Also present (lower right) is a larger discoaster, another type of nanoplanktonic alga.

component of modern deep-water carbonate oozes, particularly those of lower latitudes. They form the bulk of Tertiary and Cretaceous chalks and occur in red pelagic limestones of the Alpine Jurassic (see Fig. 4.17).

Micrite envelopes

A large proportion of skeletal fragments in modern and ancient carbonate sediments possess a dark *micrite envelope* around the grains (see Plates 7b,c,d, 8b & 11c). The envelope is produced mostly by endolithic bacteria that bore into the skeletal debris. Following vacation of the microbores (5–15 μm in diameter), they are filled with micrite. Repeated boring and filling result in a dense micrite envelope, which is the altered outer part of the skeletal grain. This process of grain degradation may eventually produce a totally micritized grain, i.e. a *peloid* (Section 4.3.1), devoid of the original skeletal structure (see Plates 7a, 8c, 12c & 13d). Some bioclasts, especially the corals, molluscs and foraminifers, are more susceptible to microboring than others, and this has implications for the preservation of grain assemblages (Perry, 1998). Inorganic recrystallization of grains on the sea floor also takes place, destroying the original skeletal structure (Reid & Macintyre, 1998).

Many other organisms bore into skeletal grains and carbonate substrates; examples include clionid sponges (in Plate 7e), bivalves (such as *Lithophaga*, see Plate 9b and Fig. 2.42d), and polychaetes, all of which produce larger borings and cavities, and fungi, which produce bores of 1–2 μm in diameter. The micrite filling the borings may be precipitated physicochemically or biochemically through decomposition of the microbes. Micrite envelopes due to endolithic cyanobacteria can be used as a depth criterion, indicating deposition within the photic zone (less than 100–200 m), but the grains can be transported to greater depths.

Stromatolites and microbialites

An important role of cyanobacteria (with other organisms such a diatoms, fungi and nematodes) is the formation of *microbial mats*, formerly called algal mats. These organic mats occur on sediment surfaces in many low- to mid-latitude, marine and non-marine environments, from moderate depth subtidal through

Fig. 4.10 Tidal-flat microbial mats from Abu Dhabi, Arabian Gulf. The microbial mats (dark areas) are desiccated into small polygons and have a thin covering of recently deposited carbonate sediment. Large polygonal desiccation cracks occur in the slightly higher, partly lithified gypsiferous carbonate sediment (light area). Knife (circled) for scale.

Fig. 4.11 Columnar stromatolites from agitated intertidal zone, Shark Bay, Western Australia. Columns increase in size upwards and join with adjacent columns to form large domal structures.

Fig. 4.12 Microbially laminated carbonate sediments from an intertidal flat. Shark Bay, Western Australia. Desiccation cracks and cavities (fenestrae) are present.

to supratidal marine areas, and fresh to hypersaline lakes and marshes. They form planar sheets, columns and domes (Figs 4.10–4.13). The cyanobacteria are mainly filamentous varieties, common mat-forming genera being *Lyngbya*, *Microcoleus*, *Schizothrix*, *Scytonema* and *Oscillatorea*, although unicellular coccoid forms such as *Endophysalis* also occur. A mat usually has a specific community that together with environmental factors produces a mat of particular morphology and structure. Areas where microbial mats of various types are developing today include the Bahamas, the Arabian Gulf and Shark Bay, Western Australia.

The cyanobacteria are mucilaginous and this, together with their filamentous nature, results in the trapping and binding of sedimentary particles to produce a laminated sediment, a *microbialite* or *stromatolite* (Figs 4.12 & 4.14). Stromatolites occur throughout the geological record but are particularly important in the Precambrian, where they have been used for stratigraphic correlation. The lamination in many modern intertidal mats consists of couplets of dark organic-rich layers alternating with light, sediment-rich laminae. Microbial filaments may be

Fig. 4.13 Stromatolite domes and columns in a few metres of water in a high-energy tidal channel. Exuma Cay, Bahamas.

Fig. 4.14 Stromatolite consisting of a large dome with internal crinkly laminae. Late Precambrian. Anti-Atlas, Morocco.

preserved (see Plate 10d). Laminae are usually less than several millimetres thick, but some sediment laminae may reach a centimetre or more. The alternating laminae reflect growth of the microbial mat (organic layer) followed by sedimentation, and then trapping and binding of the sediment particles into the mat, as the microbial filaments grow through to form a new organic layer at the surface once more. In ancient stromatolites, the laminae are usually alterations of dense micrite, perhaps dolomitized, and grains, such as peloids and fine skeletal debris (see Plate 10f). Microbially laminated sediments commonly show small corrugations and irregularities in thickness, which serve to distinguish them from laminae deposited purely by physical processes. There may be evidence of desiccation—broken laminae, intraclasts and laminoid fenestrae (see Plate 10e). The laminae may also constitute larger-scale domes and columns. A diurnal growth pattern has been demonstrated for the laminae in some subtidal mats, but in other cases, such as on tidal flats and in ephemeral lakes, mat growth is probably seasonal or related to periodic wettings, and sedimentation is erratic, being controlled largely by storm floodings.

Microbial mats give rise to a range of laminated structures (microbialites). The simplest are planar stromatolites (or microbial laminites). They typically develop on protected tidal flats and so may show desiccation polygons (Fig. 4.10) and contain *laminoid fenestrae* (elongate cavities: Fig. 4.12, Plate 10e and Section 4.6.3) and evaporite minerals or their pseudomorphs. Domal stromatolites, where the laminae are continuous from one dome to the next, occur on the scale of centimetres to metres. Columnar stromatolites are individual structures, which may be several metres high (Figs 4.13 & 4.14). Complex stromatolites, such as occur in Precambrian strata, may be combinations of domes and columns, such as a large columnar structure with linked domes internally, or they may show branching of columns. They can form build-ups many tens of metres high and laterally be very extensive.

One further type of microbial-sediment structure is a nodule (ball), or *oncoid* (Fig. 4.15). Some have an internal concentric lamination, which may be asym-

Fig. 4.15 Oncoids: spherical to irregular microbial balls. Some showing marked asymmetry, through periods of stationary growth. Carboniferous. Fife, Scotland.

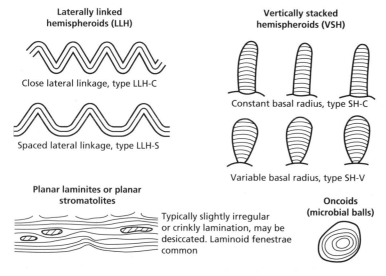

Fig. 4.16 Common types of stromatolite with terminology.

metric. Many oncoids are composed of dense micrite, whereas others are more clotted. The latter fabric also occurs in stromatolites where ones with a poor lamination and a clotted texture are called *thrombolites*. They probably have been formed by coccoid cyanobacteria.

In a notation for describing stromatolites (Fig. 4.16), domal stromatolites are referred to as laterally linked hemispheroids (LLH), columnar stromatolites are vertically stacked hemispheroids (VSH) and oncoids are spherical structures (SS). There is also a binomial classification with form genera and species, particu-

larly used for Precambrian examples. Common genera include *Conophyton*, *Collenia* and *Cryptozoon*.

The morphological variation of microbialites depends largely on environmental factors such as water depth, tidal and wave energy, frequency of exposure and sedimentation rate. For example, large columnar structures in Shark Bay (Fig. 4.11) are restricted to intertidal and subtidal areas in the vicinity of headlands; small columns and domes occur in less agitated bay waters, and low domes and planar mats dominate protected tidal flats (Fig. 4.10). In the Bahamas, large stromatolite columns (almost 'reefs') are growing in shallow, high-energy tidal channels (Fig. 4.13; Macintyre *et al.*, 1996). The microstructure of microbial mats and stromatolites is also variable; it is thought largely to be a reflection of the microbial community, as noted above.

At the present time, microbial mats are quite restricted in their occurrence, being developed extensively only in hypersaline tidal-flat and freshwater environments. Planar stromatolites are a feature of peritidal facies throughout the geological record. The dearth of domal and columnar stromatolites in the Phanerozoic, notably in shallow, normal-marine facies, is attributed to the grazing activities of organisms, especially the gastropods. The absence of such metazoans in the Precambrian and earliest Phanerozoic was one of the main factors in the widespread and diverse stromatolite development at that time, including many in subtidal and deeper-water settings.

Modern marine microbial mats are largely unlithified, whereas those of fresh and hypersaline waters may be cemented through biochemical or physicochemical precipitation of carbonate. There are particular microbes, common in the Palaeozoic, that do have calcified filaments. The group name Porostromata has been applied to these, which include *Girvanella*, *Ortonella*, *Garwoodia* and *Cayeuxia*. The structures consist of tubes or filaments with micritic walls, considered to be calcified sheaths, arranged in an irregular spaghetti-type or more ordered, radial fashion. They typically form nodules (skeletal oncoids) but also stromatolites. In some cases, the algae are associated intimately with encrusting foraminifers, as in *Osagia* and *Sphaerocodium* nodules. More bush-like, branching structures (dendrolites) of calcified microbes are common in the early Palaeozoic, commonly forming reefal structures, and include *Renalcis* and *Epiphyton* (see Plate 10c).

For a discussion of the geological history of microbial carbonates see Riding (2000), and papers in Riding & Awramik (2000).

4.3.4 Lime mud and micrite

Many grainy limestones have a fine, usually dark matrix and many others are composed entirely of fine-grained carbonate. This material is *micrite* (microcrystalline calcite), with a grain size generally less than 4 µm. Electron microscope studies have shown that the micrite is not homogeneous but has areas of finer or

Fig. 4.17 Transmission electron micrograph of a Jurassic micritic pelagic limestone showing variation in size and shape of micrite crystals, and two coccoliths consisting of elongate, radially arranged crystals. Ammonitico Rosso, Jurassic. Austrian Alps.

coarser crystals, and intercrystalline boundaries that may be planar, curved, irregular or sutured (Fig. 4.17). Micrites are susceptible to diagenetic alteration and may be replaced by coarser mosaics of *microspar* (5–15 µm) through aggrading neomorphism (Section 4.7.4).

Carbonate muds are accumulating in many modern environments, from tidal flats and shallow lagoons to the deep-sea floor. There are many possible sources of lime mud. Carbonate muds of south Florida and the Bahama Platform have been the subject of much research. They occur in the shallow subtidal, less-agitated central parts of the platform to the west of Andros Island, and in lagoons, such as the Bight of Abaco. Fine carbonate sediments also occur on tidal flats and on the slopes and in deep basins around the platform ('periplatform ooze'). The mud in many subtidal areas consists predominantly of aragonite needles and laths a few microns in length; only some 20% of the sediment is recognizably biogenic. Inorganic precipitation as a result of evaporation has been postulated for the Great Bahama Bank. The occasional 'whiting', a sudden milkiness of the sea resulting from suspended aragonite needles, may be the actual inorganic precipitation taking place, although stirring of bottom muds by shoals of fish can produce the same effect (see Shinn *et al.*, 1989). The disintegration of calcareous green algae has been widely regarded as a major process of lime-mud production, from studies in South Florida and the Bight of Abaco. When algae such as *Penicillus* (Fig. 4.8a) break down, a vast quantity of aragonite is released. Measurements of growth rates and calculations of the mass of aragonite produced have shown that sufficient mud is produced to account for all the fine-grained sediments. Indeed, it appears that there is an overproduction so that algal disintegration in lagoons could be the source of mud for neighbouring tidal flats and peri-platform areas (Fig. 4.18). Hence the shallow subtidal zone is referred to as the 'carbonate factory'. However, detailed studies with the SEM and trace elements (e.g. Milliman *et al.*, 1993) have concluded that much of the lime mud on the Great Bahama Bank is a direct precipitate, whereas that of Florida Bay is largely green algal in origin. The Sr/Mg ratio of algae is less than two, whereas it is more than four in inorganic aragonite.

Three other processes that produce lime mud, but in variable or limited quantities are:

1 bioerosion, where organisms such as boring sponges and microbes attack carbonate grains and substrates;

2 mechanical breakdown of skeletal grains through waves and currents;

3 biochemical precipitation through microbial photosynthesis and decomposition.

Carbonate mud, largely of skeletal origin, forms

subtidal banks in Florida and Shark Bay, where seagrasses and algae trap and bind the sediment (e.g. Bosence, 1995). Carbonate muds of the deep-ocean floors are oozes composed chiefly of coccoliths, with larger foraminiferal and pteropod grains.

In the lagoons along the Trucial Coast of the Arabian Gulf, inorganic precipitation is probably taking place. The aragonite needle muds contain high strontium values (9400 p.p.m.), close to the theoretical for direct precipitation from that lagoonal water. There is also a paucity of calcareous algae in the region and the possibility of other aragonitic skeletons contributing is precluded by their low strontium values.

There usually is little evidence in a limestone for the origin of the micrite. Nanofossils, in particular coccoliths, can be recognized with the electron microscope in some pelagic limestones (Fig. 4.17), but on the whole there is little to indicate a biogenic origin for the majority of ancient micrites. From the studies of modern lime muds, it is tempting to suggest that many ancient shallow-marine micrites were the product of calcareous green-algal disintegration. However, the possibility of inorganic precipitation in the past cannot be ruled out.

The original mineralogy of lime muds will have been important in their diagenesis; those with a high content of aragonite will have been more susceptible to neomorphism and microspar formation. Original aragonite-dominated precursor muds can be recognized through SEM study of lightly etched micrites, revealing minute crystal relics of aragonite (Lasemi & Sandberg, 1984). In grainstones and other coarse limestones, micrite could well be a cement, rather than a matrix (Section 4.7.1). In addition, fine carbonate sediment up to silt grade may filter into a porous limestone soon after deposition or during early diagenesis. Such geopetal or internal sediment can be recognized by its cavity-filling nature (see Plates 9a,b & 11a,c,d).

4.4 Classification of limestones

Three classification systems are currently used, each with a different emphasis, but the third, that of Dunham, based on texture, is now used more widely.

1 A very simple but often useful scheme divides limestones on the basis of grain size into calcirudite (most grains > 2 mm), calcarenite (most grains between 2 mm and 62 μm) and calcilutite (most grains less than 62 μm).

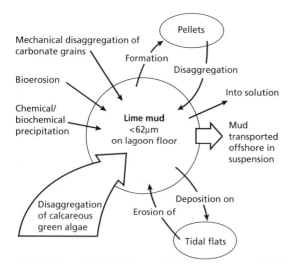

Fig. 4.18 Lime-mud budget for a lagoon in the Bahamas.

2 The classification scheme of R.L. Folk (Fig. 4.19), based mainly on composition, distinguishes three components: (a) the grains (allochems), (b) matrix, chiefly micrite and (c) cement, usually drusy sparite. An abbreviation for the grains (*bio*—skeletal grains, *oo*—ooids, *pel*—peloids, *intra*—intraclasts) is used as a prefix to micrite or sparite, whichever is dominant. Terms can be combined if two types of grain dominate, as in biopelsparite or bio-oosparite. Terms can be modified to give an indication of coarse grain size, as in biosparrudite or intramicrudite. Other categories of Folk are *biolithite*, referring to a limestone formed *in situ*, such as a stromatolite or reef-rock; and *dismicrite*, referring to a micrite with cavities (usually spar-filled), such as a birdseye limestone (Section 4.6.3).

3 The classification of R.J. Dunham (Fig. 4.20) divides limestones on the basis of texture into: *grainstone*, grains without matrix (such as a bio- or oosparite); *packstone*, grains in contact, with matrix (this could be a biomicrite); *wackestone*, coarse grains floating in a matrix (could also be a biomicrite); and a *mudstone*, micrite with few grains. Additional terms of A.F. Embry & J.E. Klovan give an indication of coarse grain size (*floatstone* and *rudstone*), and of the type of organic binding in *boundstone* during deposition (*bafflestone*, *bindstone* and *framestone*). The terms can be qualified to give information on composition, e.g. oolitic grainstone, peloidal mudstone or crinoidal rudstone.

As a result of diagenetic modifications to limestones, care must be exercised in naming the rock. For example, a homogeneous-looking micrite may be a peloidal mudstone, and micrite in a bioclastic, grain-supported rock could be cement, compacted pellets (i.e. grains), primary sediment (i.e. matrix) or internal sediment (infiltrated geopetal sediment). The second

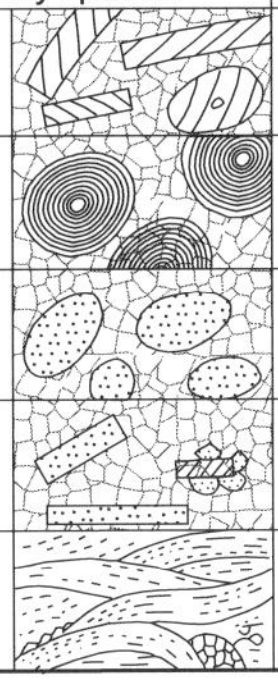

Fig. 4.19 Classification of limestones based on composition.

Principal grains in limestone	Limestone types			
	Cemented by sparite		With a micrite matrix	
Skeletal grains (bioclasts)	Biosparite		Biomicrite	
Ooids	Oosparite		Oomicrite	
Peloids	Pelsparite		Pelmicrite	
Intraclasts	Intrasparite		Intramicrite	
Limestone formed in situ	Biolithite		Fenestral limestone– dismicrite	

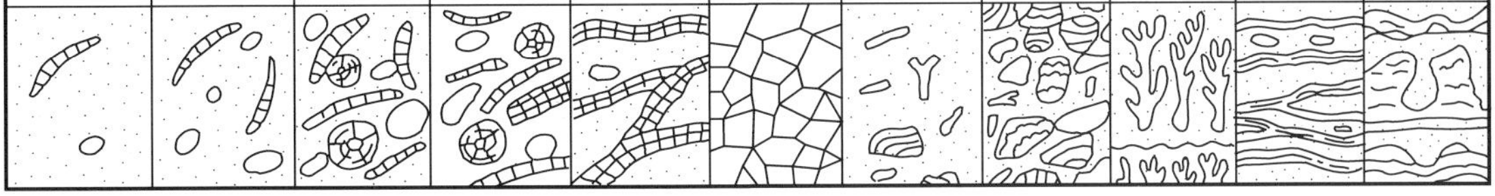

Fig. 4.20 Classification of limestones based on depositional texture.

Original components not bound together during deposition					Original compon-ents bound together	Deposit-ional texture not recogniz-able	Original components not organically bound during deposition		Original components organically bound during deposition		
Contains lime mud			Lacks mud and is grain supported				>10% grains >2mm		Organisms act as baffles	Organisms encrust and bind	Organisms build a rigid framework
Mud-supported		Grain-supported					Matrix supported	Supported by > 2mm compon-ents			
Less than 10% grains	More than 10% grains					Crystalline carbonate					
Mudstone	Wackestone	Packstone	Grainstone	Boundstone	Crystalline		Floatstone	Rudstone	Baffle stone	Bindstone	Framestone

example is basically a question of separating a grainstone from a packstone/wackestone. Many packstones are actually compacted wackestones.

The depositional environments and facies of limestones are considered in Section 4.10, where the typical limestone rock types of each environment are discussed briefly. The composition of limestones can be taken further by point-counting the different grain types present and calculating their percentages. A visual estimation of the grain percentages can be made using Fig. 2.49 in Chapter 2. Triangular diagrams can illustrate the composition for three principal components (e.g. ooids, bioclasts and peloids, or crinoids, brachiopods and bivalves). In this way different *microfacies* can be recognized.

4.5   Limestone grain size and texture

For the most part carbonate sediments are formed *in situ*. Although some may be transported from shelf to basin by turbidity currents or slumps, and from an inner to outer shelf by storms, the majority of limestones accumulated where the component grains were formed or have been subjected to only limited transport by wave and tidal currents. The skeletal grains of carbonate sediments vary greatly in size and shape. Interpretations of limestone deposition are thus to a large extent based on the types of grain present because these will often provide concise information on the depth, salinity, degree of agitation, etc. This is not to say that the grain size and degree of sorting and rounding are unimportant. Although the grain size will be a reflection largely of the size of the carbonate skeletons of the organisms living in the area and of the many biological factors involved in their breakdown, the physical factors of waves and currents will also contribute, and in cases dominate. A measure of the grain size, then, will often give useful additional information, reflecting the energy level of the environment, or energy gradient of the area. Where one is dealing with grainstones, the grain-size parameters discussed in Section 2.2.1 can be applied. It must be borne in mind, however, that carbonate particles are hydrodynamically different from quartz grains (see Kench & McClean, 1996). Apart from complications arising out of shape, carbonate grains commonly have a lower density because of pores and contained organic matter. The degree of sorting and rounding of skeletal grains (use Figs 2.4 and 2.6) can be useful in certain bioclastic

rocks, such as those of shelves and ramps where changes in these features could indicate proximity to a shoreline or zone of higher wave and tidal current activity. Some limestones of course, such as oolitic and peloidal grainstones, are very well sorted and rounded anyway. In interpreting the energy level of a depositional environment from a rock's grain-size parameters and texture, one is assuming that the sediment surface was in equilibrium with the hydrodynamic regime. With carbonate sediments this may not have been the case. In the modern shallow-marine environment, the sediment surface is commonly covered in a surficial microbial mat that stabilizes the sediment, enabling it to withstand current velocities up to five times those eroding nearby sediments lacking a microbial cover. During diagenesis, evidence of the mat could be destroyed. The probable wackestone would be interpreted as a low-energy deposit, whereas in fact periodically it was subjected to high current velocities.

Measurements of carbonate grain size in ancient limestones are made by point-counting slides or acetate peels under the microscope. For loose carbonate sands, the use of a settling chamber/sedimentation balance is recommended because this gives a better indication of the hydraulic behaviour than the grain-size distribution obtained by sieving.

In a general way the amount of micrite or lime mud in a limestone reflects the degree of agitation; lime muds tend to be deposited in quiet lagoons or on outer ramps, as well as on tidal flats and in the deep sea, in basins and periplatform areas. Increasing agitation leads to a decrease in the micrite content and increase in grain-support fabric and sparite content; sorting and rounding of grains then improves in the grainstone/biosparite. Interpretations must be made with care though because lime muds can accumulate in higher-energy environments, trapped and stabilized by sea-grass or a surficial microbial mat, which leaves no record in the sediment, and micrite can be precipitated as a cement during early diagenesis (Section 4.7.1).

4.6   Sedimentary structures of limestones

Limestones contain many of the sedimentary structures occurring in sandstones described in Section 2.3, but some structures are found only in carbonate sediments. Demicco & Hardie (1995) presented a description of sedimentary structures in limestones.

4.6.1 Bedding planes, hardgrounds, tepees and palaeokarstic surfaces

As in siliciclastic sediments, *bedding planes* generally represent a change in the conditions of sedimentation. The changes may have been subtle or short-lived. The bedding planes are mostly the result of changes in sediment grain size or composition. Thin clay seams also commonly define the bedding in limestone successions. However, with limestones it is not uncommon to find that bedding planes have been affected by dissolution as a result of overburden pressure (Section 4.7.5). Through this, originally gradational bed boundaries, such as a limestone passing up into a mudrock, or a grainstone into a lime mudstone, may become sharp. In many platform limestones, the 'bedding' planes are not primary depositional surfaces, but they have been produced by pressure dissolution during burial. This is most obvious where, for example, a 'bedding' surface occurs *within* a graded bed, or where it cross-cuts, at a low angle, a clear primary bedding surface. These *pseudobedding planes* actually account for much of the stratification in shallow and deep-water limestones (Simpson, 1985).

One particular type of bedding plane is a *hardground surface*. Hardgrounds are horizons of synsedimentary cementation, taking place at or just below the sediment surface. Where a hardground surface formed the sea floor, it was commonly encrusted by sessile benthic organisms, such as corals, serpulids, oysters, foraminifers and crinoids, and bored by polychaete annelids, certain bivalves and sponges. Hardground surfaces may cut across fossils and sedimentary structures. Two types of hardground surface can be recognized: a smooth, planar surface, formed by abrasion (Fig. 4.21), and an irregular, angular surface formed by dissolution (a corrosional hardground surface). The first type is more common in shallow subtidal sediments where waves and currents are able to move oolitic and skeletal sands across lithified sediment to produce a planar erosional surface. Corrosional hardground surfaces are more common in pelagic limestones where periods of non-sedimentation allow sea-floor cementation and dissolution. The identification of a hardground is important because it demonstrates synsedimentary submarine cementation. Hardground surfaces may become mineralized and impregnated with iron hydroxides, Fe–Mn oxides, phosphate and glauconite. Hard-

Fig. 4.21 Planar hardground surface encrusted with *Ostrea* (oysters) and penetrated by annelid and bivalve borings (seen on cut face). These features demonstrate synsedimentary sea-floor cementation of the sediment (see also Plate 11d). Jurassic. Gloucestershire, England.

grounds develop from loose sediments through firmgrounds to lithified layers, and associated with this there may be a change in the fauna, particularly of the burrowing organisms, as the sedimentation rate slowed down. Hardgrounds occur throughout the Phanerozoic and modern ones are forming at the present time off Qatar in the Arabian Gulf and on Eleuthera Bank, Bahamas (also Section 4.7.1).

One distinctive feature of peritidal limestones is the *tepee structure* (Fig. 4.22). Tepees are disruptions of the bedding into 'pseudoanticlines' and in plan view the tepee crests form a polygonal pattern. Tepees occur on the scale of tens of centimetres to several metres across. They mostly form on intertidal–supratidal flats as a result of the cementation and expansion of the surface-sediment layer. Upward movement (resurgence) of ground water, marine or meteoric, is a contributory factor in some cases. Elongate cavities (sheet cracks) commonly form beneath the uplifted slabs and in these, pisoids may form, as well as vadose cements such as dripstone and flowstone. These tepees usually are associated with planar stromatolites, desiccation cracks and intraclast conglomerates. Modern examples of these tepees, some with the spelean–pisoid association, occur on supratidal flats and around saline lakes in South and Western Australia. Tepees also form in the submarine environment, for example, where hardground surfaces have expanded through the cement precipitation. These are well developed off the

Fig. 4.22 Tepee structure in Triassic lacustrine littoral dolomite. Glamorgan, Wales.

0·1 m

Fig. 4.23 Palaeokarst. Cretaceous limestone (white) is penetrated by Eocene sands (grey), which also contain blocks of the host rock. Sub-Alpine Chains, France.

Qatar Peninsula. Tepees are prominent in many ancient peritidal sequences; classic examples occur in the back-reef facies of the Permian Capitan reef complex of Texas–New Mexico and in the Triassic of the Dolomites of northern Italy. See the review of Kendall & Warren (1987) for more information and M. Mutti (1994).

Another particular type of bedding discontinuity peculiar to limestones is a *palaeokarstic surface* (Fig. 4.23). When carbonate sediments become emergent, then dissolution through contact with meteoric water

produces an irregular, pot-holed surface. This dissolution commonly takes place beneath a thin soil cover, and the soil itself may be preserved as a discontinuous clay seam or bed immediately above the dissolution surface. The term *karst* is applied to these dissolution features, which are typical of more humid climatic areas. In a well-developed karst system, pot holes and caverns may form many tens or even hundreds of metres below the surface. Breccias form by cave collapse and deposition from subterranean streams; speleothems and flowstones are precipitated too. Important karst developed in the Lower Ordovician of Texas and Oklahoma, and in central China (Ordos Basin) and are major hydrocarbon reservoirs. Examples are given in Esteban & Klappa (1983) and James & Choquette (1988); also see Vanstone (1998), Molina *et al.* (1999) and Purdy & Waltham (1999).

Laminated crusts form upon and within uplifted carbonate sediments, as a type of calcrete or caliche (Section 4.10.1). In most cases, they are calcified root mats (Wright *et al.*, 1988), and usually they are closely associated with vadose pisoids and black pebbles. In ancient limestones, laminated crusts can be mistaken for stromatolites, but their association with palaeokarstic surfaces and palaeosoils indicates a subaerial, pedogenic origin.

Bedding surfaces and their significance in carbonates are discussed in Hillgärtner (1998).

4.6.2 Current and wave structures

All the current structures of siliciclastic rocks occur in limestones: wave and current ripples, cross-lamination, cross-bedding on all scales, planar or flat bedding, small scours to large channels, HCS in storm beds, bundled cross-beds and reactivation surfaces in tidal sands and sole structures on the bases of storm beds and turbidites. Post-depositional structures resulting from dewatering and loading are also common. For details of these sedimentary structures see Section 2.3. The lack of clay in many limestones, together with the effects of surface weathering, may make internal structures difficult to discern in the field. Careful observations, perhaps aided by polishing, etching or staining cut blocks, will often bring them to light.

The same importance is attached to current structures in limestones as in sandstones. They are essential to environmental interpretation and facies analysis, giving valuable information on depositional process, palaeocurrents, depth and water turbulence. Intraclasts of lime mud (flakes) and large fragments of skeletal debris are generally more common in limestones. They may be concentrated through current winnowing to form lag deposits, or transported by storms to give rudstones and floatstones (flakestones, Fig. 4.5). These beds may show imbrication of clasts, reverse or normal grading. Elongate fossils are commonly aligned parallel to, or normal to, the current direction, so that they too can give a palaeocurrent indication.

4.6.3 Cavity structures

Depositional and early diagenetic cavity structures are common in limestones, and there are many types. Some are partly filled with sediment that has been washed in to occupy the lower part of the cavity, with the space above occupied by sparite cement (see Plates 9a,b & 11a,c,d). Such cavity fills are known as *geopetal structures*, and they are a most useful way-up indicator (sparite at the top of course). Geopetal structures also record the horizontal at the time of sedimentation (acting as a spirit level) and in some cases show that there was an original depositional dip (as in fore-reef limestones, for example). *Umbrella structures* are simple cavities beneath convex-up bivalve and brachiopod shells and other skeletal fragments. *Intraskeletal cavities* occur in enclosed or chambered fossils, such as gastropods, foraminifers and ammonoids. *Growth cavities* are formed beneath the skeletons of frame-building organisms, corals and stromatoporoids, for example, where they build out above the sediment or enclose space within the skeletons.

Fenestral cavities or '*birdseyes*' are small cavities that occur particularly in peloidal mudstones of intertidal–supratidal environments. The majority are sparfilled only, but some may be sediment-filled. Three main types can be distinguished:

1 irregular fenestrae, the typical 'birdseyes' (Fig. 4.24), several millimetres across, equidimensional to irregular in shape;
2 laminoid fenestrae, several millimetres high and several centimetres long, parallel to bedding (Figs 4.12 & 4.25);
3 tubular fenestrae, cylindrical, vertical to subvertical in arrangement, several millimetres in diameter.

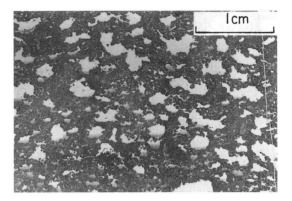

Fig. 4.24 Birdseyes (fenestrae) in lime mudstone. Fenestrae are filled with sparry calcite although some contain internal sediment. Carboniferous. Clwyd, Wales.

Fig. 4.25 Laminoid fenestrae and birdseyes filled with calcite in microbial and peloidal dolomite. The dolomite is probably penecontemporaneous in origin. Triassic. Austrian Alps.

Irregular fenestrae in abundance form the so-called birdseye limestones (my favourite). They are ascribed to gas entrapment in the sediment and desiccation and so are a characteristic intertidal-facies indicator. Similar structures also occur in subtidal grainstones associated with early cementation and hardgrounds. Beach sands also contain irregular fenestrae, called *keystone vugs*, from the movement of water and air through the sediment.

Laminoid fenestrae develop in laminated sediments, particularly planar stromatolites, from the decay of organic matter, and desiccation and parting of laminae. *Tubular fenestrae* are formed mainly by burrowing organisms, but plant rootlets produce similar tubes. Although common in tidal-flat sediments, tubular fenestrae will also occur in shallow, subtidal sediments. Fenestral limestones occur throughout the geological record; one notable example is the *loferite* of the Alpine Triassic (Fig. 4.25).

Small vugs to extensive cavern systems on the scale of tens of metres form as a result of surface and near-surface karstic dissolution of limestones (see Section 4.6.1). Dissolution also may take place during deep burial in the formation of hydrothermal karst.

A further particular type of cavity is *stromatactis* (Fig. 4.26). This is common in carbonate mud-mounds of the Palaeozoic such as the Waulsortian 'reefs' and other mounds of the European Carboniferous and Devonian, the Devonian Tully Limestone of New York, Silurian sponge 'reefs' of Quebec and Ordovician mud-mounds of Nevada. Stromatactis cavities have an irregular, unsupported roof and a flat floor, formed by internal sediment. The cement is invariably a first generation of fibrous calcite, followed by drusy sparite. Most stromatactis cavities are a few centimetres long, but they may reach tens of centimetres. The origin of the cavities has led to much discussion and speculation. Stromatactis in Upper Devonian mud-mounds has been interpreted as the product of recrystallization of algal, cyanobacterial and bacterial colonies. Fibrous calcite, which is a feature of most stromatactis cavities and normally is thought of as a

marine cement (see Section 4.7.1), also has been interpreted as microbial in origin. Various inorganic origins are available: sediment collapse and dewatering; winnowing of uncemented sediment beneath lithified crusts or beneath gelatinous microbial mats; and dissolution during deep burial. A popular mechanism now is that stromatactis formed within a sponge framework, from the decay of uncemented sponge tissue (Bourque & Boulvain, 1993; and papers in Monty *et al.*, 1995). It is likely that stromatactis structures can form through several of these processes.

Two types of cavity that form in partly lithified or cemented limestone are sheet cracks and neptunian dykes. *Sheet cracks* are cavities generally running parallel to the bedding, which have planar walls, although some may have irregular roofs. *Neptunian dykes* cut across the bedding and may penetrate down many metres from a particular bedding plane. Both sheet cracks and neptunian dykes are filled with internal sediment, in some cases with fossils a little younger than the adjacent limestone, if the cavities opened on to the sea floor. Spectacular examples of both types, in pelagic limestones and penetrating down into underlying platform carbonates, occur in the Triassic and Jurassic of the Alps and Spain (e.g. Molina *et al.*, 1995). Sheet cracks and neptunian dykes form through small tectonic movements during sedimentation and/or some slight downslope movement of sediment, causing fracturing of the lithified or partly lithified limestone mass.

Fractures and veins filled with calcite, of various types, are common in limestone. Although they can be early structures, like neptunian dykes, the majority form through later fracturing, and this is usually in response to tectonic stresses rather than purely diagenetic processes.

4.7 Carbonate diagenesis

The diagenesis of carbonates involves many different processes and takes place in near-surface marine and meteoric environments, down into the deep-burial environment. It is most important in occluding and generating porosity in the sediment. Six major processes can be distinguished: cementation, microbial micritization, neomorphism, dissolution, compaction and dolomitization. Carbonate diagenesis mostly involves the carbonate minerals, aragonite, calcite and dolomite, but other minerals such as quartz, feldspar, clays,

Fig. 4.26 Stromatactis cavities filled with fibrous calcite and drusy calcite. Mud-mound, Upper Devonian. Ardennes, Belgium.

phosphates, iron oxides and sulphides, and evaporites also may be involved. The diagenesis of carbonate sediments begins on the sea floor; in fact depositional and diagenetic processes may be going on at the same time. As a reef is growing or carbonate sand is being moved by the waves, cements may be precipitated within intraskeletal cavities and grains altered by *micritization*. The latter process has been described in Section 4.3.3, and results in the formation of micrite envelopes around bioclasts and completely micritized grains. Micrite envelopes play an important role during diagenesis by maintaining the shape of an aragonite bioclastic grain after its dissolution.

Cementation is the major diagenetic process producing a solid limestone from a loose sediment and taking place principally where there is a significant throughput of pore-fluid saturated with respect to the cement phase. The mineralogy of the cements depends on water chemistry, particularly P_{CO_2} and the Mg/Ca ratio, and carbonate supply rate (see Section 4.7.1). *Neomorphism* is used to describe replacement and recrystallization processes where there may have been a change of mineralogy. Examples include the coarsening of crystal sizes in a lime mud/micrite (aggrading neomorphism) and the replacement of aragonite shells and cements by calcite (*calcitization*). Many limestones have suffered *dissolution* as a result of the passage of pore-fluids undersaturated with respect to the carbonate phase present. This is a major process in near-surface, meteoric diagenetic environments, and may lead to the formation of karst (see Section 4.6.1), but it can also take place on the sea-floor and during deep burial. The secondary porosity created by carbonate dissolution is important in some hydrocarbon reservoirs. *Compaction* takes place during burial, resulting in a closer packing of grains, their fracture and eventual dissolution where in contact. Chemical compaction leads to stylolites and dissolution seams, when burial exceeds many hundreds of metres of overburden. *Dolomitization* is a major alteration process for many limestones and the dolomite, $CaMg(CO_3)_2$, may be precipitated in near-surface and burial environments. There are a number of models for dolomitization, but the matter is still one of great debate.

Three major diagenetic environments are distinguished: marine, near-surface meteoric and burial (see Fig. 4.27). In the marine environment, diagenesis takes place on and just below the sea-floor in both shallow and deep water, and in the intertidal–supratidal

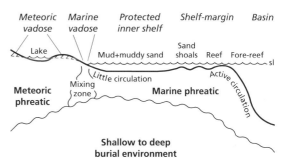

Fig. 4.27 Carbonate diagenetic environments, schematically drawn for a rimmed shelf with unconfined aquifers. Where there are confined aquifers as a result of impermeable strata, then it is possible, for example, for meteoric waters to penetrate deep beneath the marine shelf and even emerge on the sea floor as submarine springs.

zone. Meteoric diagenesis can affect a sediment soon after it is deposited if there is shoreline progradation or a slight sea-level fall, or it may operate much later when a limestone is uplifted after burial. The burial environment, the least well known, begins at a depth below the sediment surface of tens to hundreds of metres, that is, below the zone affected by surface processes, down to several kilometres where metamorphic dehydration reactions and wholesale recrystallization take over.

4.7.1 Marine diagenesis

Marine diagenesis in Recent carbonate sediments

Intertidal–supratidal diagenesis. Cementation in the intertidal zone produces cemented beach sands known as beachrock. Beachrocks are most common in the tropics and subtropics but they do occur along temperate shorelines. They are composed of the same sediment that forms the surrounding loose beach sand; this usually is calcareous but it may have a substantial or even dominant siliciclastic component. Beachrocks can form quickly, as is evidenced by the inclusion of anthropogenic objects such as beer cans. Beachrock formation probably takes place a few tens of centimetres below the surface of the beach, but in many places it is exposed through storm action, and then it can be eroded (to produce intraclasts), encrusted, and bored and grazed by intertidal organisms.

The cements in modern beachrocks are aragonite and/or high-Mg calcite. Aragonite typically occurs as fringes from 10 to 200 μm thick of acicular crystals, orientated normal to the grain surfaces (Figs 4.28 & 4.29). In many cases the cement fringes are isopachous, i.e. of equal thickness (Fig. 4.30), indicating marine phreatic (below the water table) precipitation where pores were constantly water-filled. Asymmetric cement fringes, thicker on the underside of grains, and meniscus cements, concentrated at grain contacts (Fig. 4.30), are recorded from some beachrocks and indicate precipitation in the marine vadose zone. High-Mg calcite is usually a dark micritic cement coating grains or filling pores. Micritization of grains is common in beachrocks and calcified microbial filaments are also present. If the meteoric ground-water table is high in the backshore area, then low-Mg calcite cements may be precipitated in beachrocks there in the upper intertidal zone.

Two processes are important in beachrock formation:

1 purely physico-chemical precipitation through evaporation of seawater when the tide is out and CO_2-degassing of seawater as it is pumped through the sand by waves and the rising and falling tide;

2 biochemical–microbial precipitation, involving microbial photosynthesis, bacterial calcification and the decomposition of organic matter.

Cemented surface crusts occur in the high intertidal to low supratidal zone of tidal flats in carbonate areas. Along the Trucial Coast of the Arabian Gulf, aragonite-cemented crusts are commonly brecciated or polygonally cracked, and expansion of the crusts as a result of the cementation leads to pseudoanticlines or *tepee structures* (Section 4.6.1; Kendall & Warren, 1987). The resurgence of continental ground water onto supratidal flats may contribute to crust and tepee formation. Beneath the surface crust, dripstone cements, vadose pisoids and aragonite botryoids

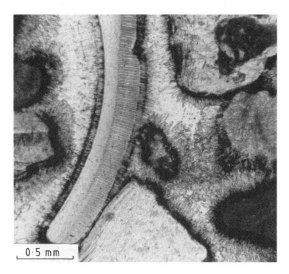

Fig. 4.28 Beachrock with cements of acicular aragonite and micritic high-Mg calcite: the latter is the dark coating around grains. The aragonite cement has grown syntaxially on the bivalve grain, which shows a two-layer shell structure. Plane polarized light. Great Barrier Reef, Australia.

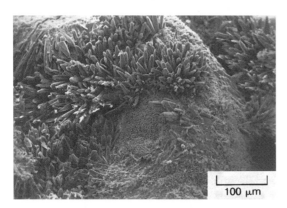

Fig. 4.29 Scanning electron micrograph of beachrock cement shown in Fig. 4.28. Where the coating of acicular aragonite crystals has come off the grain, the high-Mg calcite micrite cement is visible.

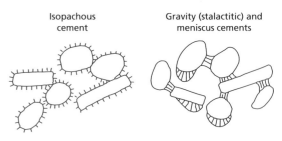

Isopachous cement

Gravity (stalactitic) and meniscus cements

Fig. 4.30 The geometry of first-generation cements: isopachous cement, indicative of precipitation in phreatic zones where all pores are filled with water (typical feature of low-intertidal and subtidal cements); and gravity (stalactitic) and meniscus cements, indicative of vadose-zone precipitation, as occurs in high-intertidal, supratidal and shallow-subsurface meteoric situations.

may develop. Supratidal crusts in the Bahamas are cemented by dolomite (Section 4.8.1, see Plate 10d).

Shallow-subtidal cementation. Sea-floor diagenesis in low-latitude, shallow-marine areas mostly involves cementation and microbial micritization. The former is more common in high-energy areas, where seawater is pumped through the sediment, whereas the latter is more common in quieter-water areas, such as back-reef lagoons. Thus an active marine phreatic diagenetic environment can be distinguished from a stagnant one. In higher-latitude, shallow-water areas, seawater is undersaturated with respect to $CaCO_3$ and so inorganic precipitation of cement does not take place. Dissolution of grains may occur, however.

In the stagnant, marine phreatic diagenetic environment, cementation is restricted to intraskeletal cavities, such as occur within gastropod and foraminiferal bioclasts. The formation of grapestones and aggregates is common in this environment, with microbial binding and filament calcification usually involved. In these areas too, micritization of grains by recrystallization is taking place on the sea floor (Reid & Macintyre, 1998).

Shallow-subtidal cementation of loose carbonate sand to produce surface crusts and lithified layers is rare but it is taking place in a few metres of water off the Qatar Peninsula, Arabian Gulf and on Eleuthera Bank, Bahamas. Off Qatar, these modern hardgrounds are being bored and encrusted by organisms, and polygonal cracks and tepee structures have formed through expansion of the cemented layer. The cements are mainly acicular aragonite, with some micritic high-Mg calcite. The acicular cements form isopachous fringes (see Plate 11b), but if well developed and completely filling the pore space, then a polygonal pattern is observed in thin-section, from the meeting of the cement fringes on adjacent grains. The precipitation of cements is probably the result of turbulent bottom conditions and the pumping of $CaCO_3$-supersaturated seawater through the sediments, particularly in areas of slow sedimentation. Microbial filaments, commonly calcified, are usually present, and they may have provided some initial stabilization of the sediment, permitting the cementation to begin.

Marine cements can be developed extensively in reefs (see papers in Schroeder & Purser, 1986). There is a wide variety of cement morphologies but the mineralogy is either aragonite or high-Mg calcite. Aragonite occurs mostly as acicular fringes (like those in Fig. 4.28), and needle meshworks, but one prominent type is the botryoid: isolated or coalesced mamelons, which reach 100 mm in diameter (see Plate 11a). They consist of fanning fibrous crystals, commonly twinned to give a pseudohexagonal cross-section. High-Mg calcite occurs as bladed cements, 20–100 μm long and less than 10 μm wide, forming isopachous fringes (see Plate 11b). They are more common than aragonite in some reefs, but rare in others. High-Mg calcite also occurs as a micritic cement. This coats grains and lines cavities, but more commonly forms peloidal structures, abundant in skeletal cavities and forming surficial crusts on corals in some instances. The peloids are 20–60 μm in diameter and are arranged in grainstone to packstone textures. The origin of the peloids has given rise to much discussion, principally over whether they are inorganic precipitates, microbial precipitates, faecal pellets, or detrital sediment. The consensus now is for a microbial origin with precipitation taking place within and around clumps of bacteria (see Chafetz, 1986).

The precipitation of cements in reefs contributes towards the generation of a solid framework, but reefs are also subject to extensive *bioerosion*. Microbial organisms, clionid sponges and lithophagid bivalves all bore into carbonate skeletons in reefs; sponges also generate much fine debris. Fish and other organisms graze on corals and generate sediment too. Internal sedimentation of this material into primary and secondary cavities is widespread, and these sediments may then become cemented.

The distribution of cements in many reefs appears to relate to the circulation of water; cementation is generally more intense along the windward margins, where seawater is constantly pumped through the reef. However, on a smaller scale, the occurrence of cements may be very patchy and varied, both in terms of extent, mineralogy and morphology. One cavity may contain acicular aragonite, an adjacent cavity high-Mg calcite peloids, whereas another close by may be empty (as in Plate 11a). One important control on the amount of cementation is local permeability, which determines the fluid-flow rates through the reef-rock (Goldsmith & King, 1987). This also may control the mineralogy itself.

Aragonite cements of intertidal to shallow-subtidal sediments usually have a high strontium content, up to

10 000 p.p.m., and Mg content of about 1000 p.p.m. or less. The high-Mg calcite cements are typically between 14 and 19 mol.% $MgCO_3$, but Sr is low at around 1000 p.p.m.

Deeper-water cementation. Cemented carbonate sediments have been recovered from the ocean floors at depths down to 3500 m, mostly from areas of negligible sedimentation, such as seamounts, banks and plateaux. The limestones consist chiefly of planktonic foraminifera, molluscs and coccoliths cemented by a micritic calcite. The limestones are commonly bored and may be impregnated with phosphate and ferromanganese oxides. On the deeper-water slopes around the Bahama Platform (700–2000 m deep), periplatform ooze is being cemented by micritic low-Mg calcite. The sediment itself consists of planktonic foraminifera, coccoliths and shallow-water material derived from the platform. Cemented crusts and hardgrounds pass downslope into patchily cemented sediment and nodules, and then there is little cementation on the deeper-water slopes. Originally aragonitic bioclasts have been leached and high-Mg calcite grains have lost their Mg (Dix & Mullins, 1988). In the Mediterranean and Red Sea, where there are and/or have been warmer bottom waters, thin surface crusts and nodules of lithified pelagic sediment are cemented by micritic high-Mg calcite. Acicular and micritic aragonite cements pteropod layers on the floor of the Red Sea.

Carbonate cement precipitation in these deeper-water environments is mainly a reflection of the very slow sedimentation rates, which allow interaction between sediment and seawater. The $CaCO_3$ is derived from seawater and the dissolution of less stable grains in the sediment; the type of cement precipitated and its Mg content are determined by the precipitation rate and the water temperature; as the latter goes down, so does the mol.% $MgCO_3$.

Marine dissolution. Dissolution of carbonate grains does take place on the shallow sea floor in higher latitudes where seawater is undersaturated with respect to the carbonate phase. The saturation state of seawater with regard to $CaCO_3$ decreases with lower temperature (and so broadly with increasing depth), and seawater becomes undersaturated in respect of aragonite and high-Mg calcite before low-Mg calcite. Dissolution of bioclasts is taking place in slope sedi-

ments off the Bahama Platform at depths of a few hundred metres, and in the geological record relatively shallow sea-floor dissolution of aragonite is recorded from hardgrounds in the Jurassic and Ordovician (Palmer *et al.,* 1988). These were times when shallow, low-latitude seawater was supersaturated with regard to calcite, but not aragonite, too, as is the case today. Sea-floor dissolution of aragonite is common in Mesozoic deeper-water pelagic limestones; this is clearly shown where, through sea-floor loss of their aragonitic shell, ammonites are preserved as casts encrusted by sessile organisms and Fe–Mn oxides. Carbonate dissolution increases with increasing depth until the CCD (carbonate compensation depth), below which limestones are not deposited (see Section 4.10.7).

Marine diagenesis in ancient limestones

In many limestones, there is abundant evidence for diagenesis taking place on the sea-floor or just below. Marine cements are common in ancient reef-rocks, and hardgrounds are well known too in the geological record (see Section 4.6.1). However, the cements themselves are variable; some were similar to those of the Recent and were composed of aragonite and high-Mg calcite, although they are now calcite (low Mg). On the other hand, other marine cements in ancient limestones do not appear to have exact modern equivalents and were precipitated as calcite with fabrics somewhat different from modern cements.

General features for the recognition of marine cements in limestones are:
1 they are the first-generation cement;
2 they usually form isopachous fringes around grains or cavity walls;
3 they usually are/were fibrous in nature;
4 they may be cut by borings or include skeletal debris;
5 they may be associated closely with internal sediments;
6 the crystals are usually non-ferroan and non-luminescent;
7 they are succeeded by clear calcite spar.

Ancient marine cements are shown in Plates 11c,d & 12a,b.

Ancient marine aragonite cements. Aragonite cements are very rarely preserved as aragonite in limestones because the mineral is metastable. They are altered to calcite in a similar way to aragonite bioclasts, either through wholesale dissolution and then later filling of the void by calcite spar, or through calcitization, whereby calcite replaces the aragonite across a thin-fluid film, with dissolution of aragonite on one side and precipitation of calcite on the other. In this process, some traces of the original cement texture may be preserved within the replacement calcite crystals by the pattern of minute relics of the aragonite or inclusions of organic matter (Fig. 4.31). The replacement calcite crystals, a type of neomorphic spar (see Section 4.7.4), are usually irregular to equant in shape, cross-cutting the original acicular pattern of the aragonite cement. This calcite may have a relatively high (several thousand p.p.m.) strontium content inherited from the aragonite. Some ancient calcitized aragonite cements have distinctive 'square-ended terminations' along the cement fringes. The gross morphology of ancient aragonite cements is similar to that of Recent cases, isopachous fringes and botryoids. Although some marine isopachous cement fringes in limestones

were calcite originally, botryoids appear to have been aragonitic only; somewhat similar structures, in calcite, do form in meteoric, spelean–vadose diagenetic environments. Descriptions of ancient aragonite are included in Tucker & Hollingworth (1986), from the Permian of northeast England, and Roylance (1990), from the Carboniferous of the Paradox Basin, USA.

Ancient marine calcite cements. The most common marine calcite cement in ancient limestones is *fibrous calcite*: elongate crystals normal to substrate, and usually cloudy or dusty in appearance relative to later calcite spar. A columnar growth form, with length-to-width ratio of more than 6 : 1 and width greater than 10 μm, can be distinguished from a more acicular variety, with a much higher length/width ratio and width of around 10 μm. Columnar fibrous calcite is abundant in reef cavities and in stromatactis structures of Palaeozoic mud-mounds (see Fig. 4.26). The more acicular type is common in grainstones but occurs in reefs too (see Plate 11c,d).

Fibrous calcite has a range of fabrics from unit-extinguishing crystals to undulose-extinguishing radiaxial fibrous calcite (RFC), with divergent optic axes, and fascicular-optic fibrous calcite (FOFC), with convergent optic axes (Fig. 4.32 and Plate 12a,b). In RFC (the most common type) the extinction swings across the crystal in the same direction as rotation of the

Fig. 4.31 Calcitized aragonite botryoid. (a) Photomicrograph of irregularly shaped neomorphic calcite cross-cutting the pattern of the original aragonite crystals. Crossed polars. (b) Scanning electron micrograph of minute aragonite crystal relics in calcite replacing an aragonite botryoid. Permian reef (Zechstein). Durham, England.

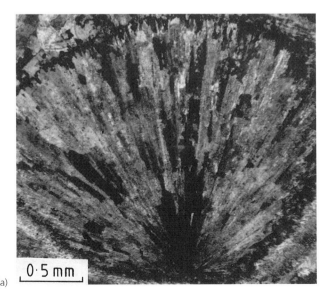

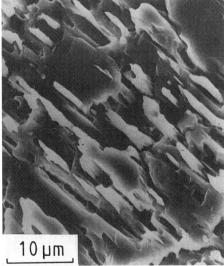

(a)   0·5 mm

(b)   10 μm

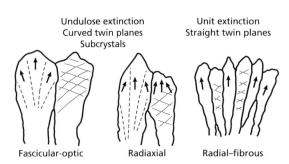

Fig. 4.32 Common types of fibrous calcite.

microscope stage. Twin planes are common in fibrous calcite and in RFC they are concave away from the substrate. There has been much discussion over the origin of fibrous calcite, with a replacement origin long popular, but it is now regarded generally as a primary precipitate, with the peculiar fabrics attributed to split-crystal growth (Kendall, 1985).

One particular point of interest with fibrous calcite is whether it was originally low- or high-Mg calcite. It is not easy to decide, as high-Mg calcite loses its Mg during diagenesis, but several lines of evidence can be used. Fibrous calcite usually is cloudy with inclusions, and some of these may be minute crystals of dolomite. Others will be fluid-filled inclusions or empty. *Microdolomites* are taken to indicate an original high-Mg calcite mineralogy; this is supported by their presence in formerly high-Mg calcite bioclasts such as echinoderms. In addition, fibrous calcite after high-Mg calcite may still contain several mol.% $MgCO_3$ after the diagenetic loss of most of the Mg. This *magnesium memory* can be detected only by geochemical means, such as with use of the electron probe. Fibrous calcites originally of high-Mg calcite may pick up some Fe^{2+}, if this is in the pore waters when the high- to low-Mg calcite transformation is taking place. Although the fabrics of fibrous calcite are mostly unchanged during diagenesis, some neomorphism may occur, especially if the crystals were originally high-Mg calcite. Dissolution may occur along twin boundaries, for example.

In addition to fibrous calcite, there is evidence in some limestones for equant sparry calcite being a marine precipitate. Although mostly of meteoric or burial origin, it does occur as the first-generation cement in some Jurassic and Ordovician hardgrounds. Syntaxial overgrowths on echinoderm debris, also generally regarded as near-surface meteoric or burial precipitates, formed early in some hardgrounds too. Early, inclusion-rich zones of crinoidal overgrowths are probably marine (see Plate 12c).

Similar textures to the micritic and peloidal high-Mg calcite precipitates in modern reefs do occur in ancient reefs, and there are the same arguments over their origin. The peloidal structures are particularly common in Triassic reefs and some Jurassic ones too (e.g. Sun & Wright, 1989).

Discussion of marine cements

The mineralogy of modern tropical shallow-marine cements is mostly either aragonite or high-Mg calcite, and in the geological record, marine cements originally of aragonite, high-Mg calcite and low-Mg calcite can be identified. In a similar manner to ooids (see Section 4.3.1), there appears to be a secular variation in marine cement mineralogy through the Phanerozoic (see Fig. 4.4). Aragonite botryoids, for example, one of the distinctive marine cement morphologies, are common in the Cenozoic and Permo-Triassic, but apparently absent from the mid-Palaeozoic and Jurassic–Cretaceous, where fibrous calcite is the dominant cement type. As with ooids, the controls on cement mineralogy are likely to be the seawater Mg/Ca ratio, Pco_2 and carbonate-supply rate (Fig. 4.33). Aragonite has a similar stability to magnesium calcite, with around 12 mol.% $MgCO_3$. It is unclear what controls whether aragonite or high-Mg calcite is precipitated. The presence of Mg^{2+} and SO_4^{2-} in seawater does result in a kinetic hindrance to calcite precipitation, so favouring aragonite. High carbonate supply appears to favour aragonite precipitation too, so that where fluid-flow rates are higher, as in very permeable reef-rocks and lime sands, aragonite will be precipitated in preference to high-Mg calcite, which will tend to occur in less permeable sediments. In some situations, substrate control is another factor, with cement crystals being the same mineralogy as the substrate, and in optical continuity (syntaxial) too. See the review of Given & Wilkinson (1985) for more information.

4.7.2 Meteoric diagenesis

Near-surface meteoric diagenesis mostly involves fresh water and the major processes are carbonate dissolution, cementation and the formation of soils. The

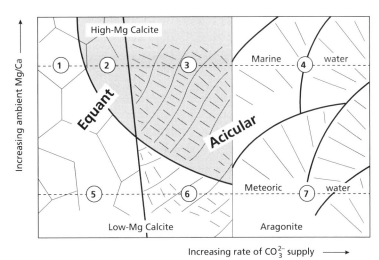

Fig. 4.33 Schematic illustration of relationship between fluid Mg/Ca ratio, rate of carbonate-ion supply, crystal morphology (equant or acicular) and mineralogy of inorganic precipitates (low-Mg calcite, high-Mg calcite or aragonite). The Mg/Ca ratio of the meteoric line is 0.3 and of the marine line 5.2. Low-Mg calcite is up to 9 mol.% $MgCO_3$ and high-Mg calcite is above. The position of modern, naturally occurring precipitates.

1　Equant calcite cement in cold, deeper-water, low-latitude sediments and shallow-water temperate sediments

2　Equant high-Mg calcite cements in reefs (rare)

3　Acicular high-Mg calcite cements in reefs and lime sands

4　Acicular aragonite cements in reefs and lime sands

5　Equant low-Mg calcite spar cements in meteoric environments

6　Acicular low-Mg calcite in speleothems and travertines

7　Acicular aragonite in speleothems (rare)

position of the ground-water table is important and the vadose zone above is distinguished from the phreatic zone below. In the vadose zone, pores periodically contain water, air or both, and an upper zone of infiltration is distinguished from a lower zone of percolation. Rainwater is undersaturated with respect to $CaCO_3$ so that dissolution is one of the main processes operating in the infiltration zone. As the water moves down through the vadose zone, it may become supersaturated with respect to $CaCO_3$ so that precipitation occurs. This is normally of calcite (low Mg), as meteoric water has a very low Mg/Ca ratio. In the phreatic zone, pores are fluid-filled all the time, and the normally fresh water gives way downwards to more saline water at depths of several hundred metres or more. In coastal regions, the phreatic meteoric ground-water realm passes into a mixing zone with seawater. It has been suggested frequently that the mixing zone is an important location for dolomitization (see Section 4.8.2), but there is now much doubt about this. Climate is a major influence on meteoric diagenesis because it controls the availability of meteoric water and

also affects the temperature and degree and nature of plant cover and soil development. Climate also determines the type and extent of karstic dissolution. Palaeokarstic surfaces (see Section 4.6.1) are now well known in the geological record and can be recognized from their morphology (an irregular pot-holed surface) and association with soil crusts and other features of subaerial exposure. Subsurface karst is the system of caves and fissures that may extend great distances down from the surface.

There have been several studies of modern to Pleistocene limestones to document the progressive alteration of marine sediments with increasing exposure to meteoric water. In the early stages, low-Mg calcite is precipitated on the surfaces of grains as an (a) isopachous (uniform thickness) fringe if precipitated in the phreatic zone, below the water table where all pores are completely filled with water, or (b) asymmetric fringe, thicker on the underside of grains (a dripstone effect) or located at grain contacts (a meniscus effect), if precipitated in the vadose zone (Fig. 4.30, and Plates 6a,b & 13a). Syntaxial overgrowths on echinoderm

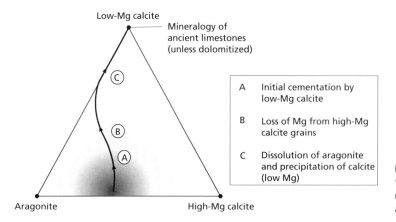

A Initial cementation by low-Mg calcite

B Loss of Mg from high-Mg calcite grains

C Dissolution of aragonite and precipitation of calcite (low Mg)

Fig. 4.34 Triangular diagram indicating the mineralogical range of modern marine carbonate sediments (stippled) and the path of meteoric diagenesis.

grains also begin to develop in this early stage. Next generally follows the loss of Mg from the high-Mg calcite, leaving a sediment of low-Mg calcite and aragonite (Fig. 4.34). The next stage is aragonite dissolution and reprecipitation of the $CaCO_3$ as drusy calcite spar. Aragonite skeletal grains are dissolved and the voids left are then filled with calcite, the shape of the grains being maintained by the earlier cement fringe or a micrite envelope (Section 4.3.3). The main feature of this dissolution–reprecipitation process is a loss of internal structure in the aragonitic skeletal grains. The reprecipitation of calcite may follow soon after aragonite dissolution, or, there may be a much longer time gap, allowing some compaction of the skeletal voids before calcite precipitation, producing broken and fractured cement fringes and micrite envelopes (seen in Plates 7c & 13d). Some aragonite grains are replaced *in situ*, i.e. calcitized (Section 4.7.4), so that the internal structure is preserved to some extent (see Plate 8a). The last stage of meteoric diagenesis would involve further precipitation of low-Mg calcite to fill all remaining voids. In most Pleistocene limestones, this final stage has rarely produced a fully cemented limestone; porosity may still reach 20%. A further phase of cementation is thus required to reduce the porosity to less than 5%, the typical value of ancient limestones.

Near-surface, meteoric vadose drusy calcite spar is generally non-ferroan, as pore waters are usually oxidizing. However, if there is decomposing organic matter, the water-flow rate is low and Fe^{2+} is available, then phreatic calcite spar may be ferroan.

Near-surface, meteoric calcite cementation takes place in many continental sediments, such as wadi gravels, scree sediments and aeolian sands, where pore waters enriched in $CaCO_3$ are evaporated. Also in the meteoric diagenetic environment, soils are formed, especially calcretes (caliches). Near-surface sediments are cemented and altered through pedogenesis, and some distinctive fabrics are produced, such as needle-fibre calcite, alveolar texture, vadose pisoids and laminated crusts from calcification of root mats (see Section 4.10.1).

Ancient meteoric calcite cements

In the geological record, the most obvious products of meteoric diagenesis are the calcareous soils (calcretes) and palaeokarsts, but the vadose cement morphologies are not that uncommon (see Plate 13a) and it is possible that some drusy calcite spar, the ubiquitous cement of ancient limestones, was precipitated in a near-surface, meteoric phreatic environment. The original mineralogy of a carbonate sediment is also a factor in the extent of meteoric diagenesis. Where there is a high percentage of metastable aragonite and high-Mg calcite, then the sediment has a much higher potential for dissolution and cementation. Sediments dominated by the more stable low-Mg calcite will have a much lower diagenetic potential. This aspect of meteoric diagenesis is explored further in Hird & Tucker (1988) and James & Bone (1989).

4.7.3 Calcite spar

The cement that occupies the majority of the original pore space in many limestones is a clear, equant calcite,

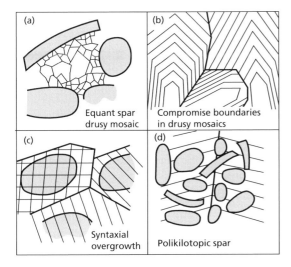

Fig. 4.35 Calcite spar. (a) Drusy calcite spar, the most common type, characterized by an increasing crystal size away from the substrate towards the cavity centre. (b) Growth zones in calcite spar showing how straight crystal boundaries (compromise boundaries) develop between adjacent crystals. (c) Syntaxial overgrowths where calcite spar cement is in optical continuity with a host grain, schematically shown here by twin planes going from cement into grain. (d) Poikilotopic calcite spar, large crystals enveloping several grains.

referred to as sparite or calcite spar (see Fig. 4.35). Sparite possesses a number of features, which, taken together, allow its cement interpretation. These are:

1 its location between grains and skeletons, and within original cavities;
2 its generally clear nature, with few inclusions;
3 the presence of planar intercrystalline boundaries;
4 a drusy fabric, i.e. an increase in crystal size away from the substrate or cavity wall;
5 crystals with a preferred orientation of optic axes normal to the substrate.

The fabric characteristics of sparite **4** and **5** are reflections of the preferred growth direction of calcite, parallel to the *c*-axis. Calcite spar is invariably precipitated after the fibrous calcite described earlier, which is mostly a marine cement. In some cases, there is a layer of internal sediment between the two cement generations. Calcite cement may also take the form of large poikilotopic crystals, several millimetres to centimetres across (as in Plate 6c). The large crystals are the result of a low nucleation rate and slow growth, perhaps because pore fluids were only just saturated with re-

spect to $CaCO_3$. Where echinoderm grains, and others composed of a single calcite crystal, are present in the limestone, then the sparite cement may precipitate syntaxially (in optical continuity) upon the grain to produce an overgrowth (see Plate 12c,d). Preferential cement growth upon such single crystal grains may envelop adjacent, small, polycrystalline grains.

Calcite-spar crystals are commonly delicately zoned as a result of subtle variations in Fe and Mn contents. The zonation can be revealed by staining with Alizarin Red S plus potassium ferricyanide, or by observing the luminescence (see Plate 13b,c and Section 4.1). Studying the zonation pattern in calcite spar in a limestone formation on a regional scale allows a cement stratigraphy to be erected from which the larger-scale hydrology of the basin can be reconstructed (e.g. papers in *Sedimentary Geology*, 65(3/4), 1989; Horbury & Robinson, 1993; Budd *et al.*, 1995).

Carbon and oxygen isotopes are being used increasingly in the study of carbonate cements. Fibrous calcite commonly has a marine signature in both $\delta^{13}C$ and $\delta^{18}O$, whereas calcite spar tends to show more negative $\delta^{18}O$, reflecting precipitation at a higher temperature during burial and/or precipitation from meteoric water (see papers in Schneidermann & Harris (1985) and Horbury & Robinson (1993); and Tucker & Wright, 1990). Fluid inclusions in calcite also provide information on the chemistry of pore fluids and the temperature of precipitation (see Emery & Robinson, 1993; Goldstein & Reynolds, 1994).

Origin of calcite spar. There has been much discussion of the environment of precipitation of sparite, the main cement in all medium- to coarse-grained limestones and the fill of most cavities in fine-grained limestones. Although drusy sparite is the typical cement of near-surface meteoric diagenesis, in many limestones there is evidence for sparite precipitation after compaction. The occurrence of broken and fractured fibrous cement fringes and micrite envelopes around formerly aragonite skeletal grains indicates mechanical compaction resulting from overburden pressure after aragonite dissolution and before sparite precipitation (see Plate 7c). In other limestones, sutured contacts between grains show that chemical compaction occurred before sparite precipitation (see Plate 13a). The pore waters for deep-burial cementation will be either connate (modified seawater buried with the sediments) or meteoric (or a mixture of the two). In fact, in many sedimentary basins it does appear that pore waters are largely of meteoric origin, although salinity is much higher than that of near-surface fresh water. There are three possible sources for the $CaCO_3$: (a) the pore water itself, (b) pressure dissolution within the limestones or at deeper levels and (c) dissolution of $CaCO_3$, mainly skeletal aragonite, in calcareous shales interbedded with the limestones. For precipitation of calcite from trapped seawater the Mg/Ca ratio would have to be lowered; the adsorption of Mg onto clay minerals could have this effect. Pressure dissolution has been considered of major importance in cementation in view of the great quantities of $CaCO_3$ that clearly have been dissolved from pressure dissolution planes, and because of the vast quantities of $CaCO_3$ that are required from somewhere to produce the fully cemented limestone formations that are seen at the Earth's surface. As noted in Section 4.7.1, in some limestones there is evidence for calcite spar having been precipitated on the sea floor.

Table 4.4 Terms for describing textures and fabrics of crystal mosaics in sedimentary rocks

For crystal shape:	anhedral—poor crystal shape
	subhedral—intermediate crystal shape
	euhedral—good crystal shape
For equigranular mosaics:	xenotopic—majority of crystals anhedral
	hypidiotopic—majority of crystals subhedral
	idiotopic—majority of crystals euhedral
For inequigranular mosaics:	porphyrotopic—where larger crystals (porphyrotopes) are enclosed in a finer-grained matrix
	poikilotopic—where larger crystals (poikilotopes) enclose smaller crystals
Size-scale:	micrometre-sized · 0–10 µm
	decimicrometre-sized · 10–100 µm
	centimicrometre-sized · 100–1000 µm
	millimetre-sized · 1–10 mm
	centimetre-sized · 10–100 mm

4.7.4 Neomorphism

Some diagenetic processes involve changes in the mineralogy and/or fabric of the sediment. For these processes of replacement, once loosely referred to as recrystallization, the term *neomorphism* is now used to include all transformations between one mineral and itself or a polymorph. There are two aspects to neomorphism: the wet polymorphic transformation of aragonite to calcite and the wet recrystallization of calcite to calcite. Both processes are wet because they take place in the presence of water, through dissolution–reprecipitation; dry, solid-state processes, such as the inversion of aragonite to calcite or recrystallization *sensu stricto* of calcite to calcite, are unlikely to occur in limestones, where diagenetic environments are always wet. Most neomorphism in limestones is of the aggrading type, leading to a coarser mosaic of crystals. Two common types are (a) microspar–pseudospar formation from micrite and (b) the calcitization of originally aragonitic skeletons, ooids and cements. Degrading neomorphism results in a finer mosaic of crystals.

A scheme for describing textures and fabrics of neomorphic limestones and dolomites, and other sediments such as evaporites that have been precipitated, crystallized or recrystallized, is given in Table 4.4.

Microspar–pseudospar; aggrading neomorphism

It is not uncommon to find that in fine-grained limestones, the micritic matrix (less than 4 µm) has been locally or even totally replaced by microspar (crystal sizes between 4 and 10 µm) and pseudospar (10–50 µm) (see Plate 13e). This *neomorphic spar* can be recognized by:
1 irregular or curved intercrystalline boundaries, commonly with embayments (contrasting with the plane intercrystalline boundaries of sparite cement, Section 4.7.3);
2 very irregular crystal-size distribution and patchy development of coarse mosaic;
3 gradational boundaries to areas of neomorphic spar;
4 presence of skeletal grains floating in coarse spar.

Aggrading neomorphic textures are prominent in micritic limestones and they can give a mottled, almost brecciated appearance, of 'clasts' of neomorphic spar/microspar in a micritic matrix. Such pseudo-breccias are common in the British Carboniferous (Solomon, 1989). Neomorphism may have taken place within an original inhomogeneous sediment, such as that resulting from bioturbation. Aggrading neomorphism involves the growth of certain crystals at the expense of others. It is likely that growth takes place in solution films and cavities between crystals, by syntaxial precipitation on pre-existing crystals. The CaCO$_3$ will be derived from dissolution of submicrometre-sized crystals and inflowing pore waters. Studies of fine-grained limestones by SEM have shown that much of the microspar is actually a cement rather than the result of aggrading neomorphism. Lasemi & Sandberg (1984) have found relics of aragonite in some ancient micrites and so were able to distinguish aragonite-dominated precursor muds (ADP) from calcite-dominated ones (CDP).

Calcitization of aragonite grains and cements

Bioclasts and ooids composed originally of aragonite are now mostly drusy sparite in limestones, through dissolution of the aragonite and later precipitation of calcite into the void (Section 4.7.1). Where calcite spar has not been precipitated, then biomoulds and oomoulds are present. In some cases, however, the grains have been replaced by calcite with no intervening void phase, a process referred to as *calcitization*. Where this has occurred, features to note are:

1 relics of the internal structure of the shell, preserved through inclusions of organic matter and minute crystals of aragonite;

2 an irregular mosaic of small and large calcite crystals, with wavy, curved or straight intercrystalline boundaries;

3 a brownish colour to the neomorphic spar, owing to residual organic matter, which imparts a pseudopleochroism to the crystals (see Plate 8a).

Aragonite cements also endure either dissolution or calcitization, but more tend to be calcitized (see Fig. 4.31).

Degrading neomorphism

Degrading neomorphism, whereby large crystals of CaCO$_3$ are replaced by smaller calcite crystals, a process of 'crystal diminution', is rare in limestones, and mostly has occurred through tectonic stress or very low-grade metamorphism. The process is most easily observed to have taken place in echinoderm grains. Micritization of skeletal grains by endolithic microbes is not a neomorphic process of course, but does result in a fine-grained mosaic.

4.7.5 Compaction

As discussed in Section 2.9.1, increasing overburden pressure leads to compaction in sediments and two categories are recognized: mechanical and chemical. Mechanical compaction may begin soon after deposition, whereas chemical compaction normally requires more than several hundred metres of burial.

Mechanical compaction in grainy sediments leads to a closer packing of the grains and a rotation of elongate bioclasts towards the plane of the bedding. As the overburden pressure increases, fracture of bioclasts may take place and micritic grains may become squashed and deformed. If there are early cements around grains, these can be spalled off, as can oolitic coatings of grains. Mechanical compaction may lead to collapse of micrite envelopes, if there has been an earlier phase of dissolution of the aragonitic bioclasts (see Plate 7c).

Lime muds generally suffer more compaction during very shallow burial as the sediments dewater. This may lead to burrows being compressed, especially if there is a high clay content in the sediment. Alternating clay-poor and clay-rich, fine-grained limestones, a common deep-water facies, usually show the differential effects of compaction well, with shell fracture and burrow flattening in the latter, but not the former. Compaction can lead to the formation of skeletal packstones from skeletal wackestones, as a result of the closer packing of grains.

Chemical compaction is the result of increased solubility at grain contacts and along sediment interfaces under an applied stress. Mostly this is the result of overburden but tectonic stresses also give rise to pressure-dissolution effects. Three common textures result from chemical compaction: fitted fabrics, stylolites and pressure-dissolution seams. In a grainstone with little or no early cement, sutured and concavo-convex contacts develop between grains (see Plate 13d). If the pressure dissolution between grains is intense, then a *fitted fabric* may be produced. This can be on a microscopic scale, as in oolitic and echinoderm grainstones, or on a macroscale, between intraclasts, fossils and early lithified diagenetic nodules and burrow fills in a muddy, compactible sediment. A *stylobreccia* texture may be produced.

Stylolites are through-going sutured surfaces that cut grains, cement and matrix indiscriminately (see

Plate 14a; see also Fig. 9.10). Clay, iron minerals and organic matter, the insoluble residue from the limestone's dissolution, are usually concentrated along the stylolites. In more argillaceous limestones, *dissolution seams* are smooth, undulose and anastomosing horizons of insoluble residue. They generally pass around grains and early diagenetic nodules, and where abundant, the term *flaser limestone* is used.

Pressure dissolution is an important process accentuating bedding planes, particularly between more muddy and more grainy sediments, and it may also lead to the development of *pseudobedding planes* (see Section 4.6.1 and Simpson, 1985). Calculations have shown that considerable amounts of $CaCO_3$ can be liberated by pressure dissolution, so that this process is often cited as one of the main sources of $CaCO_3$ for limestone cementation, especially of late-diagenetic, burial calcite spar. Burial diagenesis is reviewed in Choquette & James (1987).

4.8 Dolomitization, dedolomitization and silicification

It is not uncommon to find that ancient limestones have been partially or even completely dolomitized. The dolomite so formed can be replaced by calcite in the process of dedolomitization (Section 4.8.3). Limestones also can be silicified to various degrees (Section 4.8.4). These diagenetic processes commonly result in an obliteration of sedimentary and petrographic details.

4.8.1 Dolomites

The mineral dolomite is a rhombohedral carbonate belonging to the $\bar{3}$ trigonal/hexagonal crystal system. Ideally, it consists of an equal number of Ca^{2+} and Mg^{2+} ions arranged into separate sheets with planes of CO_3^{2-} anions between. The well-ordered nature of the dolomite lattice results in a series of superstructure reflections on X-ray diffraction (XRD), which are not present in the structurally similar calcite. The height of the ordering $d015$ peak relative to the $d110$ peak gives a measure of the degree of ordering. Most modern dolomites have a low degree of ordering compared with older dolomites. The term protodolomite was introduced for Ca–Mg carbonates made in the laboratory with no, or very weak, ordering reflections. Many natural dolomites are also not stoichiometric

($Ca:Mg$ is $50:50$), but have an excess of Ca^{2+} ions, up to $Ca:Mg$ of $58:42$. The Ca^{2+} substitution for Mg^{2+} increases the lattice spacing, and this also can be measured with XRD by the shift in the position of the $d104$ peak (see Hardy & Tucker (1988) for details of the technique). Iron substitution is common in dolomites, giving ferroan dolomite with a few mol.% $FeCO_3$; ankerite $CaMg_{0.5}Fe_{0.5}(CO_3)_2$ may be associated.

The replacement of $CaCO_3$ minerals by dolomite and the precipitation of dolomite cement may take place soon after the sediments have been deposited, i.e. penecontemporaneously and during early diagenesis, or a long time after deposition, usually after cementation, during burial. The term primary has often been applied to dolomite, implying a direct precipitate from sea or lake water. In fact the majority of dolomites have formed by replacement of pre-existing carbonate minerals, although dolomite cements are common. The word dolomite is used for both the mineral and the rock type; the term *dolostone* has been used for the latter. Carbonate rocks are divided on the basis of dolomite content into:

limestone	0–10% dolomite
dolomitic limestone	10–50% dolomite
calcitic dolomite	50–90% dolomite
dolomite (dolostone)	90–100% dolomite

To convey an indication of grain or crystal size in a dolomite, the terms *dolorudite*, *dolarenite*, *dolosparite* and *dolomicrite* can be used. In many cases, if the original structure has not been destroyed completely, the dolomites can be described in terms of Dunham's (or Folk's) classification (see Section 4.4), preceded by the word dolomitic, or prefixed by dolo-. For the description of textures and fabrics the scheme of Table 4.4 can be used.

Two common types of dolomite mosaic are *xenotopic* and *idiotopic*, the former of anhedral crystals with curved to serrated and irregular crystal boundaries, and the latter of euhedral rhombic crystals (see Fig. 4.36 and Plate 14a). The features to note when describing dolomite textures are shown in Table 4.5.

The preservation of the original limestone texture in a dolomite varies from completely fabric destructive with no obvious relics of the original sediment (see Plate 14a,c), to fabric retentive, with good to perfect preservation of original structure (see Fig. 4.3 and Plate 14b for examples). Dolomitization also may be fabric selective, only destructively replacing the matrix

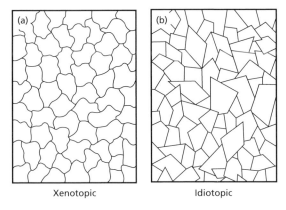

Xenotopic   Idiotopic

Fig. 4.36 Dolomite textures. (a) Xenotopic. (b) Idiotopic.

Table 4.5 Terms for describing dolomite textures

Dolomite crystal size	unimodal or polymodal
Dolomite crystal shape	anhedral, subhedral, euhedral
Dolomite mosaic	xenotopic, hypidiotopic, idiotopic
Crystal type	limpid, rhombic, baroque
Dolomite cement	isopachous, drusy
$CaCO_3$ grains	unreplaced or replaced or moulds; if replaced: partial or complete, mimic or non-mimic
$CaCO_3$ matrix	Unreplaced or replaced
Void-filling dolomite	limpid, rhombic, drusy or baroque

and not the grains, or only replacing certain bioclasts. Dolomite rhombs may be distributed randomly through the limestone (as in Plate 13f). Original crystal/grain size and mineralogy are important controls. An original micritic sediment generally is dolomitized to a fine-grained mosaic, so that primary sedimentary structures are still preserved. Grains of high-Mg calcite, such as red algae, some foraminifers and echinoderms, can be dolomitized with little fabric alteration (*mimic replacement*). By way of contrast, aragonitic grains (e.g. molluscs) are either dolomitized with much fabric alteration, or the aragonite is dissolved out and the mould filled by dolomite cement (in a similar manner to aragonite altering to calcite). Low-Mg calcite grains may resist dolomitization or be dolomitized destructively. The timing of dolomitization is also a factor, because if it is late, during burial, then it is very likely that the original, mixed-mineralogy sediment has already stabilized to low-Mg calcite, so that the dolomite is fabric destructive.

Dolomite crystals are commonly zoned; in many cases the inner part is more cloudy (from fluid inclusions or calcite relics) and the outer part is clear. Dolomite cements, as opposed to replacements, occur in primary and secondary cavities in many limestones and dolomites. They vary from cavity linings of clear, 'limpid' rhombs to drusy mosaics similar to calcite spar (as in Fig. 4.5).

One type of dolomite, which may be a replacement or a cement, is *baroque* or *saddle dolomite* (see Plate 14c). It is also known as pearlspar. The crystals generally are large (many millimetres) and have conspicuous, curved crystal faces. In thin-section, they have curved cleavage and undulose extinction. They commonly contain inclusions (fluids or mineral relics) and many are ferroan. Baroque dolomite commonly is associated with sulphide mineralization, hydrothermal activity and also hydrocarbons. It often is considered typical of burial dolomitization and the characteristic lattice distortion is attributed to variations in the concentration of Ca^{2+} ions adsorbed onto the growing crystal surface (see review of Spötl & Pitman, 1998).

The distribution of dolomites in the stratigraphic record is not even, and it has been said frequently that dolomites increase in abundance back in time. Dolomites do appear to be more common than limestones in the Precambrian, and this has led to the suggestion that seawater had a different composition then, so that dolomite could be precipitated directly, or could replace $CaCO_3$ more easily. Alternative views are that dolomitization environments were more prevalent as a result of palaeogeographical and palaeoclimatic differences, or that simply through being older, the limestones have had more time in which to become dolomitized. A survey by Given & Wilkinson (1987) suggested two maxima of dolomite occurrence in the Phanerozoic (Jurassic–Cretaceous and mid-Palaeozoic), which broadly correspond to the highstands in the first-order global sea-level curve, indicating a role of geotectonics and hydrosphere–atmosphere chemistry in dolomitization. Sun (1994) suggested that the abundance of Phanerozoic dolomites coincided with periods of extensive peritidal deposition and large-scale evaporite basins. Burns *et al.* (2000) thought that periods of more extensive dolomitization correlated with decreased oxygen levels in the atmosphere and oceans, and that this fostered more active communities of anaerobic microbes, which promoted dolomite formation.

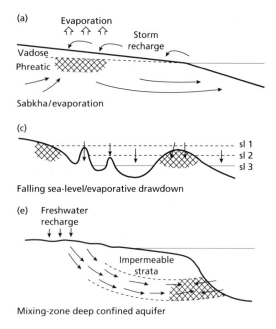

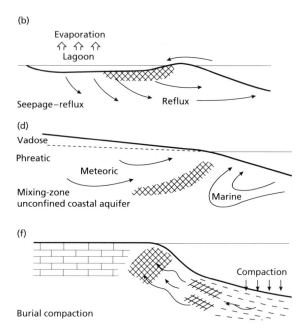

Fig. 4.37 Models of dolomitization, illustrating the variety of mechanisms for moving dolomitizing fluids through the sediments. For seawater dolomitizing models see Fig. 4.38.

4.8.2 The origin of dolomites and dolomitization models

There is still much debate and argument over the origin of dolomite, particularly concerning the pervasive dolomitization of extensive limestone platforms. One problem with dolomite is that it is difficult to manufacture in the laboratory under sedimentary–diagenetic conditions using natural waters. It is thus difficult to unravel the chemical controls on dolomite precipitation. Seawater is supersaturated with respect to dolomite, but with its highly ordered structure, dolomite appears to be inhibited from direct precipitation by various kinetic factors (the high ionic strength of sea-water, Mg^{2+} hydration and low CO_3^{2-} activity) so that aragonite and high-Mg calcite, with their simpler structure, are precipitated instead. The major considerations in the dolomitization of a limestone are the source of the Mg^{2+} and the mechanism by which the dolomitization fluids are moved through the rocks (see reviews of Tucker & Wright, 1990; Purser et al., 1994). Models for dolomitization are shown in Figs 4.37 & 4.38.

At the present time, dolomites are being precipitated in intertidal–supratidal sediments of the Bahamas, Florida and Trucial Coast (Arabian Gulf). The dolomite itself, poorly ordered and calcium-rich, consists mostly of 1–5 μm rhombs, occurring within the sediments or forming hard surface crusts. There is much evidence in the form of dolomitized gastropods and pellets for replacement, but direct precipitation also takes place (Lasemi et al., 1989). In some of these modern peritidal occurrences, it appears that the dolomite is being precipitated from seawater,

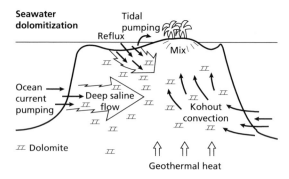

Fig. 4.38 Models for seawater dolomitization of limestones, all basically different ways of pumping seawater through a carbonate platform.

evaporated to a greater or lesser extent. The Mg/Ca ratio of the pore fluids is raised by the precipitation of aragonite and gypsum–anhydrite, and this promotes dolomite precipitation. Reduced SO_4^{2-} also helps. The Coorong in South Australia is a series of coastal lakes where dolomite is being precipitated, but not in association with evaporites. The dolomite forms where Mg^{2+}-rich continental ground water discharges into shallow lakes and is evaporated. Dolomite does precipitate directly from saline lake waters; this occurs in Victoria, Australia (Deckker & Last, 1989), where Mg/Ca ratios are high from weathering of basalts nearby. In the shallow-marine environment, dolomite is forming in Baffin Bay, a shallow, periodically hypersaline lagoon with restricted connection to the Gulf of Mexico and in a lagoon near Kuwait.

In the geological record, many fine-grained dolomites with evidence of peritidal deposition (planar stromatolites, birdseyes, polygonal cracks, tepees and intraclasts) have probably formed through evaporitic dolomitization and direct dolomite precipitation. Such dolomites generally show good preservation of the sedimentary structures, as in Plate 10e.

A model for dolomitizing shallow subtidal and reef limestones is *seepage–reflux*, whereby high Mg/Ca ratio fluids are generated in lagoons and beneath tidal flats and sabkhas (Section 5.2.1) by evaporation and these descend into the subsurface through density contrasts with marine pore waters. Unfortunately, there are no large-scale modern examples, but the model is applied frequently to dolomite formations closely associated with evaporites, such as those in the Upper Permian (Zechstein) basin of northwest Europe. Dolomitization of platforms and reefs around evaporite basins could well occur during drawdown (Fig. 4.37c, see Kendall, 1989).

Another model that has been very popular, especially for relatively early, near-surface dolomitization of limestone where there are no associated evaporites, is the *meteoric–marine, mixing-zone* model. As noted above, seawater is supersaturated with respect to dolomite but because of kinetic factors, dolomite normally does not precipitate unless the Mg/Ca ratio is raised (as in the evaporative model), and dolomite is forced out of solution. Therefore, it has been argued, dolomite precipitation is more likely to take place from dilute solutions (where there would be fewer interfering ions present) and at slow crystallization rates. Calculations have suggested that the mixing of meteoric ground waters with up to 30% sea water would cause undersaturation with respect to calcite but increasing saturation for dolomite. Dolomitization has thus been predicted for seawater–meteoric-water mixing zones, where salinities are reduced but Mg/Ca ratios are maintained. In the mixing model, the Mg^{2+} is derived from seawater, the Mg/Ca ratio is maintained above unity and ground-water movement pumps the dolomitizing fluid through the limestones. Although a most attractive hypothesis and applied to many pervasively dolomitized limestones without associated evaporites in the 1970s and 1980s, the mixing-zone model has come in for much criticism and is now abandoned. In essence, the very slow rate of dolomite precipitation (because of the ordered nature of the crystals) relative to calcite dissolution would mean that dolomite could not precipitate in significant quantities in a mixing zone. See Machel & Mountjoy (1986) and Hardie (1987) and reviews in Tucker & Wright (1990) and Purser *et al.* (1994).

Much dolomite is precipitated during the burial of a limestone formation, but there has been much debate as to whether this could result in the pervasive dolomitization of whole carbonate platforms. The driving force in this model is considered to be compaction of basinal mudrocks and the expulsion of Mg^{2+}-rich fluids into adjacent platform-margin limestones. The Mg^{2+} is thought to be derived from clay minerals, marine pore waters and high-Mg calcite, but mass-balance calculations have suggested that there is insufficient Mg^{2+} for large-scale dolomitization. The higher temperature of the burial environment should mean that some of the kinetic obstacles to dolomite precipitation are overcome. However, another problem appears to be that there is not a continuous source of fluid in the burial environment. Hydrothermal fluids may circulate, especially through fractures and faults and lead to local veins of baroque dolomite. Scattered rhombs, dolomite crystals along stylolites and pressure-dissolution seams, and late cavity-filling dolomite cements are common forms of burial dolomite in many limestones. Burial dolomites generally are coarse and fabric destructive; baroque dolomite is an especially common form. Burial dolomites have been well described from the Cambrian Bonneterre of Missouri, the Upper Devonian of Alberta (e.g. Dix, 1993; Drivet & Mountjoy, 1997), the Carboniferous of England and the Triassic of the

Dolomites (see papers in Purser *et al.,* 1994; Kupecz *et al.,* 1997; and Yoo *et al.,* 2000).

In recent years, attention has focused on the possibility that sea water alone, perhaps with a little modification, is the dolomitizing fluid, and a number of mechanisms have been put forward to drive seawater through the sediment (see Fig. 4.38). Discoveries of modern dolomite in the marine environment are significant here. Beneath Enewetak atoll in the Pacific, Eocene strata at a depth of 1250–1400 m are being dolomitized. As a result of the cooler water at this depth, seawater is just undersaturated with respect to calcite, but still supersaturated with regard to dolomite. Seawater is pumped through the atoll by oceanic tides and upwards by thermal convection as a result of the high heat flow out of underlying volcanic basement. Within the Bahama Platform, there also is a strong circulation of sea water as a result of ocean currents such as the Gulf Stream impinging on the Bahama escarpment. This *ocean current/tidal pumping* could play a major role in dolomitization because it drives huge volumes of seawater through carbonate platforms. The high geothermal gradient beneath carbonate platforms is thought to generate a large-scale convection system, which draws cold seawater into the platform. Diagenetic changes, including dolomite precipitation, in periplatform ooze on the Bahama slopes have been attributed to this *Kohout convection.* Lagoonal or platform-interior seawater is frequently a little more saline (40–45‰) than open-ocean water (35‰) and this is sufficient to cause *reflux* into the underlying sediments. Restudy of tidal-flat dolomites in the Florida Keys has suggested that they are forming through *tidal pumping* of Florida Bay water through the sediments, rather than through simple evaporation of seawater. Experiments have suggested that SO_4^{2-} in seawater is an important kinetic inhibitor of dolomitization. The SO_4^{2-} is reduced by microbial activity and this takes place in sediments containing organic matter. Dolomite being precipitated from marine pore waters within anoxic pelagic sediments in the Gulf of California and off Peru, and in lagoonal muds of Brazil, could well be the result of microbial mediation and sulphate reduction. Indeed, there is a growing body of evidence that microbes play a major role in dolomite precipitation generally; see review of Burns *et al.* (2000), and Teal *et al.* (2000). One other mechanism that will move seawater through sediments is relative sea-level change. There also is much fluid movement within the meteoric–marine mixing zone, which will generate pore-water circulation within the adjacent marine phreatic zone. Thus, now, seawater dolomitization is very popular; it provides the necessary Mg ions and there are several powerful processes for circulating seawater through a carbonate platform.

The origin of dolomites remains an enigma but careful field, petrographic and geochemical study does permit an elimination of some dolomitization models and implication of others. An analysis of the stable isotope and trace-element geochemistry of dolomites, as well as fluid inclusion data, can help to distinguish different generations and to make inferences on the nature of the pore waters involved: hypersaline/marine/mixed/connate. [87/86]Sr isotope data can be useful too, to identify the origin of the fluids, and even date the time of dolomitization.

An important consequence of many styles of dolomitization is that porosity is increased (see Plate 14a,b). Dolomite has a more compact crystal structure than calcite so that theoretically the complete dolomitization of a limestone results in a porosity increase of 13%, as long as there is no subsequent compaction or cementation. In addition, dissolution of relict calcite in dolomitized limestone or increasing calcite dissolution relative to dolomite precipitation during the dolomitization process creates extra porosity. Thus, dolomitization is important for hydrocarbon reservoir potential. Many oilfields of western Canada, for example, are in dolomitized Devonian reef limestones (see papers in Roehl & Choquette (1985) for case histories). Many papers on dolomite are contained in Shukla & Baker (1988) and Purser *et al.* (1994).

4.8.3 Dedolomitization

Dolomite may be replaced by calcite to produce limestone again. This calcitization process is referred to as dedolomitization and predominantly takes place through contact with meteoric waters. Calcite replacement of dolomite commonly is associated with the dissolution of gypsum–anhydrite, a near-surface phenomenon as well (Section 5.5). Burial dedolomitization also may occur. Recognition of 'dedolomites' is similar to that of replaced evaporites, a question of noting dolomite crystal shapes (rhombohedra) occupied by calcite (pseudomorphs), or calcite crystals with replacement fabrics (see neomorphic spar,

Section 4.7.4) containing small relict inclusions of dolomite (see Plate 14d). In some cases the original limestone texture is partially regenerated on de-dolomitization; in other instances layers and concretions of fibrous calcite randomly and completely replace the dolomite. The dissolution of dolomite rhombs may lead to a mouldic porosity (see Plate 14e).

4.8.4 Silicification

Silicification, like dolomitization, can take place during early or late diagenesis. It takes the form of selective replacement of fossils or the development of chert nodules and layers (Section 9.4). Silica also occurs as a cement in some limestones. The main types of diagenetic silica in limestones are: (a) euhedral quartz crystals, (b) microquartz, (c) megaquartz and (d) chalcedonic quartz. They are described in Section 9.2 and shown in Figs 9.1, 9.2 & 9.11. Both length-fast and length-slow chalcedonic quartz occur and the latter may indicate the former presence of evaporites (Section 5.5). Sponge spicules are the main source of silica, together with diatoms and radiolarians (Section 9.4).

4.9 Porosity in carbonate sediments

The porosity of carbonate sediments shortly after deposition is very high: sand-sized sediments around 50%, lime mud around 80%. Porosity is lost or reduced through cementation, compaction and pressure dissolution, and gained through dissolution, dolomitization and tectonic fracturing.

Porosity in limestones can be divided into two main types: primary (depositional) and secondary (diagenetic–tectonic). Three common types of primary porosity are:

1 framework porosity, formed by rigid carbonate skeletons such as corals, stromatoporoids and algae, especially in reef environments;
2 interparticle porosity in carbonate sands, dependent on grain-size distribution and shape;
3 porosity in carbonate muds provided by fenestrae (birdseyes) and stromatactis (Section 4.6.3).

Secondary porosity includes:
1 moulds, vugs and caverns formed by dissolution of grains and rock, commonly through leaching by meteoric ground waters, but also by basinal (connate) waters;

2 intercrystalline porosity produced through dolomitization;
3 fracture porosity, formed through tectonic pressures, and through collapse and brecciation of limestone as a result of dissolution, such as of interbedded evaporites, or the limestone itself in karstification.

Primary porosity, and also secondary, is commonly facies controlled. Certain facies, such as reefs, fore-reefs and oolites have high primary porosities, whereas others have low porosities, lagoonal micrites and outer-ramp carbonates, for example, unless affected by the diagenetic–tectonic processes leading to porosity creation. Studies of carbonate facies distributions, cementation patterns and diagenesis, in particular dolomitization, coupled with porosity–permeability measurements are thus all required to detect any reservoir potential. Examples of carbonate hydrocarbon reservoirs are: the Upper Jurassic Arab Formation of Saudi Arabia, with a primary intergranular and dolomite porosity; Middle and Upper Devonian reef and fore-reef limestones of western Canada and the Ordovician Trenton Limestone of northeastern USA, both with a porosity, at least in part, as a result of dolomitization; the Ellenburger of Oklahoma and Texas, with karstic porosity, and the Upper Cretaceous Chalk of the North Sea and the Tertiary Asmari Limestone of Iran, both with fracture porosity. Carbonate reservoirs are described in Roehl & Choquette (1985), Kupecz *et al.* (1997) and Harris *et al.* (1999), and porosity is reviewed by Moore (1989) and Lucia (1995, 1999).

4.10 Carbonate depositional environments and facies

4.10.1 Non-marine carbonate sediments

Lacustrine limestones

Lacustrine carbonates are of three principal types: (a) inorganic precipitates, (b) algal/microbial sediments and (c) skeletal sands. Inorganic precipitation, producing lime muds, mostly takes place through evaporation, but CO_2 loss, as a result of plant photosynthesis or pressure–temperature changes, and mixing of fresh stream or spring water with saline lake water, also causes carbonate precipitation. Tufa mounds are spectacularly developed in some lakes (e.g. Mono Lake, California) as a result of sublacustrine springs. Precipitation in shallow, agitated zones may produce ooids, as in the Great Salt Lake, Utah and Pyramid Lake, Nevada. The mineralogy of the carbonate mud and ooids precipitated depends largely on the Mg/Ca ratio of the water. Aragonite, calcite (high and low Mg) and dolomite (as mud not ooids) may all be precipitated.

Lime muds also may be produced through the activities of algae, cyanobacteria and microbes, and from phytoplankton blooms. The main role of cyanobacteria, however, is in the formation of stromatolites, common in modern lakes (e.g. Great Salt Lake) and in many ancient lake formations (e.g. the Green River Formation, Wyoming and Utah and the Pliocene Ridge Basin of California). Oncoids also occur, those from Lake Constance, Switzerland, being especially well known. Skeletal sands contain fragments of calcareous algae, such as *Chara*, as well as bivalves and gastropods.

Lacustrine carbonates are arranged in a similar facies pattern to their marine counterparts. Stromatolite 'reefs' and ooid shoals occur in more agitated, shallow waters, with lime muds occurring shoreward on littoral flats and in protected bays, and in the central deeper parts of lakes. One characteristic feature of lake-basin deposits is a rhythmic lamination, consisting of carbonate–organic matter couplets, often interpreted as seasonal in origin.

Two broad types of lake can be distinguished:
1 hydrologically open—these have an outlet and so are relatively stable;
2 hydrologically closed—these have no outflow and so are subject to rapid changes in lake level through fluctuations in rainfall and outflow.

Evaporation may exceed inflow so that a saline lake develops, where evaporites may be precipitated (see Section 5.4.2). Perennial and ephemeral closed lakes are recognized. In the geological record, both open- and closed-lake carbonates occur in the Green River Formation, western USA, and perennial–ephemeral saline-lake carbonates, mostly dolomites, occur in the Cambrian Officer Basin, South Australia (Southgate *et al.*, 1989). Stromatolites, laminites, fish beds and dolomites were formed in the thermally stratified Devonian Orcadian lake of northeast Scotland (Trewin & Davidson, 1999).

Shallow-water, muddy lacustrine facies are commonly modified by pedogenesis and the term *palustrine* is used for these deposits, which usually are nodular and mottled. They are widely developed in the Tertiary of the Mediterranean area, Cretaceous of Spain (Platt, 1989), and the Devonian Catskills of New York (Dunagan & Driese, 1999). The Everglades of Florida are a good modern analogue (Platt & Wright, 1992).

Reviews and descriptions of lake carbonates can be found in Matter & Tucker (1978), Dean & Fouch (1983), Anadon *et al.* (1991) and Talbot & Allen (1996).

Calcrete or caliche

In many parts of the world where rainfall is between 200 and 600 mm year^{-1}, and evaporation exceeds this precipitation, calcareous soils are formed. They are typically seen in river floodplain sediments but they also develop in other continental sediments (aeolian, lacustrine and colluvial deposits), and in marine sediments too, should they become subaerially exposed. Many terms are applied to these pedogenic carbonates but calcrete and caliche, the latter chiefly in the USA, are widely used. Calcrete occurs in several forms, from nodules to continuous layers, with massive, laminated and pisolitic textures. Many calcretes form in the upper vadose zone through a *per descensum* process of dissolution of carbonate particles in the upper A horizon of the soil profile and reprecipitation in the lower B horizon (see Fig. 3.7). They also may form within or just below the capillary-rise zone, where they are referred to as phreatic or *ground-water calcretes*, well known from southwest Australia. Some of these are composed of dolomite (*dolocrete*; e.g. Colson & Cojan, 1996). Calcretes develop in time from scattered to packed nodules (Fig. 4.39), to a massive limestone layer. The amount of time involved varies but is of the order of several to tens of thousands of years.

The characteristic fabric of calcretes is a fine-grained equigranular calcite mosaic with floating quartz grains through displacive growth (Fig. 4.40). Grains and pebbles may be split in this displacive process. Replacement of some grains also takes place. Circumgranular cracks and spar-filled veins also are common. Many calcretes possess spar-filled tubules, formerly occupied by rootlets, and lacey, micritic 'septa' within these, formed by calcification of fungal filaments, give the so-called alveolar texture. Rhizocretions (rootlet encrustations) and microbial grain coatings are common in some calcretes. Pisoids

Fig. 4.39 Calcrete, consisting of closely packed elongate nodules that have grown in a river floodplain sediment. Old Red Sandstone, Devonian. Gloucestershire, England.

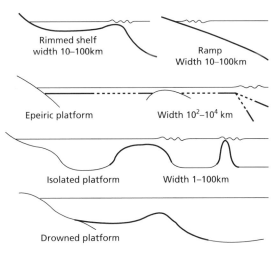

Fig. 4.41 Different types of carbonate platform.

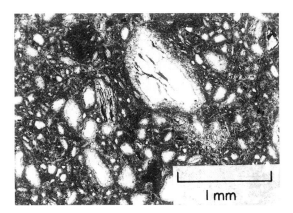

Fig. 4.40 Photomicrograph of modern calcrete showing dense micritic calcite, which has displaced the quartz grains so that the latter are now not in contact. There is also an envelope of coarser calcite crystals around each quartz grain; this is a typical feature of calcretes. Plane-polarized light. Recent calcrete. Almeria, Spain.

('vadoids') form in some calcareous soils from calcification of fungal–bacterial clusters. Some of these show evidence for *in situ* growth: dominant downward-directed laminae, reverse-grading and a fitted, almost polygonal arrangement of pisoids. Others accrete during imperceptible downslope movement of the soil. Laminated crusts are associated with calcretes and pisoids and some form from calcification of root mats. Black pebbles may be present (Section 4.3.1). Reviews on calcretes have been presented by Wright & Tucker (1991) and Retallack (1997). Specific descriptions include Williams & Krause (1998) for Devonian cal-

cretes in floodplain mudrocks, and Beier (1987) for modern Bahamian calcretes.

4.10.2 Marine carbonates and carbonate platforms

Thick successions of mostly shallow-marine limestone are a feature of the geological record. *Carbonate platform* is a general term widely used for these and they typically develop along passive continental margins, in intracratonic basins, failed rifts and foreland basins. Five major categories of carbonate platform are recognized: the rimmed shelf, ramp, epeiric, isolated and drowned platforms (Fig. 4.41). Each platform type has a particular pattern of facies and facies succession. The *rimmed shelf* (Fig. 4.42) is a shallow-water platform with a distinct break-of-slope into deeper water. Reefs and carbonate sand bodies occur along the high-energy shelf margin, restricting the circulation in the shelf lagoon behind to a greater or lesser extent. Along the shoreline, depending on the energy level and tidal range, tidal flats or a beach-barrier complex will be present. Debris from the rimmed-shelf margin is shed onto the adjacent slope and into the basin. Modern rimmed shelves occur off South Florida, Belize and Queensland (Great Barrier Reef). The *carbonate ramp* (Fig. 4.43) is a gently sloping surface with a generally high-energy, inner-ramp shoreline passing offshore to a quiet, deeper-water outer ramp, affected periodically by storms. Along the shoreline there may be a beach-barrier–tidal-delta complex, with lagoons and tidal

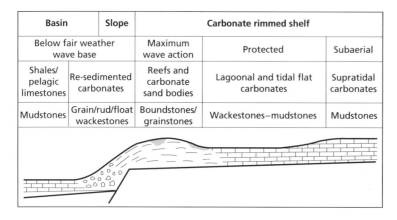

Basin	Slope	Carbonate rimmed shelf		
Below fair weather wave base		Maximum wave action	Protected	Subaerial
Shales/ pelagic limestones	Re-sedimented carbonates	Reefs and carbonate sand bodies	Lagoonal and tidal flat carbonates	Supratidal carbonates
Mudstones	Grain/rud/float wackestones	Boundstones/ grainstones	Wackestones–mudstones	Mudstones

Fig. 4.42 General facies model and limestone types for a rimmed shelf.

Basin	Carbonate ramp		
	Deep ramp	Shallow ramp	Back ramp
Below fair weather wave base		Wave-dominated	Protected/subaerial
Shale/ pelagic limestone	Thin bedded limestones Storm deposits mud mounds	Beach–barrier/ strandplain/ sand shoals patch reefs	Lagoonal–tidal flat– supratidal carbonates, evaporites palaeosols, palaeokarsts
Mudstones	Grain/wacke/ mudstones	Grainstones	Wackestones–mudstones

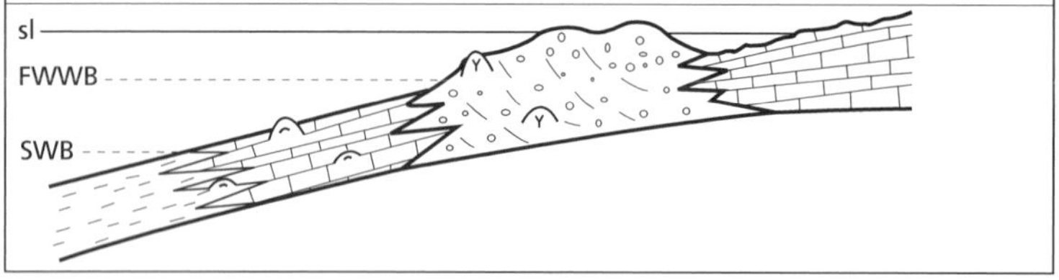

Fig. 4.43 General facies model and limestone types for a carbonate ramp.

flats behind, or a beach-ridge/strandplain system. Large reefs generally are not present on ramps, but patch reefs may occur on the inner ramp, and pinnacle reefs and mud-mounds in deeper water on the outer ramp. In effect, the carbonate ramp is equivalent to a siliciclastic open shelf. Modern carbonate ramps are located off the Yucatan Coast of Mexico, the Trucial Coast of the Arabian Gulf and Shark Bay of Western Australia. The *epeiric platform* is a very extensive (100–10 000 km across), relatively flat cratonic area covered by shallow sea. Along the margin, there may be a gentle (ramp-like) or steep (shelf-like) slope into the adjoining basin, but the margin is not a major feature of the platform. Within the platform itself, there may be deeper-water basins, surrounded by ramps and rimmed shelves. Epeiric platforms are dominated by low-energy, shallow subtidal–intertidal sediments.

Storms and tidal currents may be important. There are no good modern examples of epeiric platforms of the size that existed in the past, but the interior of the Great Bahama Bank and Florida Bay may be close analogues.

Isolated or detached platforms are surrounded by deep water and so are very much affected by prevailing wind and storm directions. They vary in size from a few kilometres to a few hundred kilometres across. The Bahamas is a large, modern example; smaller ones occur off the Belize shelf and in the Red Sea. Detached platforms consist of pure carbonate sediment because terrigenous material, apart from wind-blown dust, is excluded. The margins of these platforms may be shelf or ramp in character. *Drowned platforms* are ones that have suffered a relatively rapid sea-level rise so that deeper-water facies are deposited over shallow-water facies. Many pelagic limestones were deposited in these situations.

The type of carbonate platform developed is determined largely by tectonics and relative sea-level change, although one type commonly evolves into another. For discussions, reviews and case histories of carbonate platforms see Wilson (1975), Crevello *et al.* (1989), Tucker & Wright (1990), Tucker *et al.* (1990), Loucks & Sarg (1993), Wright & Burchette (1996), Harris *et al.* (1999) and Insalaco *et al.* (2000). The carbonate sediments of the major depositional environments are now briefly described.

4.10.3 Intertidal–supratidal carbonates

Tidal flats are areas regularly to rarely covered by water, dominated by weak currents and wave action. They are developed extensively upon epeiric platforms and they occur along the shorelines of low-energy shelves and ramps, typically behind beach-barriers and around lagoons. Tidal-flat carbonates are dominantly lime mudstones, commonly peloidal, although local lenses of coarser sediment (grainstone) may represent tidal-channel fills. Fenestrae (Section 4.6.3) are the characteristic structure, giving rise to the distinctive birdseye limestone (Fig. 4.24). The fauna may be restricted in diversity; gastropods in particular may abound, together with ostracods, foraminifers and bivalves. Thin, coarse layers of subtidal skeletal grains may occur, transported onto the tidal flat by storms. Microbial mats and stromatolites (Section 4.3.3) are typical of tidal-flat deposits. Many are simple planar varieties, showing desiccation cracks and laminoid fenestrae (e.g. Figs 4.12 & 4.25). Small domes may develop and in higher-energy areas, columnar stromatolites. Bioturbation is common, and rootlets may occur. Synsedimentary cementation of tidal-flat sediments can produce surface crusts, which may expand to form tepee structures (see Fig. 4.22) and megapolygons (Section 4.6.1). Crusts may break up to give intraclasts, reworked into edgewise conglomerates and flakestones (Fig. 4.5). Penecontemporaneous dolomitization may take place, giving fine-grained dolomite mosaics (Section 4.8.1). In arid climatic areas, the evaporite minerals gypsum–anhydrite, and possibly halite, will develop in the sediment. They may be preserved as pseudomorphs (Section 5.5). Slight uplift and contact with meteoric waters may result in palaeokarstic surfaces, laminated crusts, calcretes,

vadose pisoids and black pebbles. Tidal-flat sediments commonly occur at the top of shallowing-upward cycles (see Section 4.10.9 and Fig. 4.54). Modern carbonate tidal flats and their ancient equivalents have been described by Ginsburg (1975), Hardie (1977, 1986), Shinn (1983), Tucker & Wright (1990) and Demicco & Hardie (1995).

4.10.4 Lagoonal limestones

Lagoons are subtidal areas located behind barriers, which may be reefs or carbonate sand shoals. They occur on rimmed shelves (shelf lagoon) and along the inner part of carbonate ramps. Protected, lagoonal-type environments are widely developed on epeiric platforms too. Organisms living in lagoons, and therefore the sediments accumulating in these predominantly quiet-water areas, depend largely on the degree of restriction and permanence of the barrier. Lagoons may be normal in terms of salinity, as in the lagoons of atolls; brackish where there is much freshwater runoff, as in the inner part of Florida Bay, for example; or hypersaline to a greater or lesser extent, such as the inner part of the Great Bahama Bank, Shark Bay (Western Australia), and the lagoons of the Trucial Coast (Arabian Gulf). The sediments are variable in grain size, although many are carbonate muds, rich in peloids. Aggregates are common. Towards the barrier the muds pass into coarser sediments, and coarse skeletal debris may be derived from small, coral patch reefs, which commonly grow in lagoons of normal salinity. The lagoon floor is dominated by molluscs, green algae and foraminifers. The green algae in particular are a major sediment contributor (Section 4.3.3), and microbes play a significant role in skeletal breakdown and production of micritized grains. Surficial microbial mats and sea-grasses may cover the lagoon floor, as in the Bahamas and Shark Bay (Section 4.3.3). Bioturbation is intensive, largely through the activities of crustaceans and bivalves. Sedimentary structures may be poorly developed, although thin, vaguely graded beds of coarser grains and shell lags may be formed through periodic storm reworking.

Lagoonal limestones are common in the geological record, particularly in back-reef and back-barrier situations passing shorewards into tidal-flat facies. Many are wackestones and mudstones, in some cases with fossils preserved in growth position. Examples are the *Amphipora* wackestones of Devonian back-reefs in western Canada, Europe and Australia, the calcisphere-rich *porcellanites* of the Carboniferous, and the thick-shelled megalodont bivalve facies of Triassic back-reefs.

4.10.5 Intertidal–subtidal carbonate sand bodies

These sand bodies occur in areas of high tidal current and wave activity and include barriers, beaches, the shoreface and tidal deltas along ramp shorelines, and shoals and banks along exposed, rimmed-shelf margins. Depths of deposition are mostly less than 5–10 m. The sediments are carbonate sands and grainstones composed of ooids and rounded and sorted skeletal grains. The latter are fragments of normal-marine organisms: corals, bivalves, foraminifers and algae, and echinoderms and brachiopods in the Mesozoic–Palaeozoic. Sedimentary structures are ubiquitous, chiefly cross-bedding on all scales, perhaps with herringbone cross-bedding through tidal-current reversals (Fig. 2.23), reactivation surfaces, HCS + SCS, also planar bedding, scours and channels (see Section 2.3).

Carbonate shoreline sands are similar to their siliciclastic equivalents (see Section 2.11.5), in terms of their facies succession and sedimentary structures. The northeast Yucatan coast of Mexico is a modern example of a carbonate strandplain and the Trucial Coast of the Arabian Gulf is a barrier–lagoon–tidal-delta system. The Jurassic Lincolnshire Limestone of eastern England was deposited in such a barrier-island–lagoon system and the Jurassic Great Oolite of southern England was also deposited along an inner-ramp shoreline. Oolites in the South Wales Carboniferous were deposited in strandplains and barriers of a carbonate ramp (Burchette et al., 1990). The Jurassic Smackover Formation of the Gulf Coast Rim was deposited in similar inner-ramp environments, as were Mississippian grainstones in the Illinois Basin (see papers in Wright & Burchette (1998)).

Shelf-margin sand bodies are best developed along windward shelf-margins, where skeletal debris also may be derived from nearby reefs. The shoals are elongate parallel to the shelf-break and traversed by tidal channels that have lobes of sediment (spillovers) located at their ends. The shoals are covered in sand-waves and dunes that mostly are orientated lagoon-wards. Lily Bank on the northeast margin of the Little Bahama Bank is a good example of this sand-body type. Through time, windward sand shoals may prograde into the lagoon to generate a muddy to grainy, coarsening-upward unit. Continued growth of a sand shoal may lead to the development of a sand flat, with islands. This has happened at Joulters Cay in the Bahamas over the past 5000 years. In areas where tidal currents are strong, sand bodies are ridges orientated more normal to the shelf break, with grass-covered muddy sands in the depressions between. Such oolite ridges are prominent on the platform margins at the heads of the deep channels cutting into the Bahama Banks (e.g. Schooner and Exuma Cays, see Gonzalez & Eberli, 1997). Lime sand also accumulates along more leeward shelf-margins, and here off-shelf transport of sediment to the adjoining foreslope is a major process in platform progradation. This has occurred along the western side of the Great Bahama Bank, generating clinoforms (see Section 4.10.8 and Eberli & Ginsburg, 1989).

Occurring in high-energy locations, where water is continuously pumped through the sands, these grainstones commonly have marine cements, and hardgrounds may be formed. Ancient examples of shelf-margin, carbonate sand bodies include Zechstein oolites in northeast England (Kaldi, 1986). See papers in Harris (1984), the review of Handford (1988) and Mutti *et al.* (1996).

Oolitic and skeletal grainstones are important hydrocarbon reservoirs. The Jurassic Smackover Formation of the Gulf Coast Rim has substantial oil reserves. The type of porosity varies considerably over the region, from primary intergranular to dissolutional to intercrystalline where dolomitized. The original mineralogy of the ooids, aragonite to the north (landwards) and calcite to the south (basinwards) is a major factor in porosity evolution (Heydari & Moore, 1994).

Carbonate sands are also deposited in deeper waters on carbonate ramps by the action of storms. The sediments are grainstones and packstones, especially where depths are close to wave-base, passing into wackestones and lime mudstones in deeper water. Storm waves and currents generate HCS in above storm-wave-base lime sands, and shell lags and graded beds below storm wave-base (see Section 2.3.2). Bioturbation is widespread. Common skeletal components are molluscs, chiefly bivalves, foraminifers and coralline algae, and brachiopods and echinoderm debris in the Mesozoic and Palaeozoic. Terrigenous silt and clay are an important constituent of many outer-ramp and deep-shelf limestones. Storm-deposited limestones ('tempestites') are described by Aigner (1984) from the Triassic in Germany, Handford (1986) from the Mississippian of Arkansas and Jennette & Pryor (1993) from the Ordovician of Ohio.

4.10.6 Reefs and carbonate build ups

Although coral reefs are one of the most familiar and most studied of modern carbonate environments, there are many other types of reef developing at the present time and preserved in the geological record. Such carbonate *build ups*, a term widely used for locally formed limestone bodies that had original topographic relief, are a common feature of many carbonate formations going back to the Precambrian. The literature on modern and ancient reefs is vast; reviews and compilations include Toomey (1981), James (1983), Fagerstrom (1987), Geldsetzer (1989), Loucks & Sarg (1993), Monty *et al.* (1995), Harris *et al.* (1999) and Wood (1999).

The term reef itself is best restricted to a carbonate build up that possesses(ed) a wave-resistant framework constructed by organisms, but to be clear the term ecological reef or organic framework reef can be used for this. Specific types of such ecological reef, shown in Fig. 4.44, are *patch reef*, small and circular in shape; *pinnacle reef*, conical; *barrier reef*, separated from the coast by a lagoon; *fringing reef*, attached to the coast; and *atoll*, enclosing a lagoon. Other terms frequently used are *bioherm* for local *in situ* organic growth with or without framework; *biostrome* for laterally extensive *in situ* growth with or without framework; *organic bank* or *loose skeletal build up* for an accumulation of mostly skeletal sediment, chiefly through trapping or baffling; *mud-mound* or *mud bank* (formerly reef knoll) for an accumulation of mostly lime mud (micrite), probably by trapping and baffling.

Many different organisms can be and have been involved in the construction of reefs. At the present time, the main reef builders are corals and coralline algae; others of limited importance are sponges, serpulids, oysters and vermetid gastropods. In the past, practically all invertebrate groups have at one time or another contributed to reef growth. Special mention can be made of cyanobacteria and microbes (formerly blue–green algae) constructing stromatolite bioherms/biostromes in the Precambrian and Cambrian (locally in younger rocks too), stromatoporoids in the Ordovician to Devonian, rugose corals in the Silurian to

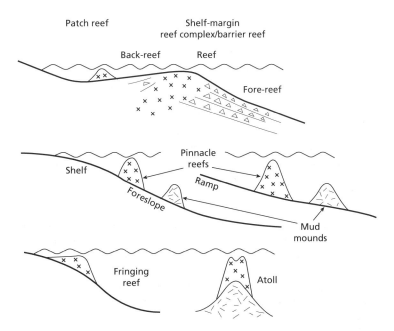

Fig. 4.44 The main types of reef and their location.

Carboniferous, scleractinian corals from the Triassic to Recent, phylloid algae in the Carboniferous and Permian, sponges in the Triassic–Jurassic and rudistid bivalves in the Cretaceous (see Kiessling *et al.* (1999) for a review).

Organisms in reefs take three roles: (i) the frame-builders, those providing a skeletal framework (corals at the present time); (ii) the frame-binders and en-crusters, organisms which consolidate the framework, such as calcareous algae and bryozoans; and (iii) the reef-users, such as boring bivalves and sponges, preda-tory fish and echinoderms. With many ancient reefs it is clear that there was no true solid framework, but much *in situ* organic growth. This last feature gives rise

to the two typical features of reef limestones, a massive appearance with no stratification (Fig. 4.45) and the presence of organisms in growth position (Fig. 4.46). Many reef limestones would be classified as bound-stones, with the terms framestone, bindstone and baf-flestone used to describe specific types (see Section 4.4); some reef-rocks do not have an obvious reef structure on the hand-specimen or thin-section scale and are a wackestone, lime mudstone or even a grain-stone. Primary cavities are a feature of many reefs, al-though they are commonly filled with skeletal debris and cement. Synsedimentary cementation is a feature of many modern and ancient reefs (Section 4.7.1).

There are many factors controlling the growth of

Fig. 4.45 Devonian reef limestones in the Canning Basin, Western Australia. The reef consists of massive unbedded limestone (central part of cliff); in front of the reef (to the left) occurs a talus slope of reef debris, these fore-reef limestones have an original dip; behind the reef (to the right) are flat-bedded limestones of the back-reef lagoon.

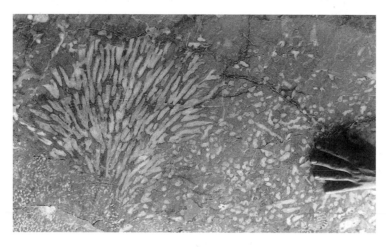

Fig. 4.46 *In situ* coral colony (*Thecosmilia*) from Triassic reef. Adnet, Austria.

modern coral reefs and it is likely that these same factors exerted an influence on coral and other reefs in the past. For coral-reef growth, these factors are:
1 water temperature—optimum growth occurs around 25°C;
2 water depth—most growth takes place within 10 m of the surface;
3 salinity—corals cannot tolerate great fluctuations;
4 turbidity and wave action—coral growth is favoured by intense wave action and an absence of terrigenous silt and clay.

The majority of reefs occur along shelf margins, an agitated zone where waves and currents of the open sea first impinge on the sea floor. Smaller patch reefs develop in open lagoons behind shelf-margin reefs, on carbonate ramps and epeiric platforms. Reefs, usually atolls, are also developed on submerged volcanic islands within the ocean basins. The configuration and morphology of some present-day reefs is a reflection of karstic dissolution of earlier reefs during glacial lowstands of sea-level.

Many modern reefs along shelf margins show a characteristic threefold division into fore-reef (reef front/slope), the reef itself (reef-crest, reef-flat) and

back-reef. The reef front is a steep slope, vertical in places, with organisms constructing reef in the upper part, passing down to a talus slope of coarse reef debris. Reef-derived carbonate breccias, debrites and turbidites may be present in the adjoining basin. A prominent system of surge channels gives a spur and groove morphology along the reef front, extending up to the reef flat, in some cases. The reef crest, covered by no more than 1–2 m of water, is the site of prolific organic growth, of corals and algae on modern reefs. Behind the crest is the reef-flat, a pavement of mostly dead coral. The back-reef area consists of reef debris adjacent to the reef-flat, passing shoreward to a quiet-water lagoon, where there may be patch reefs. A broadly similar facies pattern is seen in many ancient build ups (Fig. 4.45). Classic examples include the Permian Capitan Reef in Texas (see Saller *et al.*, 2000), the Devonian reefs of western Canada, Europe and Australia, and the Triassic reefs of the Northern Calcareous Alps (Austria) and Dolomites (Italy).

Mud-mounds are massive accumulations of lime mudstone and wackestone, tens to hundreds of metres across, which pass laterally into well-bedded limestone. The mud-mounds contain only scattered fossils; those of some possible significance are crinoids, bryozoans, sponges and algae. Much of the mud is peloidal. Mud-mounds typically occur in deeper water on carbonate ramps and shelf slopes. The best developed mud-mounds occur in the Palaeozoic, in the Carboniferous of northwest Europe, where some are referred to as Waulsortian reefs, the Carboniferous of New Mexico and Montana, and the Devonian of New York State and Belgium (Fig. 4.47).

The origin of mud-mounds is still a problem. Processes that could be involved are (a) the precipitation of lime mud by microbes, (b) entrapment of mud through the baffling action of bryozoans and crinoids and (c) concentration of mud into mounds by currents. Comparisons have been made with the mud banks of Florida and Shark Bay, where lime mud and skeletal debris is baffled and trapped by sea-grass and algae (Bosence, 1995). The sediment itself is derived from green-algal disintegration (Section 4.3.4) and the breakdown of larger skeletal grains. Many papers on mud-mounds can be found in Monty *et al.* (1995). Perhaps the most spectacular examples occur in the Algerian Sahara (Wendt *et al.*, 1997).

Reefs are important hydrocarbon reservoirs with the most porous facies generally occurring in the upper

Fig. 4.47 Lower Carboniferous mud-mound (muleshoe mound), Sacremento Mountains, New Mexico. Photograph courtesy of Dave Hunt.

fore-reef and reef-framework facies. However, marine cementation is most prevalent in this zone and so porosities can be reduced from high primary values. The talus at the toe of reef slopes also makes good reservoirs. Many reefs are dolomitized, and also this enhances their reservoir qualities. Examples include the Devonian reefs of western Canada, such as the Leduc and Golden Spike, and the Cretaceous rudist reefs of the Gulf Coast Rim, Mexico and the Middle East. Rudist bivalves have very porous skeletons but much of this intraskeletal porosity is not connected. The best reservoirs are thus located in rudist debris beds, rather than in the reefs themselves (which tend to be muddy) or in rudist limestones that have suffered much mechanical compaction during burial.

4.10.7 Pelagic limestones

Where water depth is too great for benthic organisms to flourish, in excess of some 50–100 m, then carbonate sediments composed of pelagic organisms will accumulate in the absence of clay. The maximum depth of accumulation is controlled by the rate of carbonate dissolution. In low latitudes, the ocean is saturated with respect to $CaCO_3$ in the upper few hundred metres and below this it becomes undersaturated, first with respect to aragonite, and then calcite. Below a few hundred metres $CaCO_3$ begins to dissolve, but it is not until greater depths that the rate of $CaCO_3$ dissolution increases substantially (this depth is the *lysocline*) (see Fig. 4.48). The depth at which the rate of dissolution is balanced by the rate of supply is known as the *carbonate compensation depth* (the CCD). This depth varies in the oceans, its position being controlled by calcareous plankton productivity, which itself depends largely on nutrient supply, and water temperature. In the tropical regions of the world's oceans, the CCD for calcite is between 4500 and 5000 m; the CCD for aragonite is about 2000 m less. The CCD shallows into higher latitudes and seawater is undersaturated in respect of $CaCO_3$ in temperate and polar waters. Calcareous oozes can accumulate on the sea floor, which is shallower than the CCD; siliceous oozes and red clays are present below this depth. Fluctuations in the CCD back into the Cenozoic and Mesozoic are now well documented.

Modern pelagic carbonates are composed of pteropods (aragonitic), coccoliths (Fig. 4.9) and foraminifers (both calcitic), and are found on outer

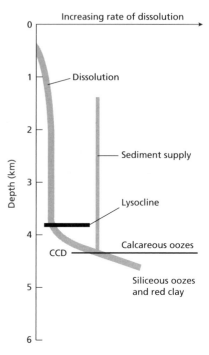

Fig. 4.48 Carbonate dissolution in the ocean. Below the carbonate compensation depth (CCD), carbonate sediment does not accumulate.

continental shelves, continental slopes and ocean floors starved of terrigenous clay, and on submarine rises, drowned reefs and volcanoes (seamounts and guyots) rising from the ocean floor. Ancient pelagic limestones occur in the Mesozoic of the Alpine region, the Ammonitico Rosso, Maiolica, Biancone and Scaglia, for example, and in the Devonian and Carboniferous of Hercynian Europe, the Cephalopodenkalk and Griotte. Characteristic features of pelagic limestones, apart from a dominantly pelagic fauna (e.g. Fig. 4.17), are their condensed nature and evidence for synsedimentary cementation in the form of hardgrounds, lithoclasts, sheet cracks and neptunian dykes. Many pelagic limestones are nodular (Fig. 4.49) and some contain ferromanganese nodules and crusts (Section 6.7). Resedimentation of pelagic sediment is common too, particularly in slumps and debrites.

The Cretaceous chalks of northwest Europe and southern USA are composed largely of coccoliths and are thus pelagic limestones. Deposition took place at depths of around 50–150 m, so that there is a signifi-

Fig. 4.49 Pelagic limestone: a micritic, nodular limestone that contains a pelagic fauna. Pressure dissolution effects are common and result in the clay seams (flasers) between nodules. Devonian. Montagne Noire, France.

Fig. 4.50 Block of reef-rock in deep-water hemipelagic mudstones, which also contain limestone turbidites of shallow-water skeletal debris. Triassic. The Dolomites, Italy.

cant benthic macrofauna of echinoids, bivalves and brachiopods. Hardgrounds are common within the chalks and these are mineralized with phosphate and glauconite. In the Chalk of the North Sea, resedimentation is widespread, with major slides and slumps, debrites and turbidites, all greatly expanding the thickness of the formation. The North Sea Chalk is a major hydrocarbon reservoir and depositional as well as diagenetic factors were important in the acquisition of reservoir qualities. The most porous horizons are commonly those that were resedimented, rather than the pelagic chalks. Fracturing of the chalk through up-doming of Zechstein evaporites has greatly improved the permeability. Early oil entry and overpressuring inhibited burial compaction and cementation. See Glennie (1998) for more information on North Sea petroleum geology.

Papers on pelagic limestones are contained in Hsü & Jenkyns (1974); also see the reviews of Scholle *et al.* (1983), Tucker & Wright (1990) and Clari & Martire (1996).

4.10.8 Resedimented deep-water limestones

Shallow-water carbonate sediment is transported into the deep sea by the same processes as discussed for siliciclastics (see Section 2.11.7): slides, slumps, debris flows, turbidity currents and modified grain flows. In the slope area between shelf and basin, slump-folded limestones and slides are common, especially in pelagic–hemipelagic slope facies. Megabreccias, including very large blocks of shallow-water limestone,

are common off major shelf-margin reefs (e.g. Fig. 4.50), and many have formed through sea-level falls (Spence & Tucker, 1997). Carbonate debrites are very variable in texture, commonly forming a spectrum from unsorted and chaotic breccias, to inversely and normally graded breccias, perhaps with associated graded grainstones (A to E in Fig. 2.78). Such beds are prominent in the Devonian of the Canadian Rockies and Cambro-Ordovician Cow Head Group of Newfoundland. In the basins themselves, limestone turbidites are usually interbedded with hemipelagic dark shales (Fig. 4.51). Sole structures, graded and planar bedding and cross-lamination are all developed in limestone turbidites, just as in siliciclastic examples, although in many, good Bouma divisions (see Section 2.11.7) are not present.

Carbonate turbidite basins are usually fed from a line source, i.e. the whole platform margin, rather than one (or several) point sources(s) as in the case of many siliciclastic turbidite basins. Thus the submarine fan model is not appropriate for most limestone turbidite formations, and in fact very few ancient, carbonate submarine fans have been described. Resedimented limestones are usually ascribed to either the slope apron (see Fig. 4.52), or base-of-slope apron models, depending on whether the resedimented limestones interfinger with the shallow-water shelf limestones (the first model) or whether the upper slope is bypassed and sediment accumulates at the toe of the slope (the second model). Ancient examples of slope aprons are well developed in the Triassic of the Dolomites (Bosellini, 1984), the Permian Capitan Reef of Texas

Fig. 4.51 Limestone turbidites composed of crinoidal and other skeletal grains interbedded with hemipelagic mudrocks. Succession is inverted. This is the typical appearance of turbidites (both carbonate and siliciclastic) in laterally continuous beds with interbedded deep-water mudrocks. Devonian. Cornwall, England.

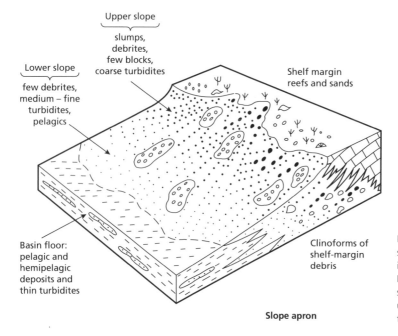

Slope apron

Fig. 4.52 Facies model for a carbonate slope apron where slope facies interdigitate with platform margin facies. In the base-of-slope apron model, not shown here, sediment mostly bypasses the upper slope and is deposited at the toe of the slope. From Tucker & Wright (1990).

and the Devonian Canning Basin reefs of Western Australia (Fig. 4.45), where large-scale, quite steeply dipping sheets of shallow-water debris extend up to the platform margin. These megascale cross-beds are called *clinoforms*, a term coming from seismic stratigraphy for dipping reflectors. Base-of-slope apron deposits are known from the Lower Palaeozoic of the Appalachians and Western Cordillera in North

America (Mullins & Cook, 1986). Resedimented Cretaceous rudist limestones are important oil reservoirs in the Golden Lane Atoll of Mexico (Enos & Stephens, 1993). They were deposited in a base-of-slope apron with debris coming from shelf-margin rudist reefs. Basin-margin, resedimented carbonates are commonly the sites of syngenetic mineralization. Hydrothermal fluids ascending along faults have precipitated base-metal sulphides in the porous limestones in a number of cases (see Anderson & McQueen, 1982).

4.10.9 Carbonate sequences

Carbonate sedimentation responds readily to relative changes in sea-level and within many carbonate formations, large-scale and small-scale cycles can be identified. As noted in Chapter 1, relative sea-level changes operate on many scales. For carbonates, the long time-scale second- (10^7 years) and third- (10^6 years) order sea-level changes are responsible for the development of the carbonate platforms themselves and the deposition of hundreds of metres of limestone. Within these platforms, depositional sequences can be identified, by recognizing the key surfaces, and their component systems tracts, from the large-scale arrangement of facies packages, in terms of their onlap (retrogradation), aggradation and offlap (progradation) or from the facies successions, reflecting relative sea-level changes. Rimmed shelf margins, where sedimentation rates are high, are strongly affected by relative sea-level changes and stationary (aggradational), offlapping, onlapping, back-stepping, drowned and emergent types have all been described. In the concepts of sequence stratigraphy, most carbonate deposition takes place within the transgressive and highstand systems tracts, especially the latter when large quantities of sediment are deposited in the basin (*highstand shedding*) and

clinoforms are well developed. During sea-level lowstands, carbonate platforms are usually emergent and so subject to extensive meteoric diagenesis and karstification. However, collapse of the margin may take place at this time to generate megabreccias. Figure 4.53 shows a sequence stratigraphic model for a rimmed shelf. For papers on carbonate sequence stratigraphy see Loucks & Sarg (1993) and Harris *et al.* (1999).

Metre-scale cycles (or parasequences) are common within platform carbonate sequences. These are the result of depositional processes and fourth- and fifth-order relative sea-level changes, operating on a 10^4–10^5 years scale. Many of these cycles display a shallowing-upward trend. There are many types but a common one consists of shallow subtidal sediments passing up into tidal-flat facies, with evidence of emergence at the top (Fig. 4.54). The latter may take the form of a palaeokarstic surface or palaeosoil. Mixed siliciclastic–carbonate cycles (see Fig. 2.68) and evaporite–carbonate cycles (see Figs 5.16 & 5.17) also occur. The parasequences may be repeated many times

Fig. 4.53 Sequence stratigraphic model for a rimmed shelf. There is much possible variation in the response of carbonate sedimentation to relative sea-level change. Here, the first sequence shows the case of reefal sedimentation keeping up with the sea-level rise during the TST so that the sediments aggrade, whereas in the second sequence the sea-level rise exceeds carbonate production in the TST and the shelf-margin sands back-step, i.e., retrograde. Abbreviations as for Fig. 2.86.

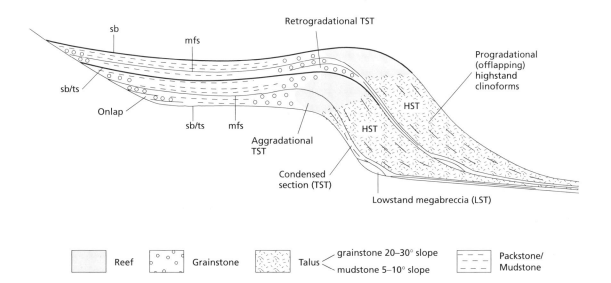

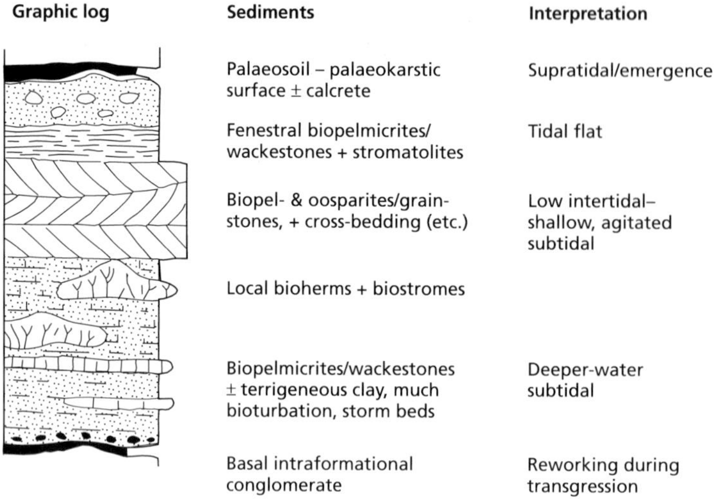

Graphic log	Sediments	Interpretation

	Palaeosoil – palaeokarstic surface ± calcrete	Supratidal/emergence
	Fenestral biopelmicrites/ wackestones + stromatolites	Tidal flat
	Biopel- & oosparites/grain-stones, + cross-bedding (etc.)	Low intertidal– shallow, agitated subtidal
	Local bioherms + biostromes	
	Biopelmicrites/wackestones ± terrigeneous clay, much bioturbation, storm beds	Deeper-water subtidal
	Basal intraformational conglomerate	Reworking during transgression

Fig. 4.54 Shallowing-upward limestone cycle (a parasequence): there are many variations on this general theme, depending largely on the energy level of the shoreline and on the climate. Such cycles are typically 1–10 m thick.

in a succession, and they can show systematic variations in thickness and character upwards. Upward-increasing thickness and proportion of subtidal facies in each successive cycle generally indicate an overall transgressive trend, and upward-decreasing thickness of cycles and increasing proportion of tidal-flat facies reflect an overall regressive trend. Many parasequences have formed through tidal-flat progradation and the lateral, seaward migration of facies belts. However, there has been much argument over the causes of the repetition of the cycles. Purely sedimentary processes have been invoked in a migrating tidal-island model and loss of carbonate source area. A tectonic mechanism of periodic fault movement or jerky subsidence has also been suggested (e.g. Satterley, 1996). However, it is currently very popular to invoke sea-level changes though ocean-water volume changes, especially via ice-caps, brought about by orbital forcing in the Milankovitch band (20 000–400 000 years). Statistical analysis and computer modelling of carbonate cycles have lent support to an astronomic control. See the reviews and papers by Tucker & Wright (1990), Wilkinson *et al.* (1997), Lehrmann & Goldhammer (1999) and Lehmann *et al.* (2000).

Relative sea-level changes are a major factor in the diagenesis of limestones as well as their sedimentation. Long periods of relative sea-level rise lead to the pumping of seawater through carbonate sediments, and this can result in extensive marine cementation and dolomitization. Extended periods of relatively low sea-level allow meteoric water to penetrate a carbonate formation, and, under a humid climate, this may generate dissolutional porosity, surface and subsurface karst, calcite cementation and mixing-zone-related dolomitization. If the climate is more arid, then evaporite precipitation and reflux dolomitization are likely early diagenetic processes. See the reviews of Read & Horbury (1993) and Tucker (1993).

Further reading

Crevello, P., Sarg, J.F., Read, J.F. & Wilson, J.L. (Eds) (1989) *Controls on Carbonate Platform and Basin Development*. Special Publication 44, Society of Economic Paleontologists and Mineralogists, Tulsa, OK, 405 pp.

Harris, P.M., Saller, A.H. & Simo, J.A. (Eds) (1999) *Advances in Carbonate Sequence Stratigraphy: Application to Reservoirs, Outcrops and Models*. Special Publication 63, Society of Economic Paleontologists and Mineralogists, Tulsa, OK, 421 pp.

Insalaco, E., Skelton, P.W. & Palmer, T.J. (Eds) (2000) *Carbonate Platform Systems: Components and Interactions*. Special Publication 178, Geological Society of London, Bath, 240 pp.

Loucks, R.G. & Sarg, J.F. (Eds) (1993) *Carbonate Sequence Stratigraphy*. Memoir 57, American Association of Petroleum Geologists, Tulsa, OK, 545 pp.

Moore, C.H. (1989) *Carbonate Facies, Diagenesis and Porosity*. Elsevier, Amsterdam, 338 pp.

Morse, J.W. & Mackenzie, F.T. (Eds) (1990) *Geo-

chemistry of Sedimentary Carbonates. Elsevier, Amsterdam, 696 pp.

Purser, B.H., Tucker, M.E. & Zenger, D.H. (Eds) (1994) *Dolomites: a Volume in Honour of Dolomieu*. Special Publication 21, International Association of Sedimentologists, Blackwell Science, Oxford, 451 pp.

Scoffin, T.P. (1987) *Carbonate Sediments and Rocks*. Blackie, Glasgow, 274 pp.

Scholle, P.A., Bebout, D.G. & Moore, C.H. (Eds) (1983) *Carbonate Depositional Environments*. Memoir 33, American Association of Petroleum Geologists, Tulsa, OK, 708 pp.

Tucker, M.E. & Bathurst, R.G.C. (1990) *Carbonate Diagenesis*. Reprint Series 1, International Association of Sedimentologists, Blackwell Science, Oxford, 312 pp.

Tucker, M.E. & Wright, V.P. (1990) *Carbonate Sedimentology*. Blackwell Science, Oxford, 482 pp.

Tucker, M.E., Wilson, J.L., Crevello, J.R., Sarg, J.R. & Read, J.F. (Eds) (1990) *Carbonate Platforms: Facies, Evolution and Sequences*. Special Publication 9, International Association of Sedimentologists, Blackwell Science, Oxford, 328 pp.

5 Evaporites

5.1 Introduction

Evaporites are mainly chemical sediments that have been precipitated from water following the evaporative concentration of dissolved salts. The principal evaporite minerals are gypsum ($CaSO_4.2H_2O$), anhydrite ($CaSO_4$) and halite ($NaCl$). There are many other naturally occurring evaporite minerals and of these the potassium and magnesium salts—sylvite, carnallite, polyhalite, kainite and kieserite—are important constituents of some marine salt deposits (formulae given in Table 5.1).

Evaporites are of great economic importance and have a wide range of uses and applications. Evaporite beds are an essential component of many oilfields of the world, commonly being the cap rocks to carbonate reservoir rocks (as in the Middle East and western Canada) or effecting structural traps through salt diapirism. Geologically, evaporites are useful in the studies of palaeoclimatology because generally they are restricted to arid areas of low latitude (see Fig. 1.1), where temperatures are very high, relative humidity is low and evaporation far exceeds any rainfall. The locations of the world's major evaporite deposits are shown in Fig. 5.1. At the present time, evaporites are not being precipitated to the same extent as they were at certain intervals in the geological past.

5.1.1 Early geochemical work

Studies of evaporites really started in the nineteenth century when chemists such as Usiglio and Van't Hoff undertook experiments to evaporate seawater and to synthesize evaporite minerals with a view to determining stability fields and precipitational controls. Starting with seawater (composition given in Table 5.2), Usiglio established that gypsum was precipitated when seawater had been evaporated to about 19% of its original volume, and that halite appeared when the volume was reduced to 9.5%. Continued evaporation produced magnesium and potassium minerals, but some naturally occurring salts, such as kieserite and polyhalite, were not obtained. Van't Hoff and co-workers showed that the minerals precipitated depended on the rate of evaporation and on the composition of the water, in particular whether equilibrium between metastable salts and solution was maintained. Following such experimental studies, it was soon noted that there were discrepancies between volumes of the various evaporite minerals found in marine salt deposits, such as the Permian Zechstein of northwest Europe, and those predicted theoretically from the evaporation of seawater (Table 5.3). The discrepancies were explained by early replenishment of the barred-basin or lagoon with normal seawater and incomplete evaporation. Details of the physical chemistry of salt precipitation and reviews of earlier work can be found in Warren (1999).

5.1.2 Evaporite deposits and depositional environments

Thick evaporite successions, in some cases reaching 1000 m or more, fill many of the world's large intracratonic sedimentary basins. Examples include the Permian Zechstein Basin of northwest Europe and the Michigan (Salina), Elk Point–Williston, Paradox and Delaware basins of North America (Fig. 5.1). Other evaporite formations interdigitate with non-evaporitic sediments, limestones and marls especially, and occur on stable platforms and shelves and in subsiding basins. Evaporites deposited in lakes or marine embayments may be located in pull-apart basins, intracratonic rifts and extensional basins, which in a few cases then developed into intercratonic rifts (continental margins) through sea-floor spreading. The Dead Sea, a well-known site of modern salt precipitation, the Triassic salt basins of Britain and Tertiary salt deposits of France and Germany are examples of evaporites formed in intracratonic rifts. The Tertiary evaporites on either side of the Red Sea and the Cretaceous evaporites along the eastern and west-

Table 5.1 The common marine and non-marine evaporite minerals

Marine evaporite minerals		Non-marine evaporite minerals	
halite	NaCl	halite, gypsum, anhydrite	
sylvite	KCl	epsomite	$MgSO_4. 7H_2O$
carnallite	$KMgCl_3. 6H_2O$	trona	$Na_2CO_3. NaHCO_3. 2H_2O$
kainite	$KMgClSO_4. 3H_2O$	mirabilite	$Na_2SO_4. 10H_2O$
anhydrite	$CaSO_4$	thenardite	$NaSO_4$
gypsum	$CaSO_4. 2H_2O$	bloedite	$Na_2SO_4. MgSO_4. 4H_2O$
polyhalite	$K_2MgCa_2(SO_4)_4.2H_2O$	gaylussite	$Na_2CO_3. CaCO_3. 5H_2O$
kieserite	$MgSO_4. H_2O$	glauberite	$CaSO_4. Na_2SO_4$

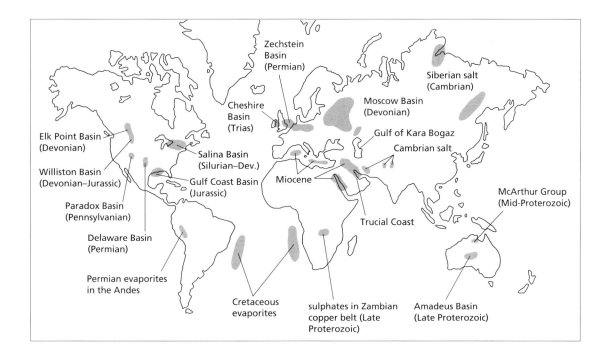

Fig. 5.1 Map showing location (and age) of the major evaporite deposits of the world. Also shown is the location of the Trucial Coast, Arabian Gulf, where sabkha sulphates are forming today, and the Gulf of Kara Bogaz, on the eastern side of the Caspian Sea, the nearest modern analogue to a barred basin.

ern continental margins of the Atlantic were deposited in rifts that subsequently became the sites of ocean crust formation.

Evaporite deposits are commonly cyclic. Some consist of numerous thin evaporite beds, a few to tens of metres thick, typically of gypsum–anhydrite with little or no halite, alternating with limestone and marl. The very thick salt deposits of intracratonic basins general-

ly consist of gypsum–anhydrite passing up into the more soluble halite, with thin beds of the highly soluble bittern salts (potassium and magnesium chlorides and sulphates) at the top. This cycle may be repeated several times, as in the Permian Zechstein of northwest Europe.

There has been much discussion on the depositional environments of salt deposits, partly because there are so few modern sites of evaporite precipitation and none on the scale of those that clearly existed in the past. Evaporite depositional environments are shown diagrammatically in Fig. 5.2, and are discussed in Kendall (1992), Kendall & Harwood (1996), Warren (1999) and Schreiber & El-Tabakh (2000).

Two principal modes of evaporite deposition are recognized:

1 *subaqueous precipitation*, from a shallow- to deep-water body, on a small (lake/lagoon) to large (intracratonic/rift basin) scale;

2 *subaerial precipitation*, taking place within sediment (sabkhas) or in very shallow to desiccated saline pans.

The subaqueous precipitation of evaporites is basically a simple 'evaporating-dish' process, with evaporite minerals forming close to the water–air interface

Table 5.2 The composition of seawater expressed in parts per million and percentage of total dissolved species. For comparison, the composition of world average river water, with a salinity of around 120 p.p.m., is also given

Dissolved species	Sea water		River water
	p.p.m.	% of total	
Cl^-	18000	55.05	7.8
Na^+	10770	30.61	6.3
SO_4^{2-}	2715	7.68	11.2
Mg^{2+}	1290	3.69	4.1
Ca^{2+}	412	1.16	15.0
K^+	380	1.10	2.3
HCO_3^-	140	0.41	58.4
Br^-	67	0.19	0.02
H_3BO_3	26	0.07	0.1
Sr^{2+}	8	0.03	0.09
F^-	1.3	0.005	0.09
H_4SiO_4	1	0.004	13.1

Saline pan—coastal or continental

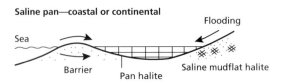

Sabkha—coastal or continental

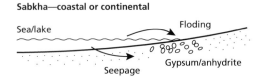

Marine marginal shelf/lagoon

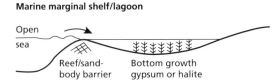

Barred-basin: deep to shallow

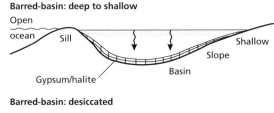

Barred-basin: desiccated

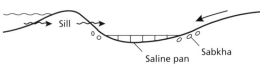

Fig. 5.2 Principal depositional environments of evaporites.

Table 5.3 The theoretical thickness of salts precipitated from seawater compared with the thickness of these salts in the Permian Zechstein of Germany, expressed as 100 m of precipitated evaporite. Note that there is much more $CaSO_4$ and much less Mg and K salt in the Zechstein deposits compared with the theoretical. Also shown is the approximate thickness of the various salts produced by the evaporation of a column of seawater 1000 m high

Component	Mineral	Thickness in 100 m of evaporite		Salt thickness from 1000 m of seawater
		From seawater	Permian Zechstein	
$MgCl_2$	in bischofite and carnallite	9.4	0.5	1.5
KCl	sylvite and in carnallite	2.6	1.5	0.4
$MgSO_4$	in kieserite	5.7	1.0	1.0
NaCl	halite	78	78	12.9
$CaSO_4$	anhydrite	3.6	16	0.6
$CaCO_3$	calcite ⎱	0.4	3	0.1
$CaMg(CO_3)_2$	dolomite ⎰			

and also nucleating on the sediment surface (bottom growth). In marine settings, some barrier is required to allow the water to evaporate to high salinities, but periodic replenishment of the brine body is necessary too. In the past, evaporites were precipitated subaqueously in shelf lagoons behind sand barriers and reefs, around barred-basin margins to form evaporite platforms, and on deep, barred-basin floors. At the present time, however, there are no good modern examples of marine barred-basins where evaporites are accumulating. An oft-quoted case is the Gulf of Kara Bogaz on the eastern side of the Caspian Sea, where halite, epsomite ($MgSO_4.7H_2O$) and astrakhanite ($Na_2Mg(SO_4)_2.2H_2O$) are being precipitated on the bay floor. Marine-fed lagoons where gypsum is precipitating occur along the coast of South Australia. Subaqueously precipitated evaporites have distinctive crystal forms, textures and bedding features. In shallow waters, the evaporite minerals may be reworked by waves and storms, and in deeper water they may be resedimented off slopes by slumps, debris flows and turbidity currents.

For many years, evaporation of standing bodies of seawater in relatively deep barred-basins was considered the only viable mechanism for the formation of the world's major evaporite bodies. However, an important alternative mode of formation was provided by the discovery in the early 1960s of gypsum–anhydrite forming *within* sediments of high intertidal–supratidal flats, called *sabkhas*, along the Trucial Coast of the Arabian Gulf. The evaporites are precipitated from sediment pore waters in the vadose and upper phreatic zones. Gypsum–anhydrite is also forming within the exposed sediment around salt lakes, oases, and near dry rivers in deserts; these are inland or continental sabkhas. A related environment is the *saline pan* (also salina), where mostly halite is precipitated. Here periodic flooding leads to shallow salt lakes that quickly evaporate away to precipitate halite, and they are then desiccated and subjected to subaerial processes. Coastal saline pans, such as the salinas along the Gulf of Aqaba, and Lake MacLeod, Western Australia, are fed by marine waters seeping through a sandy barrier; inland saline pans and playas, such as occur in Death Valley and elsewhere in the southwestern USA, North Africa and western China, are fed by fresh water from river floods and rain storms (Lowenstein & Hardie, 1985; Schubel & Lowenstein, 1997).

The subaerial precipitation of evaporites has taken place on the floor of deep-marine basins following evaporative drawdown and extreme desiccation. This occurred during the late Miocene when the Mediterranean dried up after being cut off from the Atlantic (Section 5.6).

Evaporites may precipitate in soils of desert areas to give crusts and indurated horizons. *Gypcrete* or gypsite is not uncommon in North Africa, the Middle East and India (see Watson (1985) for a review).

5.2 Gypsum and anhydrite

Rocks of gypsum–anhydrite possess distinctive structures and textures and are susceptible to replacement, recrystallization and dissolution (Fig. 5.3). Geological evidence and present-day occurrences show that both gypsum and anhydrite may be precipitated at the Earth's surface, subaqueously in shallow and deep water, and subaerially in coastal and inland sabkhas. On burial to depths greater than several hundred metres, however, all $CaSO_4$ is present as anhydrite, and on uplift anhydrite is normally converted to gypsum (secondary gypsum). The many studies of gypsum–anhydrite have shown that the stable phase is determined by the activity of water (related to salinity) and temperature.

Petrographically, gypsum and anhydrite are easily distinguished on the basis of their optical properties (see Plate 15a,b). Gypsum has low relief and weak birefringence and belongs to the monoclinic crystal system; anhydrite has moderate birefringence, higher relief and is orthorhombic. Both may show a prominent cleavage. The textures of gypsum–anhydrite vary considerably, depending on their precipitational environment and diagenetic history (discussed in the following sections). The 'gypsum–anhydrite cycle' is summarized in Fig. 5.3.

5.2.1 Sabkha sulphate and nodular anhydrite

The main site of marine sulphate precipitation today, where the early part of the gypsum–anhydrite cycle can be observed, is in the high intertidal and supratidal zones of such regions as Texas, the Trucial Coast of the Arabian Gulf and the southern Mediterranean coasts of Tunisia and Egypt. Gypsum is being precipitated displacively within the sediments as discoidal, rosette, selenite and twinned crystals from less than 1 mm to

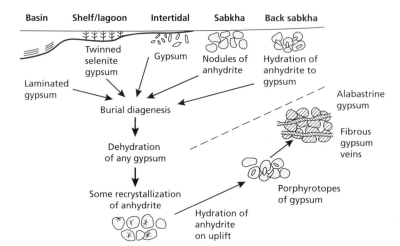

Fig. 5.3 The 'gypsum–anhydrite cycle' showing mineral and textural changes, from the surface, into the subsurface and on uplift.

more than 25 cm in size. Sediment pore waters are largely derived from surface flooding of sea water, a process referred to as *flood recharge*. Dolomitization of carbonate particles is commonly associated with gypsum precipitation, as a result of the increased Mg/Ca ratio (Section 4.8.2), and this releases calcium ions for further gypsum precipitation. Gypsum is the most common precipitate within the sediments of inland sabkhas, where it forms the familiar *desert roses*.

If the evaporation is sufficiently intense, as along the Trucial Coast, then with increasing concentration of pore fluids across the sabkhas, the gypsum crystals are replaced by a fine mush of equant and lath-shaped anhydrite crystals. This takes place where chlorinities are in excess of 145‰, i.e. a concentration of sea water by a factor of 7.5 to a salinity of around 260‰. The shape of the original gypsum crystals may be retained or pseudomorphed by the anhydrite if the host sediment is cohesive (Fig. 5.4). Continued displacive precipitation of anhydrite results in closely packed nodules with host sediment restricted to thin stringers. The nodular texture produced is referred to as *chicken-wire anhydrite* and this is the typical texture of many ancient sulphate deposits (Fig. 5.5). Anhydrite also is precipitated as thin beds or layers of coalesced nodules in the more landward parts of the sabkhas. These beds are commonly irregularly contorted and buckled, forming the so-called *enterolithic texture*, also common in ancient sulphate formations. In the most landward part of the sabkha, some rehydration of anhydrite to gypsum may

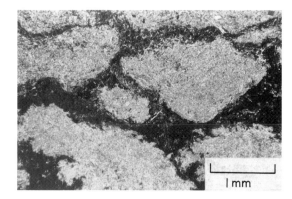

Fig. 5.4 Nodules of aphanitic anhydrite, partly retaining shape of original gypsum crystals. Dolomitic host sediment between nodules. Crossed polars. Permian. Cumbria, England.

occur from contact with fresh continental ground waters.

The formation of anhydrite requires an arid climate with high mean annual temperatures (above 22°C) and with seasonal temperatures in excess of 35°C. Where the climate is less arid, then primary nodules of gypsum crystals may develop within the sediment. This is happening in sabkhas along the Mediterranean coast of Egypt.

Primary textures exhibited by anhydrite in both modern and ancient nodular deposits include fine equant mosaics (aphanitic), and felted and parallel–subparallel arrangements of laths (Figs 5.4 & 5.6 and Plate 15a,b). Recrystallization of equant and lath

anhydrite may take place to produce coarse granular mosaics, large fibrous crystals and fibro-radiating aggregates (see Plate 15a).

5.2.2 Bottom-growth gypsum

Gypsum crystals can be precipitated on the floor of lagoons, lakes and shallow shelves around evaporite basins in a variety of crystal forms, some spectacular and very large (up to 7 m!). This *selenitic gypsum* commonly grows vertically, almost as grass, as single (prismatic), twinned (swallow-tail) and split crystals. Curved crystal faces are common, too, induced by the presence of organic matter. Beds of these gypsum crystals commonly have a distinctive palmate (palm-frond shape) texture. Gypsum fans, cabbage-like structures and metre-scale domes also occur. Thin laminae of

Fig. 5.5 Chicken-wire anhydrite. Closely packed nodules of anhydrite with thin stringers of sediment between. Upper Permian (Zechstein). Durham, England.

Fig. 5.6 Bottom-growth gypsum. (a) Twinned, selenitic gypsum (15 cm high) from the Messinian of Cyprus. (b) Palmate shapes in anhydrite after selenitic gypsum. Zechstein (Permian), North Sea.

(a)

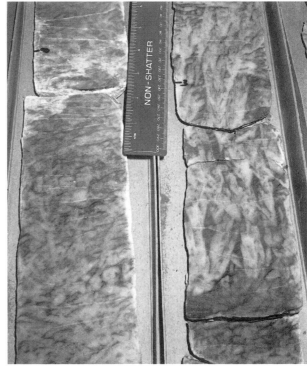

(b)

clay, detrital gypsum silt and peloids are common within bottom-nucleated gypsum beds, and thin stromatolite horizons also may be present. This type of gypsum is forming in Marion Lake, South Australia and is a feature of the Miocene Messinian evaporites of the Mediterranean area too (Fig. 5.6 and Testa & Lugli, 2000). Palmate gypsum (now anhydrite) also occurs in the Lower Cretaceous Ferry Lake Anhydrite of the Texas subsurface (Loucks & Longman, 1982), and it forms much of the gypsum platforms that developed around the Zechstein Basin in the Upper Permian. Palmate gypsum may occur as beds within nodular (sabkha) gypsum–anhydrite, from precipitation within ponds on a supratidal flat. This has been described from the Jurassic Buckner Formation of the Alabama subsurface (Lowenstein, 1987). Bottom-growth gypsum also may occur within laminated gypsum–anhydrite (next section), as a result of shallowing of the brine-filled basin. Examples of this are found in the Permian Castile Formation of the Delaware Basin, Texas (Kendall & Harwood, 1989).

5.2.3 Laminated gypsum–anhydrite

Laminated gypsum–anhydrite usually consists of thin sulphate laminae alternating with laminae of a different composition, calcite, organic-rich calcite or organic matter (Fig. 5.7). Thin, gypsum–anhydrite laminae in halite is one type of layered halite (Section 5.3). Anhydrite–calcite and anhydrite–organic matter couplets are typically less than a few millimetres in thickness, but they may form units hundreds of metres thick. The sulphate crystals are usually very fine (10–100 μm) and they probably precipitated in the surface layers of the evaporating water-body. These laminated anhydrites occur in the lower sections of thick basinal evaporite sequences, as in the Middle Devonian of the Elk Point Basin, or they constitute most of the basin fill, as in the Permian Castile Formation of the Delaware Basin. Although the anhydrite laminae are chiefly planar (Fig. 5.7), they may be contorted and buckled. The latter in fact superficially resemble sabkha-type anhydrite. However, one very important feature of laminated anhydrite is the lateral persistence of the laminae; frequently individual laminae can be correlated over vast distances, of tens to hundreds of kilometres. This feature, indicating uniform conditions over a wide area, necessitates direct precipitation of the sulphate from water in a relatively deep basin, at

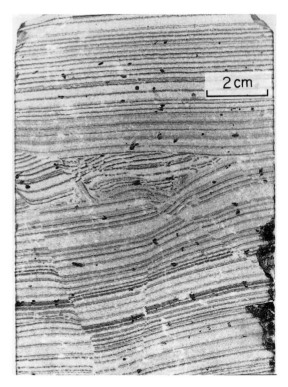

Fig. 5.7 Laminated anhydrite. Laminae of microcrystalline anhydrite alternate with organic-rich laminae. Synsedimentary faults and discontinuities (lower part) indicate downslope mass movement of sediment. Some hydration of anhydrite to gypsum has occurred (scattered dark porphyrotopes). Castile Formation, Permian. Texas, USA.

least below wave base. Seasonal (annual) changes in water chemistry and temperature have been invoked to account for the lamination, and the term 'varved' has been applied to the couplets. The organic matter is probably sapropelic in origin, derived from seasonal phytoplankton blooms within the photic zone.

5.2.4 Resedimented gypsum–anhydrite

Once gypsum–anhydrite has been precipitated, either subaqueously or subaerially, it may be transported and resedimented, or reworked by wind, wave and current processes, to produce clastic deposits. Interbedded with basinal laminated gypsum–anhydrite, for example, there occur locally thicker graded units of clastic sulphate with sole structures, which are interpreted as turbidites. Horizons of small- and large-scale folded, contorted and brecciated gypsum–anhydrite also occur (Fig. 5.7) and these are the result of slumps, slides and debris flows. Examples of this resedimented sulphate occur in the Devonian of the Elk Point Basin, the Permian Zechstein of Poland (Peryt *et al.*, 1993), the Messinian (upper Miocene) of the northern Apennines and Sicily, and the Miocene of the Red Sea area (Rouchy *et al.*, 1995). Downslope movement of evaporites into a basin is clearly indicated and this further demonstrates the subaqueous, relatively deep-water origin of the interbedded anhydrite laminites.

Reworking of intertidal–supratidal gypsum produces cross-laminated and rippled sulphates, intraformational conglomerates of gypsum crystal fragments and gypsum stromatolites where detrital grains are trapped by microbial mats. Such shallow-water and tidal-flat clastic gypsum beds (gypsarenites) are common in the Messinian of Sicily and Cyprus.

5.2.5 Secondary and fibrous gypsum

Uplift of anhydrite deposits, perhaps a long time after their formation and burial, results in the generation of secondary gypsum, as the anhydrite comes into contact with fresh near-surface ground water. Secondary gypsum consists of two varieties, porphyrotopes and alabastrine gypsum. *Gypsum porphyrotopes* are large crystals, typically several millimetres across or larger, which occur scattered through the anhydrite (see Plate 15b). *Alabastrine gypsum* consists of small to large, commonly poorly defined interlocking crystals, many with irregular extinction (Fig. 5.8). In spite of gypsifi-

cation, the original nodular or laminated texture of anhydrite normally is preserved.

Veins of fibrous gypsum (*satin spar*) are commonly parallel or subparallel to the bedding, having a displacive (intrusive) relationship. Usually they are a few millimetres to centimetres in thickness and consist of vertically arranged fibres (Fig. 5.9). It is thought that the fibrous gypsum grew under pressure in brine-filled veins induced by hydraulic fracture or unloading and exhumation. The gypsum probably has been derived from dissolution of near-surface gypsum or rehydration of more deeply buried anhydrite. See El-Tabakh *et al.* (1998) for a recent description.

5.2.6 Burial anhydrite

Anhydrite can be precipitated in the deep-burial environment as a cement (e.g. Dworkin & Land, 1994)

Fig. 5.8 Alabastrine gypsum, formed by replacement of anhydrite. Crossed polars. Gachsaran Formation, Miocene. Iraq.

Fig. 5.9 Veins of fibrous gypsum (satin spar) in gypsiferous mudrock. Crossed polars. Permian. Cumbria, England.

and as replacement crystals and nodules (e.g. Machel, 1993) in sandstones, limestones and dolomites. In these cases the anhydrite is not strictly an evaporite mineral, although the sulphate is usually derived from dissolution of evaporites within the basin. On uplift, the anhydrite would be dissolved out to leave vugs or replaced, probably by calcite or quartz.

5.3 Halite

Halite is the major component of large evaporite-basin fills, and it is the main evaporite mineral of modern salt lakes and saline pans.

There is much variation in the textural and bedding features of halite, depending largely on the environment of deposition, whether it formed subaqueously in a near-permanent water body, or in a saline pan subject to repeated flooding–desiccation cycles. Halite deposited in relatively deep water (below wave-base) is typically well bedded and laminated. This may simply be a bedding on the scale of 5–10 cm as a result of slight colour changes through variations in the clay content. Such bedded halite dominates the Triassic salt deposits of western Europe. A more particular type, laminated halite, consists of beds up to several centimetres thick alternating with laminae of anhydrite a few millimetres thick. This facies occurs in some cycles of the Permian Zechstein of northwest Europe, in the Devonian Prairie Formation of western Canada, and in the Permian Salado Formation of the Delaware Basin, Texas. Bedded halite also occurs in the Permian San Andreas Formation of Texas (Hovorka, 1987).

Halite precipitation in modern saline pans and saline lakes can be studied relatively easily, and Lowenstein & Hardie (1985) and Schubel & Lowenstein (1997) have recognized three stages to the saline-pan cycle: flooding, evaporative concentration and desiccation. Flooding planes off the irregular surface of the halite crust of the previous cycle, dissolves halite to form vugs and pipes, and deposits a layer of mud over the pan surface. Evaporation of the shallow water leads to the formation of thin halite rafts on the water surface and bottom nucleation of halite crystals on the settled-out rafts. These crystals grow most rapidly from their coigns, so developing a chevron texture (see Fig. 5.10). A thin layer of gypsum may precipitate out first, if the water contains SO_4^{2-}. Complete desiccation of the pan causes the halite crust to break up into polygons and tepees. Further halite is precipitated from

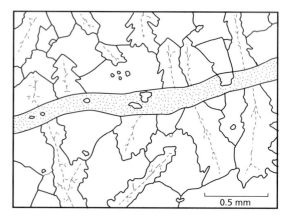

Fig. 5.10 Sketch of chevron and clear halite with an anhydrite–clay layer. Plane-polarized light.

evaporating ground water as a clear cement in vugs in the salt crust and as displacive cubes in associated muds. The dissolution surfaces produced by flooding and the chevron/clear crystals are characteristic features of saline-pan halite, as is a close association with clastics, particularly of distal, alluvial-fan–playa origin. Saline-pan halite occurs in the Permian Salado and Rutler Formations of New Mexico, above the laminated anhydrite of the Castile Formation (Casas & Lowenstein, 1989) and in the Permian of the Williston Basin, North Dakota (Benison & Goldstein, 2000).

Within many bedded-halite formations, there are horizons of megapolygons (up to 15 m across), best observed in salt-mine roofs (Fig. 5.11a), from which vertical wedges descend filled with mudrock and halite (Fig. 5.11b). These structures result from desiccation of the salt basin, and Tucker (1981) ascribed them to contraction–expansion of the superficial halite beds as a result of temperature changes, which in an arid, desert area are considerable. These megapolygons are common in the Triassic of western Europe, but they also occur in the Eocene Green River evaporites of Wyoming and in the Miocene (Messinian) of Sicily (Lugli *et al.,* 1999). In shallow water (above wave-base), it is possible for halite to be reworked into ripples and cross-lamination, and it may even form ooids.

In thin-section, halite is isotropic (as it is cubic); it usually shows strong cleavage and fracture planes, and possesses fluid inclusions (Fig. 5.12). Laminated halite and saline-pan halite may both consist of a more

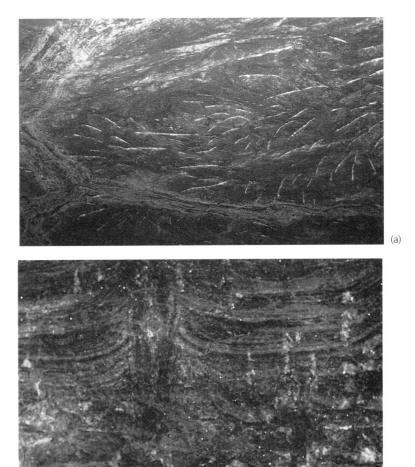

(a)

(b)

Fig. 5.11 Bedded halite. (a) Megapolygons, which reach 15 m across, on the roof of a salt mine. (b) Bedded halite in the wall of a salt mine consisting of clay-poor and clay-rich halite. The bedding, disrupted into saucer shapes, is truncated by wedges that descend from the megapolygons seen on the mine roof. Vertically laminated halite–mudrock fills the wedges. These structures formed as a result of complete and prolonged exposure of the salt basin. Triassic (Keuper). Cheshire Basin, England.

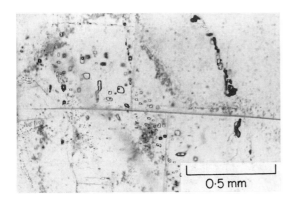

Fig. 5.12 Halite with fluid-filled inclusions. Plane-polarized light. Cheshire Halite, Triassic. Northwest England.

opaque variety with abundant liquid inclusions (chevron halite from the arrangement of the inclusions, Fig. 5.10) and a clear transparent variety. The chevron halite is the original salt precipitate, whereas the clear halite is a later precipitate filling dissolution hollows and cavities. Recrystallization of halite can take place relatively easily during diagenesis and through deformation. Halite–anhydrite alternations have been related to seasonal changes in the temperature and/or composition of the basin waters in a similar way to laminated anhydrite. The term *Jahresringe*, implying an annual rhythmicity, has been applied to halite–anhydrite couplets in the Permian Zechstein. An annual interpretation requires high rates of salt precipitation, in the Zechstein some 500 m in 8000 years. It also requires deposition in deep basins, deeper than the thickness of the salt, because the subsidence rates for a shallow-water origin are improbably great. With a deep basin, periodic influxes of brine or seawater into the basin are necessary to give the great thickness of salt (the evaporation of 1000 m of seawater only gives 12.9 m of halite, see Table 5.3).

The NaCl for most halite deposits is derived from the evaporation of seawater, which of course contains a vast reserve of NaCl. NaCl also can be concentrated from fresh continental waters, as in many salt lakes, or derived from the dissolution and recycling of older evaporites. The bromine content of halite provides useful information on the evaporation–replenishment pattern in an evaporite basin, and indicates whether the waters were marine or continental. Bromine does not form its own minerals during the crystallization of salt from seawater; instead it substitutes for chlorine in halite, and the other chloride minerals. The Br content of seawater is 65 p.p.m. and this increases as seawater evaporates to about 500 p.p.m., when halite begins to precipitate. When the first potash minerals precipitate, the Br value of the brine is around 2300 p.p.m. The Br content of the first halite precipitated theoretically should be about 75 p.p.m. and this rises to 270 p.p.m. before K^+ salt precipitation. Non-marine halite generally has a low Br value, less than 50 p.p.m., but diagenetic recrystallization of marine halite in low-Br pore waters will lower an originally high Br content (as in the Castile halite of Texas with 20–40 p.p.m.). The brine in fluid inclusions in halite also can be analysed to determine the origin of the NaCl, and Sr isotopes are particularly useful in this regard. Such studies in the Castile halite of Texas have confirmed a marine origin (Kirkland *et al.*, 2000).

Bromine profiles from ancient halite formations show a variety of trends, which can be interpreted in terms of the balance between evaporation and replenishment. An upward trend of increasing Br in the halite (Fig. 5.13a) indicates increasing salinity through evaporation within the basin water, little influx of normal seawater and no reflux of brine out of the basin. An upward trend of decreasing Br (Fig. 5.13b) suggests significant influx of normal seawater, but it is more common to find that influxes were episodic rather than continuous, and these are represented by kick-backs in the Br profile (Fig. 5.13c). A fairly uniform Br content up through a halite formation (Fig. 5.13d) indicates a near-constant salinity in the basin and to achieve this,

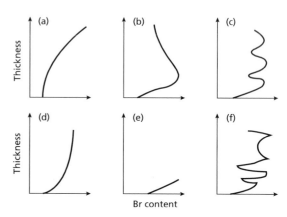

Fig. 5.13 Very schematic graphs illustrating the possible trends of bromine content in halite up through a formation as a result of variation in rates of influx, reflux and evaporation. See text for explanation.

not only a regular input of normal seawater is required, but also a reflux of bittern brines out of the basin. If an evaporating basin were to dry up completely, this would be represented by extremely high Br values in the halite (Fig. 5.13e), and there may be K^+ and Mg^{2+} salts at such an horizon. Occasionally a very irregular profile is recorded (Fig. 5.13f), and this could indicate wild fluctuations in brine salinity, or more likely the effects of diagenetic dissolution and reprecipitation of halite in low-Br pore waters.

One major feature of thick halite deposits is that commonly they are involved in subsurface mass flow to generate salt domes or diapirs. These structures generally are thought to be the result of buoyancy effects, but there are other factors involved. Below a depth of around 900–1200 m, halite has a lower density than other sediments and the overburden pressure is sufficient to cause the salt to flow. Salt domes are important in generating hydrocarbon traps, and they also can improve the reservoir qualities of overlying rocks by fracturing (papers in Alsop *et al.*, 1995). This is the case in the Gulf of Mexico and North Sea (e.g. chalk reservoirs of the Central Graben, see Glennie, 1998).

5.4 Other evaporite minerals and their occurrence

5.4.1 Potassium and magnesium salts

Potassium and magnesium salts occur in some marine

evaporite deposits and include the sulphates kieserite and kainite, and the chlorides sylvite and carnallite. These highly soluble salts are the last to form in the evaporation of seawater, hence they tend to occur in the uppermost parts of evaporite formations and rarely attain a great thickness. In view of their highly soluble nature, diagenetic mineral changes through contact with residual brines, fresh seawater and meteoric ground water are inevitable. It is probable that the mineral assemblages of these final precipitates are not original and some minerals may be entirely secondary in origin. Sylvite, for example, probably has formed through incongruent dissolution of carnallite, and some polyhalite probably has formed through alteration of kainite by solutions rich in calcium ions or replacement of anhydrite by halite-precipitating brines containing K^+ and Mg^{2+} (Peryt *et al.*, 1998). In the potash beds of the Salado Formation, Delaware Basin, the primary kainite, kieserite and carnallite are apparently replaced by langbeinite, sylvite and loewite through contact with fresh seawater (Harville & Fritz, 1986). Lowenstein & Spencer (1990), on the other hand, provided evidence for primary sylvite precipitation in potash deposits of the Oligocene Rhine Graben.

Polyhalite has been reported from supratidal flats in Baja California formed by reaction of gypsum with *bitterns*, solutions enriched in K^+ and Mg^{2+}. This modern occurrence and the association of ancient potash salts with intertidal and shallow-water sediments (as in the Permian Zechstein of Yorkshire and New Mexico) suggest that most of these deposits formed in salinas and sabkhas.

5.4.2 Lacustrine evaporites

Apart from halite and gypsum–anhydrite, there are a number of evaporite minerals that in large concentration are found only in saline lakes (Table 5.1).

The evaporite mineralogy of salt lakes can vary considerably from one region to the next, because the geochemistry of the lake waters, which determines the evaporites precipitated, is itself very variable (Table 5.4), being dependent largely on the local geology. There commonly is a zonation of evaporite minerals within a salt lake, with the least soluble occurring around the edge and the most soluble being precipitated in the lake centre, giving a bull's-eye pattern (see Fig. 5.18).

Evaporites deposited subaqueously in lakes will have similar sedimentary features to marine subaqueous evaporites: fine lamination in lake centres, slumped and turbidite beds near slopes, and reworked evaporites and evaporitic stromatolites in shallow water (see earlier sections). Salt lakes commonly are surrounded by inland sabkhas and saline mudflats. Saline pans are common in deserts, and have been described earlier too. In terms of evaporite mineralogy, *salt lakes*, such as the Dead Sea and Great Salt Lake (Utah), with dominant halite, are distinguished from *bitter lakes* such as Mono Lake (California) and Carson Lake (Nevada), where sodium carbonate and sulphate minerals dominate. Ancient lacustrine evaporites are well developed in the Eocene Green River Formation of Wyoming and Utah.

Table 5.4 The water chemistry of five salt lakes (in p.p.m.). The main point to note is the great variability in the relative concentrations of the constituents between lakes

	Dead Sea	Great Salt Lake, Utah	Mono Lake, California	Borax Lake, California	Gulf of Kara Bogaz
Cl^-	208 020	112 900	15 100	5 945	142 500
SO_4^{2-}	540	13 590	7 530	22	46 900
HCO_3^-	240	180	26 430	6 668	—
Na^+	34 940	67 500	21 400	6 199	81 200
K^+	7 560	3 380	1 120	322	
Ca^{2+}	15 800	330	11	nil	4 900
Mg^{2+}	41 960	5 620	32	31	19 900
Total salinity	315 040	203 490	71 900	>19 400	>293 000

5.5 Evaporite dissolution and replacement

As evaporites are composed of moderately to highly soluble minerals, it is not uncommon to find that at outcrop and in the shallow subsurface, evaporite beds have been dissolved away or replaced by other minerals. Evaporite dissolution leads to the collapse and brecciation of overlying strata (e.g. Swennen *et al.*, 1990). Where evaporite dissolution has taken place in the subsurface, then it may complicate the local stratigraphy and make correlation difficult. Kendall (1988) discussed such problems with respect to the Paradox Basin of Colorado and Utah. Accompanying sulphate dissolution, associated dolomites are commonly replaced by calcite (dedolomitization: Section 4.8.3).

The recognition of replaced evaporites is based on several lines of evidence. The replacement crystals may possess relics of the original evaporite mineral. This particularly is the case where silica has replaced gypsum–anhydrite. The crystal shape of the original evaporite mineral, or the nodular shape in the case of anhydrite, usually is retained on replacement to produce a *pseudomorph*. Pseudomorphs of halite are easily recognized by their cubic, usually hopper shapes, and pseudomorphs of gypsum have a characteristic lozenge (Fig. 5.14) and swallow-tail shape. The fabrics of the replacement mineral also may be a guide. Two forms of silica, length-slow chalcedonic quartz and lutecite, are commonly associated with former evaporite beds.

In the replacement of gypsum–anhydrite by calcite,

isotopic evidence suggests that one of the main processes is sulphate reduction by bacteria, particularly where organic matter or hydrocarbons are available (Pierre & Rouchy, 1988). Replaced evaporites are common in many intertidal–supratidal carbonate formations and have been described by Buick & Dunlop (1990) from the early Archaean of Western Australia, where barite is a common replacement mineral, and the Permian of Wyoming (Ulmer-Scholle & Scholle, 1994). See review in Warren (1999).

5.6 Evaporite sequences and discussion

Following the recognition of sabkhas and supratidal flats as important environments of evaporite (particularly sulphate) precipitation, many ancient examples have been described. The typical facies are nodular (chicken-wire) and enterolithic anhydrite, although as noted in Section 5.2.1, these textures can develop in subaqueous anhydrites by replacement. The key features for the identification of sabkha evaporites are the shallow-water and intertidal sedimentary structures contained within associated carbonates (Section 4.10.3). As a result of net deposition upon the sabkha surface, the sabkha gradually progrades seawards over the intertidal sediments. A sabkha cycle of supratidal evaporites overlying intertidal and subtidal carbonates is produced (Fig. 5.15), which may be repeated many times in an evaporite formation. Ancient sabkha formations have been described by Elliot & Warren (1989), and others, from the Permian of the Delaware Basin, Texas and New Mexico. See also Warren & Kendall (1985) and papers in Handford *et al.* (1982).

Sabkha successions are developed along the shorelines of carbonate ramps and rimmed shelves at times of arid climate, and they also may develop upon epeiric carbonate platforms. The very extensive Hith Anhydrite of the Upper Jurassic is an example of the latter, covering many hundreds of thousands of square kilometres of the Arabian craton.

Cycles of lagoonal evaporite passing up into intertidal–supratidal carbonate (Fig. 5.16) are not common in the geological record, but they do occur in the Ferry Lake Anhydrite of the Texas subsurface (Loucks & Longman, 1982). The evaporite beds are palmate and nodular anhydrite, replacement of twinned selenitic gypsum, but such deposits can easily be mistaken for original sabkha nodular anhydrite.

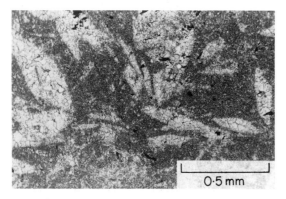

0·5 mm

Fig. 5.14 Calcite pseudomorphs after gypsum in a micritic limestone; typical of high-intertidal facies. Plane-polarized light. Carboniferous. Eastern Scotland.

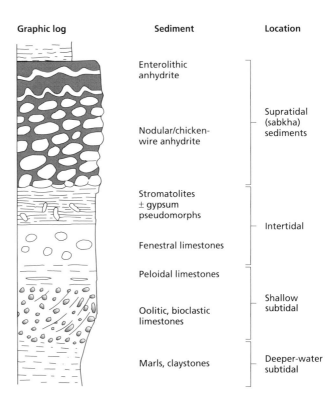

Graphic log   Sediment   Location

Enterolithic
anhydrite

Nodular/chicken-
wire anhydrite

Supratidal
(sabkha)
sediments

Stromatolites
± gypsum
pseudomorphs

Fenestral limestones

Intertidal

Peloidal limestones

Oolitic, bioclastic
limestones

Shallow
subtidal

Marls, claystones

Deeper-water
subtidal

Fig. 5.15 A sabkha cycle. Such cycles typically range from several to several tens of metres in thickness.

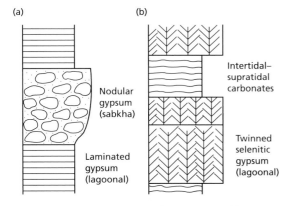

(a)   (b)

Nodular
gypsum
(sabkha)

Laminated
gypsum
(lagoonal)

Intertidal–
supratidal
carbonates

Twinned
selenitic
gypsum
(lagoonal)

Fig. 5.16 Two less common types of evaporite cycle. (a) Laminated gypsum passing up into nodular gypsum. (b) Lagoonal, selenitic gypsum passing up into intertidal–supratidal carbonates.

Metre-scale cycles (Fig. 5.16) of laminated sulphate passing up into nodular sulphate (both now secondary gypsum) are common in Triassic Keuper and Muschelkalk evaporites in western Europe and represent shallow lagoons developing into sabkhas. Coastal lagoonal gypsum occurs in the Upper Silurian Salina Formation of the Michigan Basin (Haynes *et al.*, 1989).

For the precipitation of evaporites within deep, water-filled basins, the most important factor, apart from an extremely arid climate and periodic replenishment of sea water, is the barrier that gives near-complete isolation from the main mass of sea water. The barrier may be structural, such as a fault-bounded ridge, or sedimentary, such as a carbonate reef or sand bar. In many cases, carbonate sediments and reefs deposited before evaporite precipitation contributed towards an isolation of the basins, and defined sub-basins. To account for the deficiencies of halite and/or potassium salts in some thick evaporite formations, reflux of dense, bottom-flowing brines out of the basin has been postulated.

Although many evaporite formations and textures have been re-examined and reinterpreted in the light of recent studies on sabkhas and salinas, there are several features that can be explained only in terms of the classic, relatively deep-water, barred-basin model. These are: laminated evaporites with individual laminae

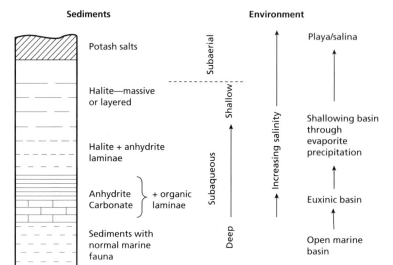

Sediments   Environment

Potash salts — Subaerial — Playa/salina

Halite—massive or layered — Shallow — Shallowing basin through evaporite precipitation

Halite + anhydrite laminae

Anhydrite / Carbonate } + organic laminae — Subaqueous — Euxinic basin

Sediments with normal marine fauna — Deep — Open marine basin

Increasing salinity

Fig. 5.17 Evaporite sequence formed in an initially deep marine barred-basin, with periodic replenishment.

correlatable over distances of many kilometres; graded beds of anhydrite clasts, interpreted as turbidites, and horizons of contorted and brecciated evaporites, interpreted as slumps. The vertical rock succession developed through evaporation of a relatively deep, water-filled basin should begin with sapropelic deposits, representing an initial stagnation of the water as the barrier became effective, overlain by laminated anhydrite as evaporation proceeded (Fig. 5.17). When evaporation is well advanced, halite would be deposited, perhaps with anhydrite laminae, until the basin was nearly filled. Final evaporation of brines and bitterns in salinas and sabkhas might then produce K^+ and Mg^{2+} salts.

The distribution of evaporites in a large barred-basin with near-continuous replenishment can be of the tear-drop type, with carbonates located near the barrier, gypsum precipitated in the central part, and halite–potash salts in the most distal, hypersaline peripheral part. This arrangement contrasts with a basin totally cut off from the open ocean, where a bull's-eye pattern develops, with the most soluble evaporites in the basin centre (Fig. 5.18).

Saline giants occurring in large intracratonic basins commonly consist of several evaporite cycles, broadly of the type shown in Fig. 5.17. One important additional control that can lead to the repetition of the cycles is eustatic sea-level change (see Fig. 5.19). When global sea-level is high relative to the intracratonic basin, open-marine carbonates are deposited there in

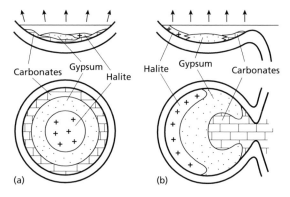

Carbonates   Gypsum   Halite   Halite   Gypsum   Carbonates

(a)   (b)

Fig. 5.18 Broad patterns of evaporite facies. (a) Bull's-eye pattern, with the most soluble salts in the basin centre, typical of completely enclosed basins. (b) Tear-drop pattern, typical of restricted basins with near-permanent connection to the open ocean, where the most soluble salts occur farthest away from the basin entrance.

marginal platforms and on the basin floor. With a moderate fall so that there is still some connection with the open ocean, evaporites will be precipitated in the restricted basin, mainly around the basin margins. Shallow, selenitic gypsum platforms with sabkhas behind are likely to form at this stage and euxinic carbonates will be deposited in the basin centre. With a more drastic sea-level fall, so that the basin is cut off, laminated anhydrite and halite will be precipitated in the basin centre. Periodic replenishment of the water

(a) High sea-level: normal marine, carbonate deposition

**(b) Intermediate sea-level:
restricted marine, gypsum deposition**

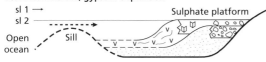

(c) Low sea-level: salt lake–saline pan, halite precipitation

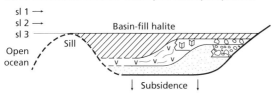

Fig. 5.19 Model for evaporite deposition in an intracratonic basin, where eustatic sea-level changes are a major control.

body will allow a thick succession to develop, but if this does not occur, saline pans and sabkhas will be present on the desiccated basin floor. A subsequent major sea-level rise will re-establish the open-marine conditions in the basin and start a new cycle. The application of sequence stratigraphy to evaporite successions has been considered by Tucker (1991) and applied to the Zechstein, and reviewed in Warren (1999). Major sea-level changes and drawdown also can account for the dolomitization of marginal carbonate platforms around evaporite basins (see Section 4.8).

The interpretation of some evaporite deposits in deep basins has been controversial. Late Miocene (Messinian) evaporites discovered beneath the floor of

the Mediterranean have been interpreted as deep-water and as subaerial. Convincing evidence now shows that the Mediterranean was subjected to substantial drawdown, causing salts to be precipitated in sabkhas, saline pans and salt lakes on the floor of this deep, desiccated basin. This event is referred to as the *Messinian salinity crisis*; see Hsü *et al.* (1978) for an early review and Clauzon *et al.* (1996) and Keogh & Butler (1999) for more recent work. Another case of controversy centres on some of the Devonian evaporites in the Elk Point Basin of western Canada. Laminated anhydrite, which passes up into thick anhydrite and halite, has been interpreted as shallow-water/intertidal and then reinterpreted as the deposits of a deep, stratified water body. The halite in this basin was deposited in a salt-pan environment (Kendall, 1989). The discovery of bottom-growth gypsum in the Castile Formation of the Delaware Basin (Kendall & Harwood, 1989) suggests that this may be more of a shallow-water deposit than previously thought.

Further reading

Kendall, A.C. (1992) Evaporites. In: *Facies Models: Responses to Sea-Level Change* (Ed. by R.G. Walker & N.P. James). Geoscience Canada, pp. 375–409.

Kendall, A.C. & Harwood, G.M. (1996) Marine evaporites: arid shorelines and basins. In: *Sedimentary Environments: Processes, Facies and Stratigraphy* (Ed. H.G. Reading). Blackwell Science, Oxford, pp. 281–324.

Schreiber, B.C. (Ed.) (1988) *Evaporites and Hydrocarbons*. Columbia University Press, New York, 475 pp.

Warren, J.K. (1989) *Evaporite Sedimentology: Importance in Hydrocarbon Accumulation*. Prentice Hall, Englewood Cliffs, NJ, 285 pp.

Warren, J.K. (1999) *Evaporites: their Evolution and Economics*. Blackwell Science, Oxford, 438 pp.

6   Sedimentary iron deposits

6.1 Introduction

Iron is present in practically all sedimentary rocks to the extent of a few per cent, but less commonly, it forms ironstones and iron-formations, where the iron content exceeds 15%. Important reserves of iron ore are contained within these sedimentary iron deposits. The element iron occurs in two valence states: a divalent form, *ferrous* iron (Fe^{2+}) and a trivalent form, *ferric* (Fe^{3+}). As a result of this, the behaviour of iron and the precipitation of its minerals are strongly controlled by the chemistry of the surface and diagenetic environments. The common iron minerals in sedimentary rocks are given in Table 6.1.

The majority of sedimentary iron deposits were formed under marine conditions and many of the Phanerozoic examples contain normal marine fossils. There are important differences between those that formed in the early middle Precambrian and those of the Phanerozoic. The former, referred to as *iron-formations* or *banded iron-formations* (BIFs), are typically thick units of various iron minerals interbedded with chert, deposited in large intracratonic basins (Section 6.5.1). The *Phanerozoic ironstones* are usually thin units, commonly oolitic in character, which were deposited in localized areas (Section 6.5.2).

One complicating factor for the interpretation of many iron-rich sedimentary rocks is that there are no good modern analogues for comparison. The only iron deposits forming to any extent at the present time are the bog-iron ores of mid- to high-latitude lakes and swamps (Section 6.6). In addition, ferromanganese nodules and crusts, and metalliferous sediments are forming on the sea floor (Section 6.7). Neither of these modern developments are very significant geologically, and they are unrelated to most ironstones and iron-formations of the geological record.

6.2 Source and transportation of iron

Traditionally, two sources of iron for ironstones have been discussed: continental weathering and contemporaneous volcanicity. Weathering and alteration of volcanic ash and lava, and associated hydrothermal activity, can supply considerable amounts of iron. Volcanic and hydrothermal sources have been proposed, notably for the Archaean iron-formations, some of which are closely associated with volcanics. Continental weathering is a major source of iron and has been proposed as the source for many Phanerozoic iron deposits, and others not associated with volcanics. Intense weathering under a humid tropical climate releases the iron from mafic and heavy minerals in igneous and other rocks and produces iron-charged ground waters and iron-rich, lateritic soils. Through erosion of these soils, the iron is carried to the sea by rivers, but the manner in which the iron travels has not been resolved. Reworking of soil through a relative sea-level rise will also make the iron available.

Iron in true solution in river water and ground water is in very low concentrations (less than 1 p.p.m.) and in seawater the concentration is around 0.003 p.p.m. The low values arise from the fact that in the pH and Eh ranges of most natural surface waters (Fig. 6.1), iron is present as the highly insoluble ferric hydroxide (also the main constituent of laterite). Three mechanisms of iron transportation have been suggested. Ferric hydroxide readily forms a colloidal suspension, which is stabilized in the presence of organic matter. Iron could be transported by rivers in this form and then precipitated in the sea through flocculation of the colloidal suspensions. Iron can be transported by adsorption and chelation onto organic matter. And in a similar manner, iron can be carried by clay minerals, either as part of the clay structure or, of greater importance, as oxide films on the surface of clays. Once deposited, the iron can be released from the clays and organic matter into the pore water if the Eh–pH conditions are appropriate, and then reprecipitated to form iron minerals.

Ironstone formation is favoured where there are low rates of sedimentation, both of siliciclastic material

Table 6.1 The iron minerals of sedimentary rocks

Oxides	hematite α-Fe_2O_3
	magnetite Fe_3O_4
	goethite α-$FeO.OH$
	limonite $FeO.OH.nH_2O$
Carbonate	siderite $FeCO_3$
Silicates	berthierine $(Fe_4^{2+}Al_2)\,(Si_2Al_2)\,O_{10}\,(OH)_8$
	chamosite $(Fe_5^{2+}Al)\,(Si_3Al)\,O_{10}(OH)_8$
	greenalite $Fe_6^{2+}Si_4\,O_{10}\,(OH)_8$
	glauconite $KMg(FeAl)\,(SiO_3)_6.3H_2O$
Sulphides	pyrite FeS_2
	marcasite FeS_2

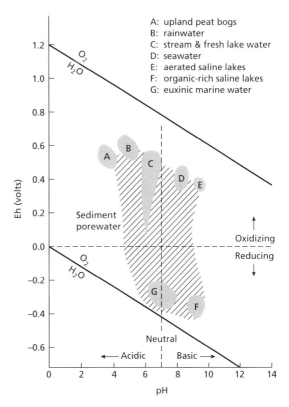

Fig. 6.1 Eh–pH diagram showing the fields of some naturally occurring waters. Redox potentials of natural solutions are limited by reactions involving water, which are dependent on pH. The upper limit of Eh is determined by the oxidation of water to oxygen (the upper diagonal line) and the lower limit of Eh is the reduction of water to hydrogen (lower diagonal). pH is the hydrogen ion concentration (pH = $-\log_{10}\alpha_{H^+}$). Eh (also E) is the potential difference of the solution relative to the hydrogen half-cell reaction, which is arbitrarily given a potential of 0.0 V at 0 pH. The redox potential also can be expressed by pe, the activity of electrons in the solution (pe = $-\log_{10}\alpha_{e^-}$).

and carbonates. Continental areas supplying the iron through deep tropical weathering are thus assumed to be low-lying with negligible relief. Many Phanerozoic ironstones were subjected to considerable reworking on the sea floor, and are part of condensed sequences and/or are associated with transgressive events.

In spite of the foregoing, it is considered by many people that the quantity of iron in early Proterozoic and Archaean iron-formations is still too high to be explained by continental weathering as occurs today, and/or volcanic–hydrothermal activities, so that an atmosphere with little oxygen and more carbon dioxide has been proposed. This would allow iron to be leached and transported more efficiently, particularly as Fe^{2+} in solution (Section 6.5.1).

6.3 The formation of the principal iron minerals

The change from one oxidation state of iron to the other is dependent on changes in the Eh and pH of the environment. Eh (also E or pe, Fig. 6.1) is a measure of the oxidizing or reducing nature of the solution, basically whether an element such as iron will gain or lose electrons; pH is a measure of the acidity or alkalinity, that is, the hydrogen ion concentration. Fe^{3+} is stable under more oxidizing and more alkaline conditions whereas Fe^{2+} is stable under more reducing and more acidic conditions. In fact, in the pH–Eh range of natural environments, Fe^{3+} is present as the highly insoluble $Fe(OH)_3$, whereas Fe^{2+} is present in solution. Apart from Eh and pH, two other factors controlling the precipitation of iron minerals are:
1 the activity (i.e. effective concentration) of carbon-

ate ions, which can be measured by the partial pressure of carbon dioxide, P_{CO_2};
2 the activity of sulphur, frequently represented by pS^{2-}, the negative logarithm of the activity of the sulphide ion.

One of the main factors affecting the Eh of natural aqueous environments is the amount of organic matter present, as its decomposition, mainly brought about by bacteria, consumes oxygen and creates reducing conditions. Normal seawater and other surface waters have a positive Eh, as do the pore waters in surficial sediments on the sea floor (Fig. 6.1). Organic matter deposited in the sediments soon decomposes through aerobic bacterial reduction, however, so that a reducing environment is formed some tens of centimetres below the sediment–water interface. Thus, an oxic, near-surface diagenetic environment passes down into an anoxic diagenetic environment. The anoxic environment can be divided into two types: sulphidic and non-sulphidic, depending on the presence or absence of S^{2-} (Table 6.2; Berner, 1981). The former is produced by the bacterial reduction of sulphate in the presence of organic matter. Anoxic non-sulphidic environments can be divided into post-oxic and methanic environments. In the post-oxic environment, there is sufficient organic matter for the consumption of all dissolved oxygen by aerobic bacteria, but not enough to bring about sulphate reduction. Further decomposition of organic matter takes place by nitrate, manganese and iron reduction (Fe^{3+} to Fe^{2+}). The lack of O_2 and H_2S may lead to high levels of Fe^{2+} and Mn^{2+} in the pore fluids. The weakly reducing, post-oxic environment is typical of sediments with low to moderate amounts of organic matter. In the non-sulphidic methanic environment, all O_2, NO_3^- and SO_4^{2-} is removed by bacterial reduction, and continued decomposition of organic matter leads to the formation of methane. This highly reducing environment forms in organic-rich sediments (see Section 3.7.2 and Chapter 8).

The stability fields of the common iron minerals are plotted on Eh–pH, Eh–pS^{2-} and Eh–Pco$_2$ diagrams in Figs 6.2 & 6.3. These diagrams are constructed from

thermodynamic data and so have their limitations when applied to natural systems. Reference should be made to the original sources for the reasoning behind the construction of these diagrams (see geochemical texts for further information).

From Figs 6.2 & 6.3, it can be seen that hematite is the stable mineral under moderately to strongly oxidizing conditions. It generally is believed that hematite forms diagenetically from a hydrated ferric oxide precursor approximating to goethite, by an ageing process involving dehydration. The formation and preservation of hematite in a sediment require a low original organic content.

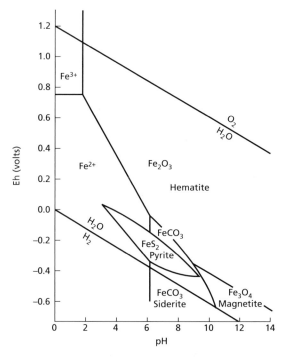

Fig. 6.2 Eh–pH diagram showing the stability fields of ferrous and ferric iron, hematite, siderite, pyrite and magnetite. The diagram shows that hematite is the stable mineral in all environments that are moderately to strongly oxidizing. For the minerals pyrite, siderite and magnetite, stable in reducing environments, the mineral stability fields are strongly dependent on the concentrations of carbonate and sulphide in the solution (see Fig. 6.3a,b), as well as pH. Figure 6.2 shows the stability fields for the condition of high carbonate and low sulphide. If sulphide is greatly in excess of carbonate, then the pyrite field expands to occupy nearly all the lower half of the diagram. When both sulphide and carbonate are in low concentrations, then the magnetite field expands into near-neutral environments.

Table 6.2 Geochemical classification of early diagenetic environments and the characteristic phases present. After Berner (1981)

Environment	Characteristic phases
Oxic	hematite, goethite, MnO_2-type minerals, no organic matter
Anoxic	
1 sulphidic	pyrite, marcasite, rhodochrosite, organic matter
2 non-sulphidic	
(a) post-oxic	glauconite, berthierine, no sulphide minerals, minor organic matter (also siderite, vivianite, rhodochrosite)
(b) methanic	siderite, vivianite, rhodochrosite, earlier formed sulphide minerals, organic matter

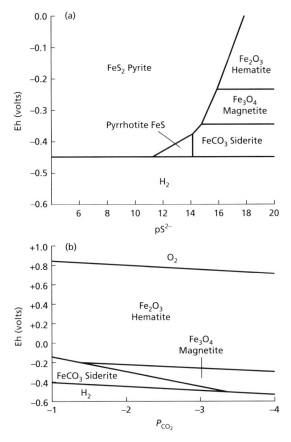

Fig. 6.3 The stability fields of iron minerals plotted on (a) an Eh–pS^{2-} diagram for a pH of 7.4 (pS^{2-} is the negative log of the activity of the sulphide ion), and (b) an Eh–log P_{CO_2} diagram (P_{CO_2} is the partial pressure of CO_2). In the latter case, the value of pS^{2-} is assumed to be so high (i.e. the activity of sulphur is very low), that pyrite and pyrrhotite do not plot on this diagram.

For the ferrous minerals, pyrite, siderite and magnetite, stable under conditions of negative Eh, the stability fields are strongly dependent on the P_{CO_2} and pS^{2-} of the solution (Fig. 6.3). As most surface environments are oxidizing these minerals usually are precipitated within sediments during early diagenesis, where reducing conditions have developed through bacterial decomposition of organic matter. The iron for these minerals is Fe^{2+} in the pore waters, mostly liberated by bacterial reduction of iron oxides/hydroxides on clays and organic matter in the sediment. The sulphide for the precipitation of pyrite comes mainly from the bac-

terial reduction of dissolved sulphate in pore waters; this produces H_2S, which reacts with the Fe^{2+} in solution. Seawater contains much dissolved sulphate and so pyrite is the typical authigenic mineral of organic-rich marine muds, forming in the anoxic sulphidic diagenetic environment. Dissolved sulphate generally is present in low concentrations in fresh water so that pyrite is less common in non-marine sediments. Pyrite also can form in association with the bacterial reduction of sulphate in gypsum. In all these cases the initial precipitates are black, finely crystalline, metastable iron sulphides: *mackinawite* and *greigite*. In a relatively short time these are transformed to pyrite. If there is excessive bacterial production of H_2S through high rates of organic-matter sedimentation and minimal circulation, H_2S will occur in waters overlying the bottom sediments. The metastable iron sulphides can then be precipitated directly.

Siderite is precipitated where the carbonate activity is high and the sulphide activity is low. The latter is rarely attained in marine sediment pore waters because of the abundant dissolved sulphate. However, if all the SO_4^{2-} is reduced, as in the anoxic non-sulphidic methanic diagenetic environment, then siderite will precipitate. Siderite is thus more common in non-marine sediments, although it does occur in many marine ironstones, mostly as a later diagenetic cement. If there is insufficient Fe^{2+} relative to Ca^{2+} and Mg^{2+} in the pore waters, then *ankerite*, $CaMg_{0.5}Fe_{0.5}(CO_3)_2$ or ferroan calcite or ferroan dolomite may form in preference to siderite.

The formation of magnetite is favoured by low activities of both sulphide and carbonate, together with negative Eh and neutral pH (Fig. 6.3). Conditions such as these are rare in nature and so magnetite precipitation is not common.

The formation of the iron silicate minerals berthierine, chamosite, greenalite and glauconite is poorly understood. Being ferrous minerals, it is to be expected from thermodynamic considerations that conditions of negative Eh are necessary for greenalite and berthierine to form, but as they are absent and poorly represented, respectively, in modern sediments, their interpretation to some extent relies on geological data (Section 6.4.4). Nevertheless, it is thought likely that they precipitate in the anoxic, non-sulphidic, postoxic diagenetic environment, where conditions may fluctuate from oxic to anoxic relatively easily, and there is no S^{2-} available.

6.4 Occurrence and petrography of the iron minerals

6.4.1 Iron oxides

Hematite is present in both Precambrian iron-formations and Phanerozoic ironstones, and of the latter it is more common in Palaeozoic occurrences. In the Precambrian cases, the hematite is chiefly present as thin beds and laminae, alternating with chert, but it also occurs as massive, peloidal and oolitic forms. In the Phanerozoic ironstones the hematite is present mainly as ooids and impregnations and replacements of fossils (see Plate 15c,d). Later diagenetic migrations and replacements of calcareous host sediment and calcite cement by hematite are not uncommon. Although the hematite itself may be a primary mineral, probably precipitated via an amorphous hydrated ferric oxide, there is often petrological evidence that the oxide has formed by replacement of berthierine. In some cases this is a synsedimentary replacement, arising from the reworking of berthierine grains into a more oxidizing environment where hematite (or its precursor) is stable.

Hematite in thin-section is opaque and typically cryptocrystalline. It can be recognized by its red colour in reflected light (quickly checked by shining a light down onto the rock slice).

Goethite is absent from Precambrian iron deposits, but it is a major constituent of Phanerozoic ones, in particular those of the Mesozoic. In some cases, the goethite appears to be a recent weathering product, having formed by oxidation and hydration of other iron minerals. However, there are ironstones where the goethite appears to be primary, or at least synsedimentary. The goethite commonly forms ooids and typically they are quite spherical. Some ooids consist of alternations of goethite and berthierine. The goethite could have formed through sea-floor oxidation of the berthierine. Ferriferous spherules largely of goethite are forming at the present time in Lake Chad, West Africa, and Fe^{3+}-rich oolitic grains occur on the Amazon inner shelf (Aller *et al.,* 1986). Sub-Recent coated grains (oncoids) of goethite occur on the Cameroon shelf (Giresse *et al.,* 1998). Goethite 'ooids' and pisoids are forming in lateritic soils of tropical regions. Goethite in section is a yellow to brown colour and generally appears isotropic.

Limonite is a poorly defined hydrated form of iron oxide, containing goethite, other materials such as clay and adsorbed water. The term is best restricted to the yellow–brown amorphous product of subaerial weathering of iron oxides and other minerals (i.e. 'rust').

Magnetite is abundant in Precambrian iron-formations, where it is interlaminated with chert. It is a minor component of Phanerozoic ironstones, but in some instances it is important. It generally occurs as small replacement crystals or granules within oolitic ironstones. It is distinguished from hematite by its steel-grey colour in reflected light. It is also magnetic of course!

6.4.2 Iron carbonates

Siderite is a major constituent of both Precambrian and Phanerozoic iron-rich sediments. It is the cement to many Phanerozoic berthierine–chamosite oolites and it can replace ooids and skeletal grains. Siderite is common in non-marine, organic-rich mudrocks, either as small disseminated crystals or as nodules and rounded masses. At the present time, siderite is forming in muds of many organic-rich intertidal marsh, delta-plain swamp and lacustrine environments, including the Mississippi Delta Plain (S. E. Moore *et al.,* 1992) and the Norfolk coast of eastern England.

Siderite crystals as seen in thin-section are of three types:
1 coarse crystals up to several millimetres across, similar to other carbonates such as calcite in terms of high birefringence and rhombohedral cleavages;
2 a very fine-grained variety of equant-rhombic crystals a few micrometres in diameter;
3 a fibrous variety that forms spherulites.

The first type occurs predominantly in the matrix of oolitic and bioclastic ironstones (as in Plate 16a) and usually can be recognized from a yellowish brown oxidation zone ('limonite') along crystal boundaries and cleavage planes. The fine-grained variety occurs interbedded with cherts in the Precambrian iron-formations and constitutes most siderite nodules. Aggregates of fibrous spherulites form the rock known as *sphaerosiderite* (Fig. 6.4), which occurs in palaeosoils. Siderite can vary considerably in chemical composition, with Ca^{2+}, Mg^{2+} and Mn^{2+} substituting for Fe^{2+} by as much as 10%. Ankerite, $CaMg_{0.5}Fe_{0.5}(CO_3)_2$ and ferroan dolomite may be

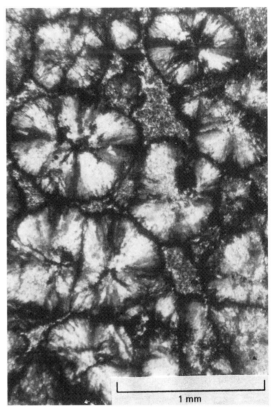

1 mm

(a)   (b)

Fig. 6.4 Sphaerosiderite: spherulites of fibrous siderite. (a) Plane-polarized light. (b) Crossed polars. Carboniferous. Yorkshire, England.

associated, but can be difficult to distinguish from siderite in thin-section.

6.4.3 Iron sulphides

The iron sulphides, in particular pyrite, are constituents of many iron-rich (and other) sediments, but they rarely form the major part. Pyrite and its metastable precursors are forming within organic-rich estuarine and tidal-flat sediments, and they are being precipitated on the floor of the Black Sea (see review of Raiswell & Canfield, 1998).

Pyrite is distinguished from other opaque iron minerals by its yellowish colour in reflected light (check by shining a light down onto the thin-section). Pyrite is present as disseminated grains and crystals (cubic); it

may replace skeletal fragments. Aggregates of spherical microconcretions of pyrite are known as framboids. *Marcasite* is a dimorph of pyrite that is rarely found in ironstones but forms nodules in chalks and coal-measure sediments.

6.4.4 Iron silicates

The important iron silicate minerals are berthierine–chamosite, greenalite and glauconite. Berthierine is an iron-rich 1 : 1-type layer silicate of the serpentine group (7 Å lattice spacing). Chamosite is a 2 : 1 trioctahedral chlorite (lattice spacing 14 Å) with Fe^{2+} as the dominant divalent octahedral cation. Berthierine is the sedimentary, early diagenetic mineral, which at a temperature of 120–160 °C, or depth in excess of 3 km, is transformed into chamosite. Thus most Jurassic ironstones contain berthierine, whereas many older, Palaeozoic ones, contain chamosite.

Berthierine–chamosite typically occurs as ooids in Phanerozoic ironstones, within a cement of siderite or calcite (see Plates 15e & 16). It also forms flakes and is finely disseminated in berthieroidal mudrocks. In many cases there are no nuclei to the ooids or they may have formed around berthierine flakes or fragments of broken ooid. One of the features of berthierine ooids that distinguishes them from the more familiar aragonite–calcite ooids is the evidence that they were soft at the time of compaction. The ooids are commonly flattened and distorted, and curious elephantine, trunk-to-tail textures can form (as in Plate 15e). In some instances, flattened ooids have formed nuclei to ooids, showing that the ooid deformation took place close to the sea floor. The term spastolith has been applied to these deformed ooids. Berthierine and chamosite are both green in colour and have a low birefringence.

There has been much discussion over the conditions of precipitation of berthierine, and of the formation of its ooids. Berthierine is a ferrous silicate and theoretical considerations of its stability field indicate that it should form under reducing conditions. In the marine environment, however, with its high SO_4^{2-} content, one would expect high S^{2-} activity in sediment with negative Eh, and berthierine is unstable, even at low sulphide activities. Some berthierine ironstones contain a rich marine benthic fauna, which requires normal oxygenated conditions on the sea floor. It has been suggested that berthierine is precipitated as a mixed gel of $Fe(OH)_3$, $Al(OH)_3$ and $SiO_2.nH_2O$, which is stable at positive Eh. Conversion of this gel to berthierine would take place after burial into the reducing zone beneath the sediment–water interface. However, it is now widely believed that berthierine is precipitated directly within the sediment in the anoxic, non-sulphidic, post-oxic diagenetic regime (see Section 6.3). Under the slow sedimentation rates and extensive reworking that typify berthierine ironstones, repeated oxic–suboxic pore-fluid changes can be expected, and these may allow the low S^{2-} activities and negative Eh conditions to develop, which are necessary for berthierine precipitation. Berthierine ooids are probably direct, intrasediment precipitates, but other suggestions have been mechanical accretion, replacement of calcareous ooids and skeletal grains, reworking of soil ooids and *in situ* microconcretions. Berthierine could form by the replacement of goethite. Current reworking of the berthierine on the sea floor could lead to its oxidation to hematite (or its precursor). In the few known marine occurrences, off the Orinoco and

Niger deltas, and on the Sarawak shelf, the berthierine is a poorly ordered ferrous-iron clay forming diagenetically within faecal pellets.

The hydrated ferrous silicate *greenalite* is interbedded with chert and constitutes beds and lenses in Precambrian sedimentary iron deposits. Greenalite occurs as rounded to subangular pellets, with little internal structure. It is a green isotropic mineral related to berthierine and chlorite. Greenalite is considered a primary mineral in iron-formations but it is not certain if it is an original precipitate. It may well have formed from an iron-silicate gel or by early diagenetic replacement of detrital iron-rich particles or berthierine.

Glauconite is a potassium–iron aluminosilicate with a high Fe^{3+}/Fe^{2+} ratio. From X-ray diffraction, several types of glauconite can be recognized, depending on the structure and degree of ordering. Well-ordered glauconite is a 10 Å illite-type clay mineral, but it commonly forms a mixed-layer clay with smectite. It typically occurs as light to dark green pellets and aggregates up to 1 mm in diameter (see Plate 16b). In thin-section, glauconite is a light green colour, usually pleochroic. Pellets are microcrystalline and show an aggregate polarization pattern.

Glauconite is present in many sandstones and may be the major constituent, forming *greensands*, well known in the Cretaceous of Britain and eastern USA. Glauconite is being formed on many modern continental shelves at depths from a few tens to hundreds of metres, but invariably it is a poorly ordered form. It tends to occur in areas with low sedimentation rates, in some cases in association with marine flooding (transgressive) surfaces and condensed sections (see Amorosi, 1995). Much modern glauconite is present within tests of foraminifera and in faecal pellets. Glauconite forms by the transformation of degraded clay minerals and by the authigenic growth of crystallites in the pores of substrates, be they clay minerals, skeletal grains or faecal pellets. Glauconite commonly is associated with organic matter, creating local reducing conditions, but the overall environment is oxic. Glauconite probably is forming in the slightly reducing, anoxic, non-sulphidic, post-oxic diagenetic environment. It appears that initially glauconitic smectites are precipitated in pores and fractures of host grains, and that in time, recrystallization and dissolution-replacement of the substrate leads to the well-ordered Fe-rich, K-rich glauconite. See papers in Odin (1988) and Kelly & Webb (1999) for an example from the Tertiary of Australia.

6.5 Precambrian iron-formations and Phanerozoic ironstones

The separation of iron-rich sedimentary rocks into the two groups Precambrian iron-formations and Phanerozoic ironstones is somewhat arbitrary and traditional, but there are differences in mineralogy, geochemistry, sedimentology and stratigraphy.

6.5.1 Precambrian iron-formations

In spite of the tremendous economic importance of these rocks and the vast amount of data available they still arouse much discussion and controversy. Iron-formations are present within the cratonic shields of all continents.

From studies of iron-formations in Canada, two groups are recognized:

1 *Algoma-type*—lenticular deposits, relatively thin and narrow across strike, usually closely associated with volcanic rocks and greywackes;

2 *Superior-type*—thicker and much more regionally extensive, deposited on stable shelves and in broad basins.

With a few problematic exceptions, Algoma types are mostly Archaean in age (2500–3000 Ma) and occur in greenstone belts, whereas Superior types are restricted to the early middle Proterozoic (1900–2500 Ma).

On the basis of the dominant early diagenetic iron mineral present, four sedimentary iron facies are recognized within iron-formations: oxide (hematite, magnetite), silicate (greenalite), carbonate (siderite) and sulphide (pyrite) facies. The primary minerals are likely to have been amorphous $Fe(OH)_3$, $FeO(OH)$, berthierine, siderite and pyrite, with the oxygen content of waters just above and within the sediments determining the minerals precipitated, as discussed in Section 6.3. Water circulation in the basin of deposition, the local energy level and degree of restriction and organic productivity are major factors here. Diagenesis ('ageing') of the oxide facies precipitates would have led to the formation of hematite, although if there were some organic matter present, then the ferric hydroxide would have been partly reduced to magnetite instead. In Archaean iron-formations of the Canadian Shield, oxide–silicate facies pass into carbonate and then sulphide facies with increasing depth and distance from the shore.

Facies in iron-formations also can be defined on sedimentary characteristics and the best known is the laminated 'banded' facies (Fig. 6.5), consisting of the fine interbedding of chert with hematite or magnetite, less commonly with siderite or greenalite. Lamination on the mesoscale (10–50 mm) and microscale (0.2–2 mm) can be distinguished, and in the Brockman Iron Formation of the Hamersley Basin, Australia, individual mesobands can be traced over $30\,000\,km^2$. Laminated iron facies were deposited in deep-water basins, on shelves below wave-base, and in lagoons, the lamination probably reflecting seasonal changes in the environment (Garrels, 1987). The absence of burrowing organisms at this time contributed to the preservation of the fine lamination. Sediments of shallow, agitated waters consist of ooids, peloids, granules and intraclasts, giving rise to iron facies exactly comparable with those of limestones (Section 4.10). Such non-laminated facies are well developed in the Biwabik and Gunflint Iron-Formations of the Lake Superior region and in the Sokoman Iron-Formation of the Labrador Trough (Fralick & Barrett, 1995). Ooids generally consist of hematite and chert, and granules are of greenalite. Sedimentary structures include cross-bedding, channels, ripples, desiccation cracks and stromatolites. These more shallow-water, grainy iron facies are generally cemented by various types of quartz (Simonson, 1987).

Points of discussion with the Precambrian iron-formations centre on the source of the iron, the mechanism of iron precipitation, the origin of the chert and the depositional environment. A volcanic source for the iron has been suggested, particularly for the Archaean Algoma types, which are clearly associated with contemporaneous volcanic rocks. The iron may derive from hydrothermal vents on the sea floor, with sulphide–carbonate facies close to the vent, and oxide facies more distal (Gross, 1983). The REE and Nd isotope data suggest a hydrothermal source for a banded iron-formation (BIF) from South Africa (Klein & Beukes, 1989). Some Archaean iron-formations are interbedded with turbidite sandstones deposited in submarine-fan–deltaic-ramp environments, and it appears that iron precipitation was a background chemical 'rainout' at times of minimal clastic influx (Barrett & Fralick, 1989).

(a)

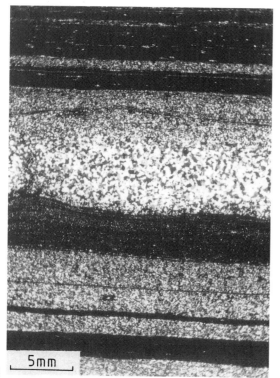

(b)

Fig. 6.5 Banded iron formation of the jaspilite variety. (a) Polished hand-specimen showing fine lamination, very persistent in the light-coloured hematitic chert alternating with darker, more structureless but locally lenticular hematite. Extensional fractures prominent in the hematite layers. Archaean Temagami Iron Formation, Ontario, Canada. (b) Graded quartz siltstone laminae, the thicker one with a scoured base, alternating with hematite laminae. Crossed polars. Early Proterozoic Transvaal Iron-Formation. Southern Africa.

For many of the Proterozoic iron-formations, there appears to have been no contemporaneous volcanicity, so that deep weathering of continental rocks probably provided the iron. To facilitate leaching of iron and its transportation, it often has been suggested that the Earth's atmosphere had a higher carbon dioxide content at that time, and little or no oxygen. A higher partial pressure of CO_2 would have the effect of lowering the pH of surface waters, leading to a greater efficiency in iron leaching and transportation. Sea water with a lower pH could itself have been a major reservoir of Fe^{2+} in solution. The large, early Proterozoic basins may well have had stratified water bodies, with an anoxic lower part containing iron in solution, and more oxic near-surface waters. Overturning and mixing of the water masses, which could happen on a seasonal scale (as in some modern lakes), could have led to precipitation of the iron. Deep-ocean, anoxic, iron-rich waters upwelling onto passive margins may have resulted in the precipitation of iron on oxic, shallow-water shelves. A biochemical mechanism of iron deposition suggested by Cloud (1983) involves precipitation of the Fe^{2+} as ferric oxides and hydroxides through reaction with oxygen derived from primitive photosynthesizing organisms. Direct chemical precipitation of silica and seasonal blooms of silica-secreting organisms have been suggested for the chert layers. The occurrence of microfossils in cherts associated with peloidal, shallow-water iron-formations (such as the Gunflint) and the presence of stromatolites indicate at least some biological activity. However, there is no evidence for any microorganisms or stromatolites in the Archaean iron-formations or in the laminated ('banded') Superior types (such as the Brockman). This could be the result of deposition in deep water, at least below the photic zone (base of photic zone now is around 120 m in the clearest ocean water).

For more information on BIFs see papers in Lepp (1975), Trendall & Morris (1983), and the reviews of H. L. James (1992) and Trendall (2000).

6.5.2 Phanerozoic ironstones

Phanerozoic ironstones vary considerably in grade, lithology and iron minerals present. The most important are the *oolitic ironstones*, which generally are hematite–chamosite in the Palaeozoic and goethite–berthierine in the Mesozoic. Of less significance are the siderite mudstones and sulphide ironstones. For a collection of papers on Phanerozoic ironstones see Young & Taylor (1989).

The oolitic ironstones show two major peaks of occurrence during the Phanerozoic, in the Jurassic and in the Ordovician. These were both times of global sea-level highstand and large areas of peneplained continental terrain. Climate generally was humid too, facilitating chemical weathering. Sea-level changes are important in controlling the amount of iron that can be supplied to the marine environment by reworking terrestrial weathering products. The Lower Ordovician Wabana Ironstone of Newfoundland is dominantly oolitic and of shallow-water aspect. Ooids consist of alternations of hematite and chamosite, with the former apparently replacing the latter. Ironstones are common in the Lower Ordovician of western Europe, and here ooids appear to have developed on marine shelves during phases of sediment starvation at times of relative sea-level rise. For some examples in Brittany, the iron is thought to be derived from leaching of continental rocks (Fernandez et al., 1998), whereas for an iron oolite on the Baltic platform, which can be traced for 1200 km, the iron is probably derived from volcanic ash, based on REE patterns (Sturesson et al., 1999).

A famous and well-documented hematitic–chamositic oolite is the Clinton Ironstone, which developed along the eastern shelf of the Central Appalachian Basin in the Middle Silurian. The ironstone consists mainly of hematite oolite, but chamosite (originally berthierine) oolite also occurs. Skeletal grains in the sediment are impregnated and replaced by hematite. Facies analysis has shown that hematite oolites were deposited in shallower water and closer to the shoreline than the chamosite facies. A glauconite facies is recognized farther offshore. These ironstones developed during periods with low rates of siliciclastic influx, and they occur at horizons of major lithological and faunal breaks in the succession, interpreted as representing periods of shallowing and reworking.

Hematite ironstones occur within limestone successions in the Devonian and Carboniferous of Britain, continental Europe and the USA. They mostly take the form of hematite-impregnated and replaced skeletal debris in a carbonate cement (see Plate 15c,d). In some cases, such as the Rhiwbina iron ore of South Wales, the ironstones are concentrates produced by the current reworking of grains impregnated with berthierine

(now chamosite), which was then converted to hematite (via goethite) on the oxic sea-floor.

The Jurassic berthierine–goethite oolites generally form thin units, of the order of 1–10 m, which are condensed in thickness relative to other contemporaneous sediments. The ironstones usually occur only in localized areas, passing laterally into typical marine sediments. Oolitic ironstones may show cross-bedding and other structures indicating shallow, agitated waters, and commonly they are interbedded with berthieroidal and sideritic mudrocks. Fossils found within berthierine oolites usually indicate normal marine conditions. Examples are the Northampton and Frodingham Ironstones of the British Jurassic (e.g. Taylor & Curtis, 1995). In these cases the ironstones probably formed upon intrabasinal highs and within local embayments away from the diluting effects of terrigenous mud and sand. A shoal within a pro-delta environment is a likely location for berthierine ooid growth, with iron-bearing waters being carried beyond the main area of terrigenous sedimentation (a 'clastic trap' mechanism). Some of the minette oolitic ironstones of Luxembourg and Lorraine were deposited in a high-energy, shallow, subtidal, sand-wave complex (Teyssen, 1984). Other oolitic ironstones contain a restricted fauna. The Westbury Ironstone (Jurassic, southern England), for example, contains an abundance of oysters. In these cases it is likely that the berthierine facies formed in coastal lagoons, at times hyposaline, separated from normal marine areas by sand bars and barriers. Cretaceous ooidal ironstones in the Western Canada foreland basin developed on a localized intrabasinal arch at a time of sediment starvation and marine transgression (Donaldson et al., 1999).

Sideritic mudrocks and nodular siderites. Mudrocks with finely disseminated siderite commonly have been deposited in deltaic, lagoonal, estuarine and other environments where iron-bearing muds were deposited and the necessary chemical conditions existed in the sediment. The sediments are usually grey or bluish grey but they take on a brown colour with weathering through oxidation of the siderite. Sideritic mudrocks are common in Jurassic ironstone formations, associated with berthierine oolites.

Mudrocks containing nodules or more continuous beds of siderite constitute the clay ironstones, blackband ironstones and sphaerosiderites. Nodular siderites are particularly common in coal-measure successions, such as the Carboniferous of Britain, the Appalachians and Russia. The clay ironstones and blackband ironstones formed mainly within fresh- and brackish-water swamp and lagoonal sediments (Fisher *et al.,* 1998), as they are forming at the present time (Moore *et al.,* 1992). Carbon isotope data indicate precipitation during microbial methanogenesis. Blackband ironstones have a higher organic content than clay ironstones. Sphaerosiderites, with their radiating–spherulitic crystal fabric (Fig. 6.4) chiefly formed in the lower horizons of fossil soils (seatearths and underclays), with the iron derived from leaching in the higher part of the soil profile.

Sulphide ironstones. Pyrite rarely makes up a large part of Phanerozoic sedimentary ironstones but it is disseminated within many black carbonaceous shales and some limestones, as framboids and crystals (e.g. Taylor & Macquaker, 2000). In some ironstones pyrite occurs as aggregates and replaces other iron minerals and skeletal fragments to form thin beds of high sulphide content. Such horizons occur within Ordovician chamosite and hematite oolites of Wabana, Newfoundland and in the Sulphur Bed of the Cleveland Ironstone and the 'Snap Band' of the Frodingham Ironstone (Jurassic), England. These beds have formed through the imposition of a sulphidic diagenetic environment after the formation of the berthierine. Berthierine is unstable in the presence of S^{2-} and is easily replaced by pyrite.

6.6 Bog iron ores

The only modern environment where significant iron ores are forming at present is in swamps and lakes of mid- to high latitudes such as northern America, Europe and Asia. The ores range from hard oolitic, pisolitic and concretionary forms to earthy and soft types. Iron minerals present frequently are difficult to identify because of their amorphous or poorly crystalline form. However, goethite is most common, siderite less so, and vivianite ($Fe_3P_2O_8.H_2O$) an unstable accessory. Manganese oxide content may reach 40% (in most ironstones Mn is less than 1%). The formation of bog ores occurs where acidic ground water seeps into oxygenated lakes, marshes and swamps. The rise in Eh and pH causes ferrous iron in solution to be precipitated, mainly as ferric hydroxides. This seepage of ground water has been suggested as a mechanism for formation of some ancient ironstones.

6.7 Ferromanganese nodules and crusts, and metalliferous sediments

Ferromanganese nodules, crusts and metalliferous sediments on the sea floor have received much attention in recent years, chiefly in view of their possible economic significance.

Nodules and encrustations of iron and manganese oxides are widely distributed on the sea floor, but particularly in parts of the Atlantic, Pacific and Indian oceans. They are especially well developed in areas of negligible sedimentation, where there are strong bottom currents, and usually at depths of several thousand metres. In oceanic settings, they commonly form on the flanks of active mid-ocean and aseismic ridges, on seamounts and on abyssal plains. They also form in sediment-starved continental-margin areas, such as on the Blake Plateau, off the eastern coast of the USA. Somewhat similar manganese nodules are found in some temperate lakes and shallow-marine areas.

The marine manganese nodules and crusts vary considerably in chemistry and mineralogy from one part of the sea floor to another. Both Mn-rich–Fe-poor and Mn-poor–Fe-rich varieties occur. One of the main points of interest is the relatively high concentrations of metals including Co, Ni, Cu, Cr and V (Table 6.3), such that the submarine 'mining' of these deposits is now considered a viability. Although the minerals constituting the nodules are frequently X-ray amorphous, todorokite $(Mn^{2+}R^+R^{2+})(Mn^{4+}Mn^{2+})O_6.3H_2O$ (where R represents other metals) is one manganese oxide that commonly occurs, together with the hydrated iron oxide, goethite. Ferromanganese nodules and pavements, being hard substrates, are commonly encrusted with organisms such as foraminifers, bryozoans, serpulids and ahermatypic corals. Phosphate nodules and phosphatization of sediment may occur in association with marine manganese deposits.

In the geological record, ferromanganese nodules are not common, but they do occur in Jurassic and Devonian condensed pelagic limestones of the Alpine and Hercynian fold belts in Europe, and also in Cretaceous red clays of Timor.

The origin of these ferromanganese deposits has given rise to much debate. A hydrothermal–volcanic origin has been favoured by some workers and this could well be the case where there is a close association with volcanism, as in active mid-ocean ridge occur-

Table 6.3 Average concentration of Fe, Mn, Cu, Co and Ni (in percentages) of shallow- and deep-water sediments and ferromanganese nodules from three sea-floor settings. From Glasby (1977)

	Nearshore sediments	Deep-sea sediments, Atlantic	Seamount nodules	Abyssal nodules	Active ridge nodules
Fe	4.83	5.74	15.81	17.27	19.15
Mn	0.0850	0.3980	14.62	16.78	15.51
Cu	0.0048	0.0115	0.058	0.37	0.08
Co	0.0013	0.0039	1.15	0.256	0.40
Ni	0.0055	0.0079	0.351	0.54	0.31
Depth			1900 m	4500 m	2900 m

rences. However, in many instances ferromanganese nodules are in areas remote from any sort of volcanic activity, so that some form of direct or indirect precipitation of the metal oxides from sea water is required. Although the rate of accretion of the nodules varies considerably, and in some cases tens of years is sufficient, the majority appear to form at incredibly slow rates of around 1 mm 10^{-6} years.

Apart from nodules and crusts, metalliferous sediments also occur in oceanic settings, especially in the vicinity of active spreading ridges, where they develop on top of ocean-floor basalts. These basal sediments are typically enriched in Fe, Mn, Cu, Pb, Zn, Ni, Co, Cr and V. Metal-rich muds have been recovered from the Mid-Atlantic Ridge and East Pacific Rise during deep-sea drilling and from the spreading ridge of the Red Sea, where sediments enriched in Zn, Cu and Pb are associated with hot, metal-rich brines (papers in Purser & Bosence, 1998). The fluids effecting these metal enrichments are thought to derive either from mantle magmatic sources or from the interaction of basalt with seawater.

Sediments enriched in metals are found in association with many ancient pillow lavas and ophiolite suites. The Cretaceous umbers (Fe- and Mn-rich) and ochres (Fe-rich–Mn-poor) of the Troodos Massif, Cyprus can be compared with the basal sediments upon present-day, active mid-ocean ridges.

Ferromanganese deposits are considered in detail in Glasby (1977), Cronan (1980) and Nicholson (1996).

Further reading

Cronan, D.S. (1980) *Underwater Minerals*. Academic Press, London, 362 pp.

Economic Geology **68**(7) (1973) for papers on iron-formations.

Glasby, G.P. (Ed.) (1977) *Marine Manganese Deposits*. Elsevier, Amsterdam, 523 pp.

Lepp, H. (Ed.) (1975) *Geochemistry of Iron*. Benchmark Papers in Geology. Dowden, Hutchinson & Ross, Stroudsburg, 464 pp.

Trendall, A.F. & Morris, R.C. (Eds) (1983) *Iron-formations, Facts and Problems*. Elsevier, Amsterdam.

Young, T.P. & Taylor, W.E.G. (Eds) (1989) *Phanerozoic Ironstones*. Special Publication 46, Geological Society of London, Bath, 251 pp.

7   Sedimentary phosphate deposits

7.1 Introduction

Sedimentary phosphate deposits or *phosphorites* are important natural resources. Phosphates are one of the chief constituents of fertilizer and they are used widely in the chemical industry. In addition, phosphorites commonly contain relatively high concentrations of useful elements such as uranium, fluorine and vanadium, and commonly are associated with organic-rich mudstones, which are potential hydrocarbon source rocks.

Phosphorus is one of the essential elements for life and is present in all living matter. Although only forming a minor part of plants and animal soft parts, phosphate is the major constituent of all vertebrate skeletons (bones) and some invertebrate hard parts. In the marine environment, phosphate is a primary nutrient so that it is a control of organic productivity. In seawater, it is present mainly as dissolved 'orthophosphate' and particulate phosphate, with the phosphate in the latter case chiefly contained within or adsorbed onto organic detritus. In the oceans, most organic productivity utilizes dissolved 'orthophosphate' and takes place in the upper levels through phytoplankton growth. The concentration of phosphate in coastal waters, such as within estuaries, is usually higher than in surface waters offshore. High phosphate concentrations also occur in the waters of anoxic basins.

Many sedimentary rocks contain a few per cent calcium phosphate in the form of grains of apatite (a heavy mineral, Section 2.5.5), bone fragments or coprolites. Rocks composed largely of phosphate on the other hand are relatively rare. Sedimentary phosphate deposits are discussed here in three categories:

1 nodular and bedded phosphorites, where upwelling and organic productivity have played a major role in their formation (Section 7.3);

2 bioclastic and pebble-bed phosphorites, where sediment reworking has been of paramount importance (Section 7.4);

3 oceanic-island phosphorites, mostly related to guano (Section 7.5).

7.2 Mineralogy

The most common sedimentary phosphate minerals are varieties of apatite. The apatite of igneous rocks is chiefly fluorapatite, $Ca_5(PO_4)_3F$, but in sedimentary apatite replacement of the phosphate by carbonate may reach several per cent; sulphate may also replace the phosphate. Fluorine may be replaced by hydroxyl or chlorine ions. In addition, the calcium ions may be replaced by sodium, magnesium, strontium, uranium and rare earths. Most sedimentary phosphates are carbonate hydroxyl fluorapatites, which can be represented by the formula $Ca_{10}(PO_4,CO_3)_6F_{2-3}$. Two specific sedimentary apatite minerals, best identified by X-ray diffraction and chemical analysis, are *francolite* with more than 1% fluorine and appreciable carbonate, and *dahllite*, a carbonate hydroxyapatite with less than 1% F. These two minerals are anisotropic. The term *collophane* is often loosely applied to sedimentary apatite of cryptocrystalline form for which the precise composition is variable. It is isotropic.

7.3 Nodular and bedded phosphorites

7.3.1 Recent–sub-Recent occurrences

Marine phosphate deposits on the sea-floor have been known since the *Challenger* oceanographic expedition of the 1870s dredged up phosphate nodules and slabs from the continental shelf and slope off southern Africa. Since then sea-floor phosphorites have been recorded off many other continents (Fig. 7.1), in particular the west coast of North and South America, the east coast of the USA, and the shelves off northwest Africa and Japan. The marine phosphorite generally occurs in areas of slow sedimentation, on outer continental shelves and slopes, particularly on the tops

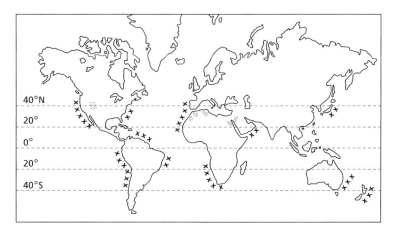

Fig. 7.1 Distribution of marine phosphorites, although in many cases the deposits are not actually forming at the present time. Also shown are the locations of the Permian Phosphoria Formation of northwestern USA and the Cretaceous–Tertiary phosphorites of the Middle East–North Africa.

and sides of local ridges and banks, on fault scarps and the flanks of submarine canyons. Phosphate nodules and crusts generally occur at depths from 60 to 300 m.

The nodules usually are several centimetres in diameter but may reach a metre or more across. They range in shape from flat slabs to spherical nodules and irregular masses. The internal structure of the nodules varies from homogeneous to concentrically laminated and conglomeratic, and many contain pellets and coated grains ('ooids'). Vertebrate skeletal debris, especially of fish, and coprolites are commonly associated. In some situations pebbles derived from local sea-floor exposures of Tertiary limestone are phosphatized, as well as the *in situ* bedrock surface itself. Drowned reefs formed upon volcanic seamounts in the oceans commonly have crusts of phosphorite upon them, and skeletal grains have been replaced by carbonate apatite. Both francolite and dahllite have been identified in these phosphorites, and glauconite (Section 6.4.4) is frequently found too.

It has been established from uranium isotope data that many marine phosphorites, such as those off California and Florida, were formed in pre-Holocene time, in some cases as far back as the Miocene. There are three areas where Recent accumulation of phosphorites has been demonstrated: the Walvis Bay region off southwestern Africa, along the continental margin off Peru and Chile and off eastern Australia. On the Namibian shelf, the phosphorite occurs within diatom oozes as dispersed phosphate, biogenic fragments, coprolites, pellets and phosphatic nodules. The nodules vary from soft and gelatinous to hard and massive forms. The phosphate content increases with increas-

ing induration. Diatoms, originally composed of opaline silica, have been replaced by cryptocrystalline phosphate within nodules. Off Peru and Chile the phosphorite is developing on the upper continental slope as replacements of benthic foraminifers. Obliteration of the foraminiferal structure by the replacement leads to the production of phosphorite pellets.

7.3.2 Origin of marine phosphorites

There is still much discussion over the origin of marine phosphorites. A major factor of course is the availability of phosphate, but beyond that there is still much uncertainty. The question is basically how to concentrate the phosphate in seawater into the sediments. One popular mechanism has been upwelling, whereby cold waters containing nutrients rise from the depths towards the surface (see Fig. 7.2). Upwelling currents lead to high organic productivities and phytoplankton growth in surface waters, which in turn results in organic-rich (and so phosphate-enriched) sediments and oxygen-deficient waters overlying the sea floor. Mass mortalities of fish occasionally take place in areas of upwelling, particularly through poisoning by phytoplankton blooms. More organic matter, with its combined phosphorus, and skeletal phosphate (bones) are thus contributed to the sea floor during these events. Upwelling is a feature of mid-latitude continental margins and it is controlled by the predominant high-pressure atmospheric systems. There are five major zones of coastal upwelling, mainly located on the western side of the landmasses. More local upwelling does occur in regions of irregular sea-floor topography in

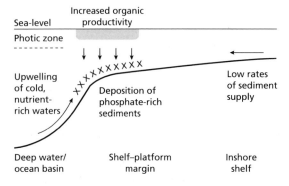

Fig. 7.2 Model for formation of marine phosphorites.

response to regional current patterns. The sea-floor where phosphorites are being deposited generally is at depths of a few hundred metres, and this commonly is within the *oxygen minimum zone*. Oxygen-depleted waters permit the deposition of organic matter, and its contained phosphate is released by bacterial reduction. If the environment becomes too reducing, then bacterial reduction of the organic matter is inhibited, and phosphorites do not form. In many instances, low sedimentation rates prevailed during phosphogenesis, and this is reflected in the associated sediments, which may be organic-rich mudrocks, cherts, pelagic limestones, hardgrounds and glauconite. Ancient phosphorites are commonly constituents of condensed successions, and in a sequence stratigraphic approach to the sedimentary record, they tend to occur in outer-shelf locations in the 'condensed section', i.e. at times of relative sea-level rise and highstand, with sediment starvation in deeper water (see Section 2.11.9).

Although it was once thought that phosphorite was precipitated directly from seawater, perhaps as some type of colloid, data from sites of active phosphorite formation indicate that much is being formed within the surficial sediments, largely by replacement and impregnation of grains. The bacterial decay of organic matter in the sediment liberates phosphate that is precipitated in pellets and coprolites, and replaces siliceous and calcareous skeletons and lime mud, eventually giving rise to nodular masses of phosphorite. Micas and detrital clays may absorb phosphate into their lattices. The role of the phytoplankton is crucial in transporting the phosphate from upwelling currents and near-surface waters to the sea floor. Microorganisms, such as bacteria and fungi, may be important in

concentrating the phosphate; SEM observations show that there is abundant evidence for microbes in ancient phosphorites (see Soudry & Southgate (1989) for an example).

A further important stage in the formation of extensive marine phosphorites is reworking. Ocean currents and severe storms remove much of the fine, unphosphatized sediment from the sea floor, leaving a concentrate of nodules, pellets and coprolites in various stages of induration and phosphate impregnation, which are then further phosphatized. Even greater reworking of bottom sediment is achieved during sea-level changes; transgressive–regressive events across the shelf off southwestern Africa have been important in the formation of the phosphorites there.

7.3.3 Ancient phosphorite sequences

Phosphate deposits, which in many cases can be related to upwelling currents and high organic productivities, are known from the Precambrian onwards. In fact, it appears that there have been a number of major phosphogenic episodes of global extent, principally in the late Precambrian to Cambrian, Permian, late Cretaceous to early Tertiary, and Miocene–Pliocene. One of the most famous deposits is the Phosphoria Formation of Permian age occurring in Idaho, Wyoming and Montana, USA. Phosphorite occurs in two members (Meade Peak and Retort) separated by chert (Rex Chert Member). An area of some $350\,000\,km^2$ is covered by these deposits, which have up to 36% P_2O_5. The phosphate units consist of interbedded carbonaceous mudrock, phosphatic mudrock, pelletal, oolitic, pisolitic, nodular and bioclastic phosphorite and phosphatic dolomite and limestone. Much of the phosphorite consists of pellets up to several millimetres in diameter composed of collophane, in a matrix of collophane with smaller pellets (See Plate 16c). The pellets could well be of faecal origin. Coated grains of phosphorite are present in some beds but their origin is less clear. Skeletal phosphate occurs in the form of fish scales and bones and inarticulate brachiopods (see Plate 16d). The Phosphoria phosphorites developed in an outer-shelf, upper-slope setting and pass laterally eastwards into nearshore carbonates, evaporites and red beds, and westwards into deeper-water sediments. The situation is thought to have been directly analogous to modern phosphorite accumulations, with phosphate derived initially from upwelling currents and delivered to the sediments via phytoplankton growth in surface waters (Fig. 7.2).

Other Palaeozoic phosphorites in North America occur in the Mississippian and Ordovician of Utah, Idaho and adjacent states. Along the western margin of North America, bedded phosphorites occur in the Miocene Monterey Formation of California and Jurassic La Caja Formation of Mexico. Extensive and valuable phosphate deposits occur in the Upper Cretaceous to Lower Tertiary of North Africa and the Middle East, from Morocco and the Spanish Sahara eastwards to Iraq and Turkey (see papers in Burnett & Riggs, 1990). These bioclastic and pelleted phosphorites accumulated on the continental margin along the southern side of Tethys and are related by many authors to upwelling currents. However, some of the phosphorus may have been supplied from rivers draining landmasses subjected to intense chemical weathering under a warm, humid climate. Bottom-current reworking of authigenic phosphorite grains produced concentrates in many of these economically important deposits.

The late Proterozoic to early Cambrian was a period of several phosphogenic events, principally in eastern Asia and Australia, but on the other continents too (see papers in Cook & Shergold, 1986). This was a time of rifting and ocean-basin formation, passive-margin development and major transgression. Upwelling, phytoplankton blooms, oceanic anoxic events (OAEs, see Section 3.7.2), expansion of the oxygen minimum zone and vigorous current circulation patterns probably caused these phosphogenic events. Such major changes within the oceans, coupled with subtle fluctuations in seawater chemistry, may have been instrumental in the Cambrian radiation event (the evolution of the metazoans) and the onset of biomineralization (Tucker, 1992; Shen et al., 2000). The late Precambrian to early Cambrian phosphorites are similar to younger ones, mainly peloidal and nodular. In the Georgina Basin of Australia, phosphorite appears to have been precipitated in microbial mats, and then reworking and exposure led to the formation of peloids and coated grains (vadoids) (Soudry & Southgate, 1989). Many of these phosphates are of economic importance, and in a number of cases they have been naturally beneficiated by more recent weathering, which has removed carbonate and led to the formation of *phoscrete* (a phosphatic soil).

Some ancient bedded phosphorites formed within confined basins, lagoons or estuaries, where high phosphate values, promoting high organic productivities, were generated by continental runoff. Examples are Miocene–Pliocene phosphorites from the Atlantic coastal plain of the USA (Florida and North Carolina especially), although topographic upwelling also may be involved. Phosphate precipitation and replacement associated with non-sequences, hardground surfaces and glauconite occur in ancient pelagic sediments, such as in the Cretaceous Chalks of western Europe (e.g. Jarvis, 1992; see Plate 16b).

Many of the phosphorite deposits of the geological record formed when sea-level was relatively high or were associated with short-lived transgressions. During these times, shallow, fertile, shelf seas promoted phytoplankton blooms, which led to poorly oxygenated shallow sea floors where organic matter (with its PO_4^{2-}) could accumulate. There also is a broad correlation of phosphogenesis with warm equable climate, probably because:

1 this induces an increase in the phosphorus flux to the oceans from increased chemical weathering on land;

2 it leads to the widespread occurrence of oxygen-depleted waters, because rates of oceanic circulation and oxygen solubility are reduced (see papers in Glenn et al., 2000).

7.4 Bioclastic and pebble-bed phosphorites

Vertebrate skeletal fragments are concentrated locally to form bone beds, commonly with fish scales. The depositional processes instrumental in their concentration are mainly current and wave reworking of sediments and winnowing of finer material, such that the phosphatic grains are left as lag deposits. Environments where these processes take place include transgressive and regressive shelf and shore zones and fluvial and intertidal channels. Bioclastic phosphorites can form in association with phosphate deposits related to upwelling (Section 7.3); the bone beds of Florida, for example. In the geological record, two classic bioclastic phosphorites are the Rhaetic Bone Bed in the Upper Triassic of southwest Britain and the Ludlow Bone Bed (Upper Silurian) of the Welsh Borderland.

In thin-section, skeletal phosphate is distinguished by its light yellow to brown colour and presence of a microstructure of regularly arranged canals (canaliculi) and growth lines (see Plate 16e). Bone phosphate is usually isotropic, or anisotropic with weak, irregular, patchy or undulose extinction.

Coprolites commonly are present in bioclastic phosphorites. They are generally spherical to elongate faecal pellets, up to several centimetres in diameter, composed of collophane. They may be homogeneous or show a vague concentric lamination. Some contain broken shell fragments or quartz silt, depending on what the animal had for dinner.

During diagenesis, bone fragments in bioclastic phosphorites are further enriched in phosphate, and their initially low fluorine content increases with age. The bone fragments may be cemented by collophane and phosphate nodules may nucleate around grains. Diagenetic phosphate can be precipitated within limestones, mudrocks and sandstones in the form of nodules, cements or replacements of calcareous skeletal grains. In most cases this phosphate would be derived from the decay of disseminated organic matter. A neutral or slightly acid pH facilitates the replacement of $CaCO_3$. The chemistry and petrology of dinosaur bones have been compared with those of modern mammal bones by Hubert *et al.* (1996).

Once formed, phosphate nodules and phosphatized limestone fragments, as well as phosphatic fossils, are very resistant to weathering. They are easily reworked into succeeding beds and may form concentrates, again of economic value, in younger formations. Examples are the phosphate pebble beds in the Pliocene Bone Valley Formation in Florida, the Palaeocene Brightseat Formation of Maryland and the Cambridge Greensand and Glauconitic Marl (Cretaceous) of England. Phosphorites on the northwest African continental shelf are largely reworked from Tertiary phosphate deposits.

7.5 Guano and ocean-island phosphorites

The excrement of birds, and to a lesser extent, bats, may in certain circumstances form thick phosphate deposits of economic significance. Leaching of the fresh guano leaves an insoluble residue composed mainly of calcium phosphate. Pedogenic processes also may be involved and produce phosphate-rich soils, *phoscrete*.

Thick accumulations of bird guano are found on some small oceanic islands in the eastern Pacific, such as Ocean Island, Nauru, Christmas Island, Makatea and the Banabans, along the Pacific coast of South America, and in the West Indies. Many of these deposits are comparatively old and are not being formed at the present day.

Geologically, guano itself is not significant. However, downward percolation of solutions derived from guano may cause phosphatization of underlying carbonate sediments and rocks, and this has happened extensively on some oceanic islands (e.g. Baker *et al.*, 1998). Atoll carbonates also may be phosphatized through accumulation of phosphorus in organic-rich sediments in lagoons and then early diagenetic release through microbial activities on subaerial exposure (e.g. Trichet & Fikri, 1997).

Further reading

Baturin, G.N. (1982) *Phosphorites on the Sea Floor: Origin, Composition and Distribution.* Elsevier, Amsterdam, 343 pp.

Burnett, W.C. & Riggs, S.R. (Eds) (1990) *Phosphate Deposits of the World*, Vol. 3. *Neogene to Modern Phosphorites.* Cambridge University Press, Cambridge, 484 pp.

Cook, P.T. & Shergold, J.H. (Eds) (1986) *Phosphate Deposits of the World,* Vol. 1. *Proterozoic and Cambrian Phosphorites.* Cambridge University Press, Cambridge, 386 pp.

Glenn, C.R., Prévôt-Lucas, L. & Lucas, J. (Eds) (2000) *Marine Authigenesis: From Global to Microbial.* Special Publication 66, Society of Economic and Paleontologists and Mineralogists, Tulsa, OK.

Iijima, A., Abed, A.M. & Garrison, R.E. (Eds) (1994) *Siliceous, Phosphatic and Glauconitic Sediments of the Tertiary and Mesozoic.* Proceedings 29th International Geological Congress, Part C. VSP, Utrecht, 450 pp.

Notholt, A.L.G. & Jarvis, I. (1990) *Phosphorite Research and Development.* Special Publication 52, Geological Society of London, Bath, 330 pp.

Nriagu, J.O. & Moore, P.B. (Eds) (1984) *Phosphate Minerals.* Springer-Verlag, Berlin, 470 pp.

8   Coal, oil shale and petroleum

8.1 Introduction

Organic matter in recent and ancient sediments is a reflection of the original organic input and subsequent processes of diagenesis and metamorphism. However, much of the organic matter that forms living organisms decays and breaks down in the presence of oxygen, eventually into carbon dioxide and water, and so disappears from the sedimentary record. Bacteria and other microbes are chiefly involved in this degradation, which takes place in most subaerial and many subaqueous environments. Where there is a deficiency of oxygen, the organic decomposition is incomplete and arrested, and quite stable organic compounds can develop and be preserved in the rock record. Organic matter is best preserved in anoxic environments, such as occur in stagnant lakes and stratified marine basins, swamps and bogs (mires), and in situations where anaerobic conditions are established early during diagenesis.

The fundamental process in the production of both plant and animal organic matter, and in fact the very basis of all life, is photosynthesis. Through this, plants manufacture carbohydrates from carbon dioxide and water, using sunlight for energy and chlorophyll as a catalyst. Although it is a complicated process, it can be represented by:

$$6CO_2 + 6H_2O \rightarrow C_6H_{12}O_6 + 6O_2$$

Decomposition of organic matter in the presence of oxygen, aerobic decay, is basically the reverse of photosynthesis, with various intermediate steps. When there is little or no oxygen present, anaerobic decay takes place to produce hydrocarbons and other more complex organic compounds.

As the majority of natural environments are oxidizing, most sediments contain only small quantities of organic matter. On average, sandstones contain 0.05% organic matter, limestones around 0.3% and mudrocks 2%. The principal organic deposits are oil shales and hard coals, brown coals (lignite) and their modern analogue, peat. These deposits and the oil and gas that are derived from them and from dispersed organic matter in sediments, make up the fossil fuels, of immense importance to humans. The now highly analytical and specialized disciplines of organic geochemistry and organic petrology are doing much to improve our understanding and exploitation of fossil fuels, all the more so in view of the expected depletion of petroleum reserves in the geologically very near future.

8.2 Modern organic deposits

The three main types of organic deposit accumulating at the present time are humus, peat and sapropel. *Humus* is fresh, decaying and decayed organic matter occurring mainly in the upper part of soil profiles. Decay products are mainly humic acids that are capable of leaching rock fragments and clays. In time, most humus is oxidized completely and so does not form significant organic deposits. It may be important, however, in contributing to the downward transportation of ions in the soil profile (eluviation) and in the formation of iron pans and other soil precipitates.

Peat is a dense mass of plant remains, which accumulates in waterlogged, boggy swamps and marshy regions, generally referred to as *mires*. The anaerobic conditions inhibit the complete breakdown of the organic material. Peat forms at all latitudes, in equatorial regions in tropical rain forests and mangrove swamps, and in temperate–polar regions in lowland and moorland bogs. In higher latitudes, the low temperatures are important in reducing the rate of bacterial decay. Vast areas of Canada, northern Europe and Russia–Siberia are covered in Pleistocene to Recent peat deposits. Two broad types of peat are the moorland variety, made up of mosses, sedges, rushes and sphagnum, and fen peat, derived from trees and leaves. Most coal comes from this second type, and much compaction and alteration takes place so that a layer of peat tens of metres thick is required to form just a single metre of coal.

Sapropel refers to organic material that accumulates subaqueously in shallow- to deep-marine basins, lagoons and lakes. The organic matter is derived largely from phytoplankton, which live in the upper water levels, but fragments of higher plants can be an important constituent. Planktonic algae, including *Botryococcus,* which contains globules of oil, proliferate in the photic zone of some lakes and give rise to hydrocarbon-rich sediments. Organic matter derived from such algae can accumulate along shorelines to form significant deposits. This occurs in the Coorong Lagoon, South Australia and in Lake Balkash, Siberia. Fine-grained terrigenous clastic sediment can be deposited along with the sapropelic organic matter so that all transitions occur from pure sapropel through organic-rich muds (see Section 3.7.2) to organic-free muds. Again, anaerobic conditions are necessary for preservation of the sapropel material, unless rates of organic production are very high; restricted water circulation, stagnation and water stratification are usually involved.

8.3 Ancient organic deposits

Organic deposits can be divided into two broad groups: those formed through *in situ* organic growth, as peat and humus, and those consisting of organic matter that has been transported or deposited from suspension, as with sapropels. Many brown and hard coals belong to the first, humic group, whereas oil shales and some coals (cannel coals) are examples of the second, sapropelic group. By definition, *brown coals and hard coals* have less than 33% inorganic material; however, most have less than 10%, consisting of clay minerals and quartz silt and sand (Section 8.5.2). *Oil shales* are mostly thin-bedded mudrocks containing more than 33% inorganic material. The organic matter is mostly kerogen (Section 8.9), and hydrocarbons are liberated on heating (Section 8.8). Organic-rich mudrocks have been discussed in Section 3.7.2.

In addition to being a primary deposit or accumulating along with carbonate or clastic debris, organic material may migrate into a sediment from a near or distant source rock. Tar sands, for example, contain solid and semi-solid hydrocarbons, which usually have permeated a porous sediment a long time after

deposition. For the organic material itself in sedimentary rocks:

1 *phytoclast* is used for a recognizable plant fragment (wood, leaf, cuticle, etc.);

2 *bitumen* is used for liquid or solid hydrocarbons, which are soluble in organic solvents, such as carbon tetrachloride or acetone;

3 *asphalt* is a solid or semi-solid bitumen, strictly one derived from an oil rich in cycloparaffin hydrocarbons;

4 *kerogen* refers to organic matter that is largely insoluble in organic solvents; it is a geopolymer consisting of long-chain hydrocarbons of high molecular weight;

5 *petroleum* consists of crude oils, chiefly short- and long-chain hydrocarbons, and gas, mainly methane, which in most cases has migrated into porous reservoir rocks from source rocks.

8.4 Coals and the coal series

Most coals are *humic coals*, formed from the *in situ* accumulation of woody plant material; *sapropelic coals* are those formed from algae, spores and comminuted plant debris. Humic coals form a natural series from peat through brown coals (lignite) and bituminous coals to anthracite. Coalification (also carbonification) refers to all the processes that take place when peat is converted into coal. As most of the changes are controlled by temperature of burial, the term *organic metamorphism* also is used.

Various microbiological, physical and chemical processes operate during coalification, all contributing towards the *rank* of the coal. The rank basically is a measure of the degree of coalification or level of organic metamorphism. With increasing rank (Table 8.1), the carbon content increases and the volatile content decreases (see Fig. 8.1; Section 8.6.2). The volatiles in coals are combustible gases (hydrogen, carbon dioxide and methane mostly), and condensible substances (such as water), which are driven off when coal is heated in the absence of air. Low-rank, volatile-rich coals burn easily with a smoky flame. Higher rank, low-volatile coals are more difficult to ignite, but burn with a smokeless flame. The carbonized residue remaining after removing the volatiles is *coke*. Humic coals are divided on their rank into a number of categories (Table 8.1): (a) peat, (b) lignite (soft brown coal), (c) sub-bituminous coal (hard brown coal), (d) bituminous coal (hard coal), (e) semi-anthracite and

Table 8.1 The rank stages of coal with approximate values of various parameters used to estimate rank

Rank stages	Carbon content as dry ash free (%)	Volatile content (%)	Calorific value (kJ g^{-1})	Vitrinite reflectance in oil
Peat	<50	>50		
Lignite	60	50	15–26	0.3
Sub-bituminous coal	75	45	25–30	0.5
Bituminous coal	85	35	31–35	1.0
Semi-anthracite	87	25	30–34	1.5
Anthracite	90	10	30–33	2.5
Graphite	>95	<5		

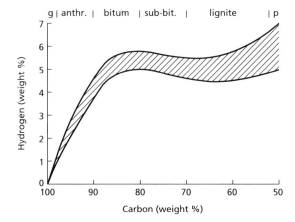

Fig. 8.1 Graph showing broad relationship between hydrogen and carbon content of coal with increasing rank.

(f) anthracite. Continued metamorphism leads to the formation of graphite. The changes in rank are gradual so that the boundaries between rank stages are somewhat arbitrary. Different terms are used in some countries. Apart from the level of organic metamorphism, coals are also divided into various types, determined by their organic constituents (Section 8.5.1).

8.4.1 Humic coals

The initial stages of coalification take place during peat formation; the processes are mainly microbial, with little alteration of the original plant material. In *soft brown coal* (lignite) many plant fragments are readily discernible, together with their original cell structure. A process of 'gelification' during the formation of *sub-bituminous coal* causes homogenization and compaction of plant cell walls, and this leads to the

formation of *vitrinite*, one of the main constituents of bituminous coals (Section 8.5.1). The lignites are relatively young, mostly occurring in Tertiary and Mesozoic strata, and important reserves occur in the Tertiary and Cretaceous of Germany, North America and the Far East.

Bituminous coals contain fewer volatiles and less moisture, and they consist of some 75–90% carbon. They are black, hard and bright in comparison with the soft, earthy and dull brown coals. They typically break into rectangular lumps, defined by prominent joint surfaces known as *cleat*. A distinctive bedding-parallel banding is usually developed, consisting of bright and dull layers of differing petrological composition. *Anthracite* contains more that 90% carbon; it burns with a smokeless flame. Anthracite is a bright shiny rock with conchoidal fracture. During the transition from bituminous coal to anthracite, much methane is liberated, plus other hydrocarbons, and these gases may accumulate in suitable reservoir rocks. This methane is the 'firedamp', which can cause explosions in mines exploiting high-rank coals. Bituminous coals and anthracites generally are older coals, widely developed in the Carboniferous of western Europe, North America and Russia and the Permo-Triassic of South Africa and Australia.

8.4.2 Sapropelic coals

Cannel coal and boghead coal are the two common sapropelic coals. Cannel is a massive unlaminated coal with an even, fine-grained texture and conchoidal fracture. It consists of altered plant debris, spores and some algae. Boghead coal is similar to cannel coal but it is derived largely from algae, particularly of the oil-bearing type such as *Botryococcus*. Many sapropelic coals formed in ponds and shallow lakes within the areas of swamp where the vegetation grew for the formation of the humic coals.

8.5 Coal petrology

8.5.1 Organic constituents

Studies of coal petrology use a polished surface of the coal sample and a reflected-light microscope, with oil-immersion objectives for increased contrast. For the microscopic constituents of coal, the term *maceral* is used, analogous to the minerals of igneous rocks. Many varieties of maceral can be recognized, placed into three groups (Table 8.2), and there are differences in the maceral content of soft and hard coals. Microscopic associations of particular macerals forming bands thicker than 50 μm are referred to as *microlithotypes*. These form different macroscopic types of coal, visible in hand specimen, known as *lithotypes*. In humic coals, the different lithotypes give rise to the millimetre- to centimetre-thick layering. The principal macerals of hard (bituminous) coals, and the various lithotypes and microlithotypes are shown in Tables 8.2 & 8.3. Photomicrographs of coal are presented in Fig. 8.2.

The vitrinite macerals (Fig. 8.2) are collinite and telinite. Telinite is derived from cell-wall material of wood and so has a cellular structure. Collinite is structureless, commonly being the organic fill of cell cavities. Vitrinite is derived from wood fragments that accumulated in stagnant, anaerobic water and were soon buried.

The inertinite macerals, fusinite and semifusinite, are derived from wood tissue so that cell structures are preserved (Fig. 8.2c). Sclerotinite forms from fungal remains, and micrinite from the thermal decomposition of resin. It is generally held that the inertinites have formed as a result of some oxidation of the woody tissue before or during the peatification stage, such as through subaerial exposure of the plant debris before burial. With fusinite, however, which possesses a charcoal-like character, formation through forest fires has been suggested, but much contested. A fusain-rich sandstone in the Carboniferous of Donegal is interpreted as the result of a palaeo-wildfire, which covered an area larger than Ireland (Nichols & Jones, 1992).

The liptinite macerals are derived from megaspores, microspores, cuticles, resins and algae, as the various names suggest. These macerals can be recognized by their shape and structure, although in many cases the original components were compacted and squashed in the peat stage (Fig. 8.2a,d). Alginite is the main constituent of boghead coals.

With increasing rank, the whole coal tends to become homogeneous and the macerals lose their identities.

The four common microlithotypes forming the microscopic bands and layers in hard coal and composed of the various macerals described above are vitrite, fusite, clarite and durite (Table 8.3). The lithotypes that can be recognized in hand specimen are composed of these microlithotypes. *Vitrain* layers are bright and glassy looking. They are brittle, have a conchoidal fracture and are largely composed of the microlithotype vitrite consisting of vitrinite macerals. Fusain is a soft charcoal-like material that gives coal its 'dirty' character. It is composed chiefly of fusite, consisting of

Table 8.2 The principal macerals of coal

Maceral group	Macerals	Origin
Vitrinite	collinite	wood
	telinite	
Inertinite	fusinite	woody tissues
	semifusinite	
	sclerotinite	fungi
	micrinite	polymerized resin
Liptinite	sporinite	spores
	cutinite	cuticle
	resinite	resin
	alginite	algae

Table 8.3 The lithotypes and microlithotypes of coal, together with the principal macerals forming the microlithotypes

Lithotypes (hand-specimen scale)	Microlithotypes (microscopic scale)	Principal macerals in microlithotype
Vitrain	vitrite	vitrinites
Fusain	fusite	inertinites, especially fusinite
Durain	durite	liptinites + inertinites
Clarain	clarite	vitrinites + liptinites

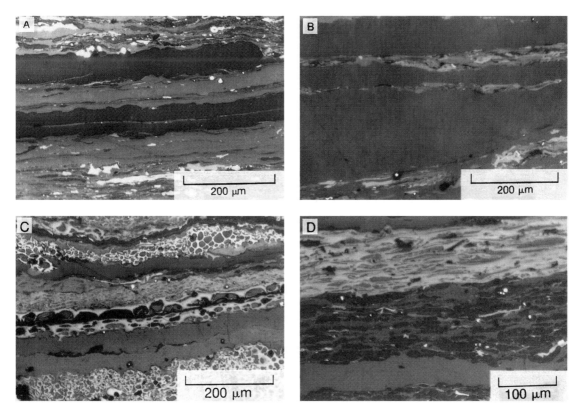

Fig. 8.2 Photomicrographs of bituminous coal showing the various macerals. Liptinite has a low reflectance and so forms the darkest areas; vitrinite has medium reflectance and so appears medium grey; inertinite has a high reflectance and so has a very bright, light-grey appearance. (A) Liptinite (the variety sporinite), vitrinite and inertinite. Liptinite comprises the two elongate dark areas, which are compressed megaspores, and the small dark elongate streaks, which are compressed microspores. (B) Vitrinite and thin streaks of inertinite (lighter grey colour). (C) Inertinite (semifusinite) and vitrinite. (D) Liptinite consisting of compressed microspores, vitrinite and inertinite (semifusinite, the very light layer). All reflected light, oil immersion. Carboniferous. Northumberland, England.

fusinite. *Durain* is a dull variety of coal with no lustre and an irregular fracture. It is composed of durite, a mixture of liptinite and inertinite macerals. *Clarain* is a laminated coal with a silky lustre and smooth fracture. It consists mainly of clarite, formed of a fine alternation of vitrinite and liptinite macerals.

Coal petrology is treated at length in Stach *et al.* (1982) Robert (1988) and Taylor *et al.* (1998), and see papers in Lyons & Alpern (1989a,b).

8.5.2 Inorganic constituents

A wide variety of inorganic minerals may be present in coal, some being sedimentary, whereas others are early or late diagenetic. Quartz grains, clay minerals and

some heavy minerals are the principal sedimentary inorganic constituents. Kaolinite is commonly the main clay mineral as many coals formed in tropical swamps, and this clay is dominant in these regions. With burial diagenesis, however, kaolinite is converted to illite, so that in higher rank coals, this clay is more abundant. Early diagenetic minerals in coal commonly occur in nodules, which may be composed of siderite, ankerite, dolomite, calcite (including coal balls, see below) and pyrite. Pyrite is widespread in coals and is largely derived from the activities of sulphate-reducing bacteria (Section 6.3). Later diagenetic minerals typically form in veins parallel to the cleat, and in cavities. These minerals include ankerite, quartz and sulphides of Fe, Pb, Zn and Cu.

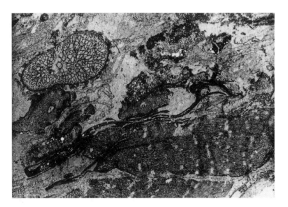

Fig. 8.3 Photomicrograph of a coal ball composed of calcite showing sections through stems and branches of various plants. Plane-polarized light. Carboniferous. Durham, England. Field of view 1.5 × 1.0 mm.

8.5.3 Coal balls

Of particular interest to the origin of coal is the occurrence of spherical concretions known as coal balls. These are composed mainly of calcite or dolomite but their significance lies in the well-preserved plant material that they contain (Fig. 8.3). The coal balls are early diagenetic nodules that formed before compaction and alteration of the organic matter that eventually formed the coal. In Carboniferous coal balls, fragments of the large trees *Sigillaria* and *Lepidodendron* are abundant, as well as *Calamites*. This indicates that the coal was derived from peat of a forest-swamp environment. Modern analogues occur in mangrove swamps of the humid tropics such as the Everglades, Florida.

8.6 Coal formation and rank

8.6.1 Chemical considerations

From the sedimentological context of coal seams, coal petrology, and coal-ball petrology, there is little doubt that humic coals are formed *in situ* as peat in forest swamps. The formation of brown coals (lignite) from peat requires burial, an absence of air, and time. The conversion of brown coals to bituminous coals in general requires more time, but also higher temperature and pressure. Changes in the brown-coal–bituminous-coal range are entirely physical and chemical processes.

Chemically, coals largely consist of humic substances derived from the biochemical and chemical breakdown of the lignin, cellulose and protein of the original plant material. The humic substances have complex structures, high molecular weights and are composed largely of carbon in combination with small amounts of hydrogen, oxygen and nitrogen. In the early stages, lignin is degraded by microorganisms to humic acids, diphenyl, benzyl and other organic compounds with ring structures and side chains. Then, polycyclic aromatic compounds are formed and with increasing rank there is an increase in the aromatic portion, so that anthracite consists principally of condensed benzene rings. The formation of bituminous coal takes place at temperatures between 100 and 200°C; at these temperatures, the processes taking place are decarboxylation, condensation with the elimination of water, and rearrangement reactions. Free carbon is not thought to be present in coal although it may occur in anthracites.

8.6.2 Rank, depth and temperature

The degree of coalification, i.e. the rank of the coal, can be measured by several parameters, including the amount of carbon, hydrogen, oxygen, volatiles and moisture present, the calorific value and the reflectance of the vitrinite. Not all parameters are useful over the whole coal-rank spectrum. The reflectance of vitrinite, which increases with increasing rank (Table 8.1), is particularly useful because measurements need not be confined to coal seams, but can be undertaken on other sediments that contain plant fragments (phytoclasts) preserved as vitrinite. Vitrinite reflectance can be used as an indication of burial history and to evaluate hydrocarbon potential because it will indicate the burial temperatures reached by the sediments; one can then assess whether the temperatures were high enough for petroleum generation (Section 8.10 and Tyson, 1995).

It has long been known that the rank of coal increases with increasing depth (Fig. 8.4). The increase in rank is not so much a function of depth as temperature, because it is the latter that effects chemical changes in coal. Rank increasing with depth has been demonstrated for numerous coal basins, mainly from borehole core studies. The depth at which coal of a particular rank is converted to one of higher rank does vary between coal basins. If the coals are of the same

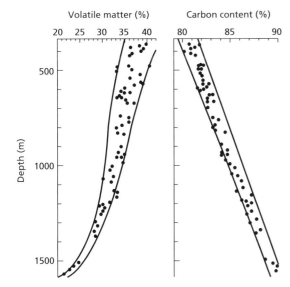

Fig. 8.4 An example from the Carboniferous Westphalian Coal Measures of Germany of the increase in coal rank with depth, on the basis of volatile matter and carbon contents (both dry-ash free). After Teichmuller (1987).

general age then this is a reflection of variations in the geothermal gradient. Apart from temperature, the length of time that a coal has been subjected to that burial temperature is also important. With coals of different ages buried to the same depths, the older coals have a higher rank. Where coals occur in areas of folding and thrusting then they too are commonly of higher rank. This is the case with anthracite in the South Wales Coalfield and high-rank coals associated with thrusts in the Coal Measures of the Ruhr, Germany. Tectonic pressure and frictional heat have been invoked to account for this, but local higher geothermal gradient and hydrothermal fluid flow also are important factors. Tectonic pressure is more physical in its effect, inducing an anisotropy in vitrinite of a polarization direction following original wood grain.

8.7 Occurrence of coal

In the geological record, coal-forming conditions were widespread in humid climatic areas from the late Devonian onwards, when land plants evolved and proliferated. Paralic and limnic coals are distin-

guished, the paralic ones developed in coastal situations, particularly in deltaic environments, usually as thin (<3 m) but relatively persistent seams. Limnic coals form in continental basins around lakes, and in rift basins rapid subsidence can give rise to very thick seams indeed (hundreds of metres). The majority of the coal seams in European and North American Carboniferous strata are of the paralic type; limnic coal basins are especially well developed in the late Carboniferous in France.

Paralic, humic coals typically occur at the top of coarsening-upward deltaic cycles (Fig. 2.68 and Section 2.11.4). In a typical coal-measures cycle, mudrocks with marine fossils (in some cases a marine limestone) give way to non-marine mudrocks then siltstone and sandstone, before the coal seam. Coal seams overlie soil horizons containing roots and rootlets but the nature of the soil varies considerably, depending on the lithology of the sediment in which the plants grew. Where the soils were sandy, they may have developed into quartz arenites known as *ganisters*, through the leaching of less stable minerals by acid pore waters (Section 2.7.1). Other soils are mudrock of mixed-clay mineralogy (Section 3.7.1), commonly with siderite nodules (Section 6.3), referred to as seatearth or underclay, or if suitable for refractory purposes, then fireclay. Many of these Carboniferous soils are vertisols. In some Carboniferous coal formations, there are prominent bands of volcanic ash known as *tonsteins* if dominated by kaolinite, or *bentonites* if dominantly smectite (e.g. Spears *et al.*, 1999; see Section 10.7). They can occur below, within and above coal seams, and are useful for correlation. Individual coal seams may be very continuous laterally, but in most cases they are restricted to a particular coal basin and cannot be correlated from one region to the next. Thick coal seams generally develop when base level (relative sea-level) is rising, but not so fast that the growth of the peat in the mire cannot keep up. Marine horizons that occur within coal-measure successions, and tonsteins, frequently can be traced over vast areas. Many papers on the sedimentology of coal deposits are contained in Rahmani & Flores (1985), Scott (1987), Whateley & Pickering (1989), Whateley & Spears (1995) and Gayer & Petek (1997). See Heckel (1995), Fielding & Webb (1996), Breyer (1997), Hampson *et al.* (1999) and Miller & Eriksson (2000) for papers on coal deposition and coal in a sequence stratigraphic context.

8.8 Oil shales

Oil shales are a diverse group of rocks which contain organic material that is mostly insoluble in organic solvents, but that can be extracted by heating (distillation). The organic matter is largely kerogen, but some bitumen may occur. The quantity of oil that can be extracted ranges from about 4% to more than 50% of the weight of the rock, i.e. between 10 and 150 gallons of oil per ton of rock or 50–700 L tonne^{-1}. Oil shales contain a substantial amount of inorganic material consisting largely of quartz silt and clay minerals. Some oil shales are really organic-rich siltstones and mudrocks (Section 3.7.2), whereas others are organic-rich limestones.

Much of the organic matter in oil shales is finely disseminated and so altered that the organisms from which it formed cannot be identified. In many oil shales the remains of algae and algal spores are common and so much of the organic matter is assumed to be of algal origin. Fine-grained higher-plant debris and megaspores also may be an important constituent. The common macerals in oil shales are thus of the liptinite group. The typical sedimentary feature of many oil shales is a distinct lamination, on a millimetre scale, of alternating clastic (or carbonate) and organic laminae. Seasonal or annual blooms of planktonic algae frequently are invoked to account for the rhythmic lamination. As with the formation of coal, anaerobic conditions are required to prevent oxidation of the organic matter and to reduce the bacterial degradation, unless the rate of organic productivity is very high, when accumulation can take place in an oxidizing environment. Many oil shales formed in stratified water bodies where oxygenated surface waters permitted plankton growth, and anoxic bottom waters allowed the preservation of the organic matter (see Fig. 3.11).

The kerogen in oil shales is largely type I, that is it has a high H/C and low O/C ratio (see Fig. 8.5 and next section) and is derived largely from algal lipid matter (fats and fatty acids), rather than carbohydrates, lignins or waxes. Some kerogen in oil shales may be type III, formed from vascular plant debris. Certain metals, such as V, Ni, U and Mo, are enriched in oil shales.

Oil shales were deposited in lacustrine and marine environments. Oil shales in the Eocene Green River Formation of the western USA contain much dolomite

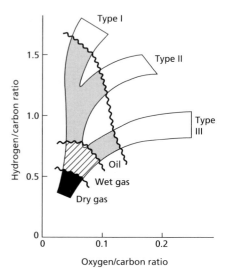

Fig. 8.5 Van Krevelen diagram showing kerogen types I, II and III and their evolution paths with increasing burial as oil, wet gas and dry gas are generated.

and calcite and are rhythmically laminated or varved. Although formerly regarded as relatively deep-water in origin, deposition is now thought to have taken place in relatively shallow ephemeral lakes, frequently subjected to desiccation (Dean & Anders, 1991). Thin cycles of oil shale passing up into evaporite (of trona and halite especially) reflect a permanent, non-saline stratified lake developing into a saline lake and saline pan. Oil shale formed from a single algal species occurs at several horizons in the Lower Carboniferous of the Midland Valley of Scotland. This formed in freshwater lakes within a deltaic complex where humic coals also developed. Marine oil shales deposited in shallow seas, on continental platforms and basins are known from the Devonian of eastern and central North America, the Jurassic of Europe and the Miocene of California. In some cases they formed in association with upwelling (Section 7.3.2). As planktonic algae are the principal source of the organic matter and these have a long geological history, Precambrian oil shales do occur. An example is the Nonesuch Shale of Michigan and Wisconsin, from around 1100 Ma.

There is considerable interest in oil shales because they are a source of fossil fuel and they may help to offset the expected exhaustion of petroleum reserves. Extensive deposits of oil shale occur in Russia, China and Brazil and low-grade deposits, which may become

economically viable, occur in many other countries of
the world. Oil shales also are potential petroleum-
source rocks (see next section), and many descriptions
of these are contained in Brooks & Fleet (1987), Fleet
et al. (1988) and Whelan & Farrington (1992).

8.9 Formation of kerogen

Kerogen is a very complex geopolymer of high molecular weight formed from the diagenesis of organic
matter. There are four major groups of organic compounds in living organisms—carbohydrates, lignin,
proteins and lipids—and their elemental composition
is shown in Table 8.4. Of note is the fact that lipids have
the lowest oxygen, and little sulphur and nitrogen
relative to the other compounds. The composition of
lipids is also closest to kerogen (Table 8.4). In oxidizing
environments, organic matter generally is broken
down through aerobic bacterial activity and oxidation
to CO_2, NH_3 and H_2O. In reducing environments,
however, anaerobic bacteria decompose organic
matter (especially carbohydrates, through fermentation and other processes) to new and residual organic
compounds, and CH_4 and CO_2. The material left over
from microbial–bacterial activity recombines by poly-
condensation and polymerization to form insoluble
organic compounds, such as fulvic and humic acids.
During shallow burial, depths of tens to hundreds of
metres, and over several millions of years, these organic acids are converted to insoluble *humin*. Further burial, with decreasing O/C and N/C ratios, leads to the
formation of kerogen. The composition is very variable and normally not determined, but the kerogen in
the oil shales of the Eocene Green River Formation is
given as $C_{215}H_{330}O_{12}N_5S$. With further burial still into
the realm of catagenesis, the composition of kerogen is
modified through decreases in the O/C and H/C ratios

as oil and gas are generated (Section 8.10.2). Kerogens
from different depths in the same formation plot along
a curve or *evolution path* on an O/C versus H/C graph
(a Van Krevelen diagram, Fig. 8.5).

On the basis of O/C and H/C ratios (Fig. 8.5), three
types of kerogen are distinguished: types I, II and III.
Type-I kerogen has a high initial H/C ratio, and low
initial O/C ratio. It is derived from lipid material, particularly that of algal origin, and also from the severe
biodegradation of other organic compounds during
early diagenesis. Type-I kerogen is common in oil
shales, such as those of the Eocene Green River Formation, and it is rare compared with types II and III. Type-
I kerogen is oil-prone. Type-II kerogen has relatively
high H/C and low O/C ratios. It is common in many
marine organic-rich mudrocks, which may be hydrocarbon source rocks if buried to sufficient depths. The
organic matter is derived mainly from phytoplankton,
zooplankton and bacteria. Type-III kerogen has relatively low initial H/C and high initial O/C ratios. It is
derived largely from the diagenesis of vascular plants
and often much plant material can be identified. This
kerogen type does not generate much oil, but it does
provide gas if buried deep enough (i.e. it is gas-prone).

Kerogen in polished section is a structureless organic material, usually occurring in bands and
stringers parallel to stratification. Immature (shallow-
burial) kerogen, as in the Green River Formation, is a
yellow–amber colour, but with increased burial and
evolution (maturation), it takes on a brown and then
black colour.

8.10 Petroleum

8.10.1 Composition and occurrence

The generation of petroleum is one of the stages in the
alteration of certain types of organic matter buried in
sediments. It is formed through increasing burial and
temperature and so is part of the general process of
organic matter diagenesis and metamorphism. Petroleum consists of crude oil and gas. Crude oils are largely carbon (average 85 wt%) and hydrogen (13 wt%),
in the ratio of 1.85 hydrogen atoms to 1 carbon atom.
Minor elements, S, N and O generally constitute less
than 3% in most oils (see Table 8.4), and phosphorus
and vanadium also may be present in significant
amounts. The S, N and O values do vary considerably,
and sulphurous oils, known as sour oils with up to 7%

Table 8.4 Average composition of main organic compounds in
organic matter compared with typical petroleum and kerogen

	C	H	S	N	O
Carbohydrates	44	6	—	—	50
Lignin	63	5	0.1	0.3	31.6
Proteins	53	7	1	17	22
Lipids	76	12	—	—	12
Petroleum	85	13	1	0.5	0.5
Kerogen	79	6	5	2	8

S, are distinguished from low-sulphur sweet oils. Hydrogen is a much lighter element than the others, so that the specific gravity of oil, which is easily measured, does indicate the H content. A higher hydrogen content gives a lower specific gravity (e.g. a 14% H oil has a specific gravity of 0.86; a 12% H oil, specific gravity 0.95). Oil is also characterized by its °API (degrees American Petroleum Institute), expressed by:

$$°API = 141.5/ρ - 131.5 \qquad (ρ = \text{specific gravity at } 60°)$$

This is an arbitrary scale with a range from 5 to 60°API. In this case a higher specific gravity oil has a lower °API. A typical North Sea crude has 37°API; a heavy Orinoco crude oil has 6–20°API.

Petroleum is composed of a great number and variety of simple and complex hydrocarbon compounds, from the smallest methane (CH_4), with a molecular weight of 16, to the largest asphaltene molecules with molecular weights in the thousands. The common hydrocarbon compounds in petroleum belong to the alkane–paraffin (C_nH_{2n+2}), naphthene–cycloalkane (C_nH_{2n}) and arene–aromatic (C_nH_{2n-6}) homologous groups. Compounds with sulphur, nitrogen and oxygen include the thiols, thiophenes, pyridines, quinolines, carboxylic acids and phenols.

Natural gas occurs as a gas cap to oil reservoirs, in solution in oil (released when pressure decreases) and as a reservoir fluid alone. Dry gas, mostly consisting of methane (CH_4) and ethane (CH_6) is distinguished from wet gas, with more than 50% propane (C_3H_8) and butane (C_4H_{10}). Wet gas is usually closely associated with oil, whereas dry gas is more associated with coal deposits and derived from deeply buried source rocks. H_2S, CO_2 and N_2 may form a significant component of natural gas and helium also is present. Water occurs in most oilfields and is typically a brine, much more saline than seawater.

Petroleum can occur in any porous rock but in fact most of the world's oil is located in Tertiary and Mesozoic sedimentary rocks. Petroleum is derived from source rocks (discussed below) and then migrates into reservoir rocks, which typically are sandstones and limestones. The porosity and permeability of the reservoir rocks are obviously very important (Sections 2.10 and 4.9). An impervious seal is required to prevent upward escape of the petroleum from the reservoir and common cap-rocks in oilfields are mudrocks and evaporites. To contain the petroleum some form of trap is necessary. Many traps are structural, involving folds (domes and anticlines, especially), faults and salt diapirs, whereas others are depositional, arising from the geometry of the reservoir sandstone body or limestone mass and its overlying cap rock.

At the present time around one-third of the world's oil comes from the Middle East: Saudi Arabia, Iran, Iraq and the UAE. There are major oilfield 'giants' in the USA, Canada, Russia, Venezuela, Nigeria, Libya, Mexico, western Europe, North Sea and Indonesia, and many other countries have smaller oilfields.

Petroleum can escape from a reservoir and reach the surface. This happens mostly where a reservoir is subsequently faulted. Gas may emerge from the ground and burn for a long time. The 'eternal fires' of the Middle East, referred to in the Bible, are a case in point. Tar pits and asphalt lakes also may form; those in Trinidad and Venezuela are well known, and in Los Angeles many prehistoric animals fell into the tar pits at La Brea (Fig. 8.6). Near-surface sediments may become impregnated with asphaltic hydrocarbons. In Alberta, Canada, tar in the Cretaceous Athabasca sands came from a fractured Devonian reefal reservoir at depth.

Fig. 8.6 Tar pit at La Brea, California, with *Mastodon* sadly trapped in the tar. Bubbles are produced by natural gas rising through the oil.

8.10.2 Formation of petroleum

Petroleum is derived largely from the maturation of organic matter deposited in fine-grained marine sediments. As discussed in Section 3.7.2, organic-rich sediments can be deposited in anoxic silled basins, on shelf margins in association with upwelling and the oxygen-minimum zone, and on the sea floor at times of oceanic anoxic events. Many marine hydrocarbon source rocks formed at times of high organic productivity of marine plankton, coinciding with transgressive events and highstands of sea-level (Fig. 8.7). Peritidal microbial mats and organic matter in reefs and ooids may be other sources. Lacustrine source rocks also are important. Diagenesis of the organic matter begins very early at shallow burial depths, and substantial quantities of methane can be produced through bacterial fermentation. This 'marsh gas' normally escapes into the atmosphere, but it can be trapped. Burial diagenesis of the deposited organic matter leads to the formation of kerogen, the type depending on the nature of the organic input (see Section 8.9).

Burial to temperatures of 50–80 °C causes thermo-catalytic reactions in the kerogen, and in types I and II, cycloalkanes and alkanes are generated, two of the main constituents of crude oil. When this process takes place, the source rock is said to be mature. With increasing temperature, more and more oil is generated until a maximum is reached, and then the quantity decreases and an increasing amount of gas is formed (see Fig. 8.8). The principal phase of oil generation takes place at temperatures around 70–100 °C; in areas of average geothermal gradient this is at depths of 2–3.5 km (this is the *oil window*). The gas produced is wet initially, but above 150 °C only methane (dry gas) is generated. Time also is a factor in source-rock maturation; higher temperatures/greater burial depths are required for oil generation from younger rocks, compared with older rocks, which can thus reach maturity at lower temperatures (Fig. 8.9). Burial of type-III kerogen and coal into the catagenesis realm leads to the generation of much gas, mostly methane, and little oil.

Some organic compounds in organic matter, the porphyrins, for example, are very resistant to diagenetic alteration and are found in hydrocarbon source rocks as well as crude oils. These *geochemical fossils* or *biomarkers* can be very useful in correlating an oil with its source rock. At higher temperatures, the

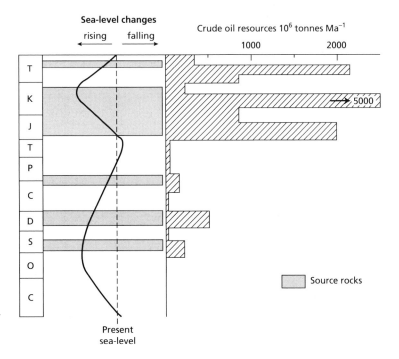

Fig. 8.7 Schematic distribution of source rocks through time, relative to the first-order global sea-level curve, and of crude-oil resources through time. After Tissot & Welte (1984).

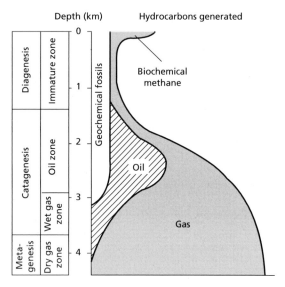

Fig. 8.8 Hydrocarbon generation with depth from organic matter, mainly kerogen contained in sediments. The precise depth at which hydrocarbons evolve depends on the geothermal gradient, the burial history, and type of kerogen present. After Tissot & Welte (1984).

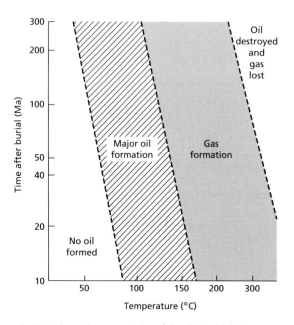

Fig. 8.9 Schematic representation of the relationship between time after burial of source rocks and the temperatures for oil and gas generation.

Table 8.5 Vitrinite reflectance, spore colour and hydrocarbon generation

Vitrinite reflectance	Spore colour	Hydrocarbon generation		
0.5	yellow orange	—		
1.0	brown	oil window	wet	—
1.5	—		gas	dry
2.0	black	—		gas
5.0		—		

biomarkers begin to break down so that they can also give an indication of the maturity of the source rock. This is important for calculating the amount of oil that has been generated.

In the search for petroleum, use can be made of the colour of pollen and spores (palynomorphs) in the source rocks to see if the stage of petroleum generation has been reached. With rising temperature and higher level of organic metamorphism, palynomorphs change colour from yellow to brown when crude oil is evolved, and to black when dry gas is generated. An indication of burial temperature also can be obtained from the vitrinite reflectance of phytoclasts and the colour of conodonts (Table 8.5). Other tests for source-rock maturity are the determination of H/C and O/C ratios, UV fluorescence and pyrolysis.

The study of petroleum is a science in itself and for further information reference should be made to petroleum geology textbooks, such as Tissot & Welte (1984), Selley (1988), Bjørlykke (1989), North (1990) and Gluyas & Swarbrick (2001).

Further reading

Brooks, J. (Ed.) (1990) *Classic Petroleum Provinces*. Special Publication 50, Geological Society of London, Bath, 568 pp.

Gayer, R. & Petek, J. (Eds) (1997) *European Coal Geology and Technology*. Special Publication 125, Geological Society of London, Bath, 448 pp.

Gluyas, J. & Swarbrick, R.E. (2001) *Petroleum Geoscience*. Blackwell Science, Oxford, 400 pp.

Lyons, P.C. & Alpern, B. (Eds) (1989a) *Coal*. Elsevier, Amsterdam, 678 pp.

Lyons, P.C. & Alpern, B. (Eds) (1989b) *Peat and Coal*. Elsevier, Amsterdam, 892 pp.

North, F.K. (1990) *Petroleum Geology*. Unwin–Hyman, London, 607 pp.

Rahmani, R.A. & Flores, R.M. (Eds) (1985) *Sedimentology of Coal and Coal-bearing Sequences*. Special Publication 7, International Association of Sedimentologists, Blackwell Scientific Publications, Oxford, 418 pp.

Robert, P. (1988) *Organic Metamorphism and Geothermal History*. Reidel, Dordrecht, 311 pp.

Scott, A.C. (Ed.) (1987) *Coal and Coal-bearing Strata: Recent Advances*. Special Publication 32, Geological Society of London, Bath, 340 pp.

Stach, E., *et al.* (1975) *Textbook of Coal Petrology*. Gebrüder Borntraeger, Stuttgart, 428 pp.

Taylor, G.H. *et al.* (1998) *Organic Petrology*. Gebrüder Borntraeger, Berlin, 704 pp.

Tissot, B.P. & Welte, D.H. (1984) *Petroleum Formation and Occurrence*. Springer-Verlag, Berlin, 699 pp.

Whateley, M.K.G. & Pickering, K.T. (Eds) (1989) *Deltas: Sites and Traps for Fossil Fuels*. Special Publication 41, Geological Society of London, Bath, 360 pp.

Whateley, M.K.G. & Spears, D.A. (Eds) (1995) *European Coal Geology*. Special Publication 82, Geological Society of London, Bath, 500 pp.

Whelan, J.K. & Farrington, J.W. (Eds) (1992) *Organic Matter: Productivity, Accumulation and Preservation in Recent and Ancient Sediments*. Columbia University Press, New York, 533 pp.

Also see recent papers in *Bulletin of the American Association of Petroleum Geologists, Bulletin of Canadian Petroleum Geology, Coal Geology, Journal of Petroleum Geology, Marine and Petroleum Geoscience, Organic Geochemistry* and the *Treatize of Petroleum Geology*, Reprint Series, American Association of Petroleum Geologists.

9 Cherts and siliceous sediments

9.1 Introduction

Chert is a very general term for fine-grained siliceous sediment, of inorganic, biochemical, biogenic, volcanic or hydrothermal origin. It usually is a dense, very hard rock, which splinters with a conchoidal fracture when struck. Most cherts are composed of fine-grained silica, and contain only small quantities of impurities. Certain types of chert have been given specific names. For example, *flint* frequently is used as a synonym for chert and more specifically for chert nodules occurring in Cretaceous chalks. *Jasper* refers to a red variety of chert, its colour being due to finely disseminated hematite. Chert of this type is interbedded with iron minerals to form jaspilite in some Precambrian iron-formations (Section 6.5.1). *Porcelanite* refers to fine-grained siliceous rock with a texture and fracture similar to unglazed porcelain. The term porcelanite is also used more specifically for an opaline claystone composed largely of opal-CT (Section 9.3.2).

Cherts in the geological record usually are divided into bedded and nodular types. Some bedded cherts are associated with volcanic rocks and the 'chert problem' has centred on a volcanic versus biogenic origin of the silica. The modern equivalents of many ancient bedded cherts, the radiolarian and diatom oozes, cover large areas of the deep-ocean floors. Nodular cherts are developed mainly in limestones and to a lesser extent in mudrocks and evaporites. Most bedded cherts are primary accumulations; many nodular cherts on the other hand are diagenetic, having formed by replacement of the host sediment, although they commonly do still reflect deposition of silica-rich sediment. Siliceous sediments are also being deposited in lakes, and they do form soils (*silcretes*, Section 9.5).

9.2 Chert petrology

Bedded and nodular cherts consist of three main types of silica: *microquartz*, *megaquartz* and *chalcedonic quartz* (Fig. 9.1). Microquartz consists of equant quartz crystals only a few microns across. Megaquartz crystals are larger, reaching 500 μm or more in size; the crystals have unit extinction and often possess good crystal shapes and terminations. Megaquartz is often referred to as drusy quartz because it commonly occurs as a pore-filling cement, just like calcite spar. Chalcedonic quartz is a fibrous variety with crystals varying from a few tens to hundreds of microns in length. They usually occur in a radiating arrangement, forming wedge-shaped, mammillated and spherulitic growth structures (Fig. 9.2). Most chalcedonic quartz is length-fast (*chalcedonite*) but a length-slow variety (*quartzine*) also occurs. The latter is rare, but where found it is commonly associated with replaced evaporites (Section 5.5).

Radiolarians (marine zooplankton with a range of Cambrian to Recent), diatoms (marine and non-marine phytoplankton, Triassic to Recent) and siliceous sponges (marine and non-marine, Cambrian to Recent) are composed of *opaline silica*. This is an isotropic amorphous variety, containing up to 10% water. Opaline silica is metastable so that it decreases in abundance back through time and is absent from Palaeozoic cherts. Radiolarians and diatoms have disc-shaped, elongate and spherical tests with spines and surface ornamentation (Fig. 9.3). They range in size from a few tens to hundreds of microns. Sponge spicules are a similar size and up to a few millimetres in length, and have a trilete or Y-shape, giving circular and elongate sections in thin-section.

9.3 Bedded cherts

9.3.1 Siliceous oozes and bedded cherts

Radiolarian and diatom oozes are accumulating on the ocean floors at the present time. They occur especially where there is high organic productivity in near-surface waters and this is controlled largely by oceanographic factors of upwelling and nutrient supply. Diatoms dominate in siliceous oozes in the

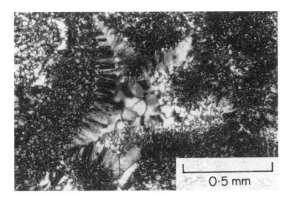

Fig. 9.1 Photomicrograph showing: microquartz—finely crystalline mosaic with pin-point extinction; megaquartz—larger quartz crystals in central part; chalcedonic quartz—in a fibrous fringe. In this case, the microquartz is a replacement of carbonate grains and the megaquartz and chalcedonic quartz are pore filling. Crossed polars. Carboniferous. Glamorgan, Wales.

Fig. 9.2 Photomicrograph of chalcedonic quartz in spherulitic growth structure, also some microquartz and megaquartz. Crossed polars. Chert in limestone, Jurassic. Yorkshire, England.

Southern Ocean around Antarctica, and in the northern Pacific. Radiolarian-rich oozes occur in the equatorial regions of the Pacific and Indian Oceans. It is likely that prior to late Mesozoic times, radiolarians occupied the niches of diatoms and so were more widely distributed than at present. The siliceous oozes preferentially accumulate in abyssal areas where depths exceed the carbonate compensation depth (CCD, Section 4.10.7), around 4500 m in the central Pacific. Siliceous oozes form at shallower depths where surface oceanic waters are fertile and there is a paucity of calcareous plankton and terrigenous detrital material. Diatomaceous sediments accumulating at depths of less than 1500 m in the Gulf of California are of this type. The depth at which silica itself dissolves rapidly, the opal compensation depth (OCD), is around 6000 m. As a result of dissolution during settling, only a small percentage of diatoms and radiolarians actually reach the ocean floor to form sediments. Laminae within siliceous oozes are reflections of seasonal variations in plankton productivity and/or seasonal influxes of detrital clay.

Ancient bedded cherts commonly occur in mountain belts and other zones of folded rocks, many having been deposited in deeper-water basins, which are then structurally deformed. The uniform rhythmic bedding, which is a characteristic feature of these cherts, is generally on the scale of several centimetres, with millimetre-thick beds or partings of shale between (Fig. 9.4). The rhythmic bedding can be related to orbital forcing and Milankovitch rhythms, which cause periodic variations in biogenic silica precipitation and/or terrigenous influx (see Tada (1991) for a Miocene example from Japan). Other chert beds are massive with no internal sedimentary structures. Although recrystallization has commonly occurred, structureless beds may well have resulted from the slow and steady deposition of the silica. Some chert beds show graded bedding, parallel and small-scale cross-lamination, and basal scour structures. Cherts with these features have been either deposited by turbidity currents derived from some nearby topographic high where the siliceous sediments were first deposited, or reworked by contour-flowing bottom currents (Section 2.11.7). Bedded cherts locally show slump folding and contemporaneous brecciation as a result of instability and mass-sediment movement during sedimentation.

Many bedded cherts consist of radiolarians to the exclusion of other biogenic material. The radiolarians generally are poorly preserved, consisting of megaquartz-filled moulds contained in a matrix of microquartz (Fig. 9.5). Sponge spicules are common in some radiolarites. Fine clastic and carbonate sediment may be present in the cherts, and with increasing concentrations the cherts pass into siliceous shales and limestones.

Some bedded cherts are associated with volcanic rocks, others are not. Where there is a volcanic association, the cherts commonly were deposited within or

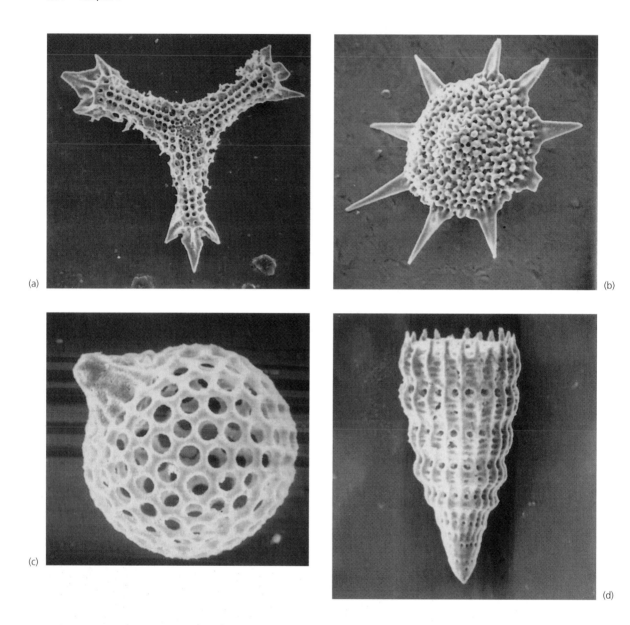

(a)

(b)

(c)

(d)

Fig. 9.3 Scanning electron micrographs of radiolarians, illustrating variety of shape. All from the Upper Cretaceous of Cyprus and between 200 and 300 μm across. (a) *Paronaella* sp., (b) *Pseudoaulophacus* sp., (c) *Cryptamphorella* sp. and (d) *Dictyomitra* sp.

above pillow lavas. Lava flows and volcaniclastic sediments may be intercalated, as well as horizons of black shale and pelagic limestone. In some cases ultramafic rocks and dyke complexes also are present, so that the whole igneous–sedimentary assemblage constitutes an *ophiolite suite*, generally accepted as a fragment of ocean floor. The chert may be derived from devitrification of volcanic ash or a biogenic source. Classic Mesozoic volcanic–chert associations occur in the Franciscan rocks of the Californian Coast Ranges (Robertson, 1990), in the Apennines of Italy (Barrett, 1982) and in the Troodos Massif, Cyprus (Robertson & Hudson, 1974). A Tertiary example occurs in Japan (Hattori *et al.*, 1996) and Cambro-Ordovician chert is found in Newfoundland and Maine (Pollock, 1987), also Girvan–Ballantrae, southwest Scotland, where REEs have helped identify the plate-tectonic setting (Armstrong *et al.*, 1999).

Bedded cherts independent of volcanics usually are associated with pelagic limestones, and siliciclastic and carbonate turbidites. Such deposits are typical of ancient passive continental-margin successions and commonly rest upon foundered shallow-water platform carbonates. Examples occur in the Mesozoic of the Tethyan–Alpine region and the Lower Carboniferous of western Europe, in southwest England and central Germany especially.

By analogy with modern siliceous oozes, many ancient radiolarian-rich bedded cherts are interpreted as very deep-water in origin, having been deposited below the carbonate compensation depth (CCD), at depths of several kilometres. Although this may be so for some cherts, they could have been deposited at much shallower depths if there was no calcareous plankton available for sedimentation. This may well have been the case during the Palaeozoic and well into the Mesozoic, because the main calcareous planktonic organisms, coccoliths and foraminifers, did not evolve

Fig. 9.4 Rhythmically bedded chert, with thin shale partings. Carboniferous. Montagne Noire, southern France.

Fig. 9.5 Photomicrographs of bedded chert, composed of microquartz, with radiolarians preserved as microquartz and megaquartz: (a) plane-polarized light; (b) crossed polars. Ordovician. Ballantrae, Scotland.

(a)

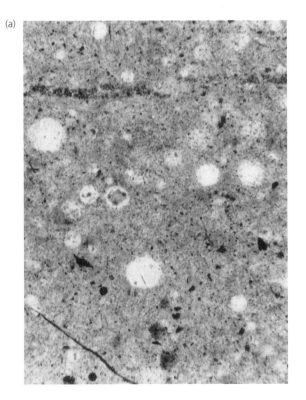

(b)

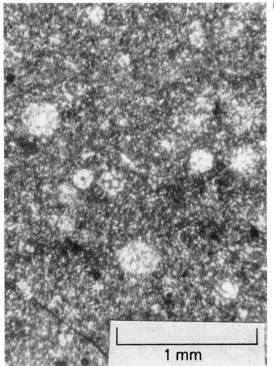

1 mm

until the Mesozoic. Variations in the CCD also could have permitted siliceous sediments to form in shallower water.

Cherts are relatively common in Precambrian successions but it is thought generally that siliceous organisms evolved later, in the Palaeozoic. The most likely sources of silica are volcanic material and hydrothermal fluids. The early Proterozoic from eastern Canada contains chert-cemented sandstones and thin chert beds, and intraclasts of these facies suggest very early precipitation of the silica (Simonson, 1985). It is thought that the silica was derived from hydrothermal waters migrating up through the sediments and discharging onto the sea floor. It is very likely that seawater had higher concentrations of silica in the early Precambrian than in the Phanerozoic, and a lower pH, so promoting primary silica precipitation (see Sugitani *et al.* (1998) for an Australian Archaean case study). See Zhou *et al.* (1994) for a description of hydrothermal cherts from China, interpreted on the basis of REE patterns.

Siliceous sediments rich in diatoms are common in the Miocene and Pliocene of the circum-Pacific area and in localized regions of the Mediterranean (papers in Iijima *et al.*, 1983; Hein & Obradovic, 1989). These *diatomites*, which locally are bituminous and phosphatic, formed in relatively small back-arc and rifted basins that were starved of terrigenous sediment but were the sites of vigorous upwelling and thus phytoplankton productivity. Diatomites of the Monterey Formation, California, are of this type (Beil & Garrison, 1994) and the diatom oozes of the Gulf of California are a modern analogue. Diatomites were deposited in small restricted basins in the Mediterranean region, just before the Messinian salinity crisis, towards the end of the Miocene (Bellanca *et al.*, 1986).

9.3.2 The origin of chert

Broadly there are two alternative views for the formation of chert:
1 that the cherts are entirely biogenic in origin, unrelated to any igneous activity;
2 that the cherts are a product of submarine volcanism, either directly through inorganic precipitation of silica derived from subaqueous magmas and hydrothermal activity or indirectly through plankton blooms induced by submarine volcanism.

A better understanding of submarine volcanism, in recent years, through plate-tectonic theory, has made a volcanic–sedimentary origin of cherts less likely. Sea-floor volcanism is restricted to oceanic ridges and localized 'hot-spots' and so is unlikely to give rise to regionally extensive cherts. In fact, hydrothermal silica is precipitated close to vents, but it is quantitatively insignificant. In addition, the occurrence of radiolarian cherts in non-volcanic successions and the dominantly biogenic origin of modern siliceous sediments, controlled by oceanographic factors, indicate that the formation of cherts is not related to contemporaneous volcanism. The situation may well have been different in the early Precambrian, however.

Cores collected during deep-sea drilling have permitted detailed studies of the siliceous ooze to chert transformation. Cores from the ocean floors of the Pacific and Atlantic have encountered well-indurated cherts in Pliocene and older sections. Chert is particularly widespread in the Eocene of the North Atlantic. Contributions to silica diagenesis also have come from studies of cherts exposed on land.

From the biogenic amorphous opal, frequently referred to as *opal-A*, the first diagenetic stage is the development of crystalline opal, identified by X-ray diffraction (Fig. 9.6) and referred to as *opal-CT*, also called disordered cristobalite, alpha-cristobalite or lussatite. Opal-CT consists of an interlayering of cristobalite and tridymite, and its disordered nature probably results from the small crystal size and incor-

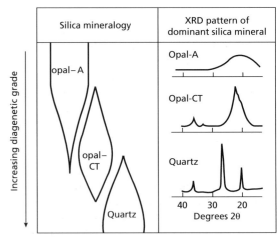

Fig. 9.6 Schematic changes in silica mineralogy with increasing diagenesis, and X-ray diffraction patterns for opal-A, opal-CT and quartz showing the increasing crystallinity. After Pisciotto (1981).

Fig. 9.7 Lepispheres of opal-CT growing in voids in silicified Eocene chalk from the Arabian Sea. Sample from 630 m below sea floor. Prismatic crystals are clinoptilolite (a zeolite). Scanning electron micrograph.

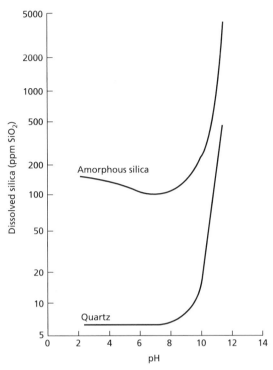

Fig. 9.8 The solubility of quartz and amorphous silica at 25 °C. At pH values less than 9, the silica is in solution as undissolved orthosilicic acid (H_4SiO_4); above pH 9, this dissociates into $H_3SiO_4^{2-}$ and $H_2SiO_4^{2-}$.

poration of cations into the crystal lattice. Opal-CT replaces radiolarian and diatom skeletons and is precipitated as bladed crystals lining cavities and forming microspherules (5–10 μm diameter) called *lepispheres* (Fig. 9.7). Further diagenesis results in the metastable opal-CT being converted to quartz chert, mostly an equant mosaic of microquartz crystals but also chalcedonic quartz. This recrystallization of opal-CT to quartz obliterates the structure of many diatom and radiolarian tests.

The driving forces behind chert formation from biogenic opal-A are the solubility differences and the chemical conditions. Biogenic silica has a solubility of 120–140 p.p.m., cristobalite of 25–30 p.p.m. and quartz of 6–10 p.p.m. in the pH ranges of marine-sediment pore water (Fig. 9.8). Once the metastable opal-A dissolves, the solution is saturated with respect to opal-CT and quartz. The precipitation of opal-CT in preference to quartz probably results from the more internally structured nature of quartz, which would require slow precipitation from less concentrated solutions. Temperature also is involved; with a rise in temperature, as through increasing depth of burial, the rate of transformation of opal-CT to quartz increases substantially. The formation of chert from opal-CT has been referred to as a 'maturation' process and in the Miocene Monterey Formation of California, the

term porcelanite or opaline claystone is used for the metastable precursor to chert. Further studies have shown that the maturation of opal-A to quartz depends on the nature of the host sediment and on the chemical conditions. The presence of excess alkalinity in the sediments, as occurs where there is much calcareous material, favours the initial opal-CT precipitation and enhances the rate of transformation of opal-CT to quartz. Where there is much clay in the sediment, opal-CT contains abundant impurities, mainly foreign cations, which retard the maturation to quartz. The end-product of these processes of silica dissolution, reprecipitation and replacement is a mosaic of microquartz and chalcedonic quartz, with relatively few of the biogenic particles identifiable, although the original ooze was wholly composed of them. The maturation of siliceous sediments also leads to a decrease in porosity. In the Monterey Formation diatomites have porosities of 50–90%, porcelanites up

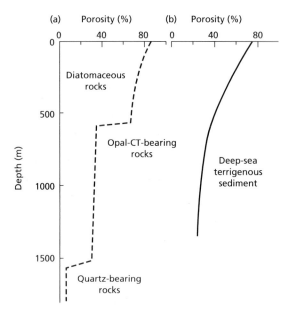

Fig. 9.9 Porosity reduction with depth for (a) diatomaceous oozes to opal-CT porcelanites to quartzose cherts, compared with (b) deep-sea terrigenous sediment, where porosity loss is gradual rather than stepwise. After Isaacs (1981).

Fig. 9.10 Chert nodules (flint) in Cretaceous chalk of Yorkshire, England. Sutured stylolites are present in the chalk.

to 30% and cherts less than 10% (Fig. 9.9). This is a result of the silica-mineral transformations rather than compaction (Isaacs, 1981).

Although most Phanerozoic bedded cherts are now regarded as biogenic, some contain minerals thought to be volcanic alteration products, e.g. montmorillonite, palygorskite, sepiolite and clinoptilolite. The devitrification of volcanic glass and clay transformations, such as montmorillonite to illite, do liberate silica. In most, if not all Phanerozoic bedded cherts, however, there is at least some preservation of siliceous microfossils. This can be taken to indicate a dominantly biogenic origin, with most microfossils destroyed through the dissolution–reprecipitation of the maturation process.

Many papers on chert diagenesis are contained in Hsü & Jenkyns (1974), Iijima *et al.* (1983, 1994), Hein & Obradovic (1989), Heaney *et al.* (1994) and Knauth (1994).

9.4 Nodular cherts

Nodular cherts occur predominantly in carbonate host rocks. They are small to large, subspherical to ir-

regular nodules, commonly concentrated along particular bedding planes; they may coalesce to form near-continuous layers, when they resemble bedded cherts. Nodular cherts are common in shelf limestones such as the Lower Carboniferous limestones of Britain and North America. They are also common in pelagic carbonates, with many examples occurring in the Cretaceous and Tertiary of the Alpine–Mediterranean–Tethys region. In the Cretaceous Chalk of western Europe and southeastern USA, nodules of chert (flint) are common (Fig. 9.10) and many have developed in burrow fills and nucleated around fossils. Flint pebbles derived from the Chalk are a major constituent of Tertiary and Quaternary gravels. Nodular cherts have been recovered from Miocene and older deep-sea chalks and pelagic limestones in cores from the ocean floors.

As with bedded cherts the origin of nodular cherts also has been much discussed. The older view involved the direct precipitation of silica from seawater to form blobs of silica gel on the sea floor that hardened into chert nodules. Nodular cherts in limestones, however, contain much evidence to demonstrate a replacement and thus diagenetic origin. Within the nodules, originally calcareous grains such as ooids and skeletal debris are preserved in silica (Fig. 9.11). Bedding structures such as lamination may be preserved in the nodules. The diagenetic processes involved in chert-nodule formation are thought to be similar to those operating in bedded cherts. Biogenic silica disseminated in the sediment dissolves and is reprecipitated in the form of opal-CT at nodule growth points.

Fig. 9.11 Silicified oolite. Ooids have been replaced by microquartz and are enclosed in a megaquartz mosaic. Crossed polars. Cambrian. Pennsylvania, USA.

Pore spaces are first filled with opal-CT lepispheres and then carbonate-skeletal and matrix replacement by opal-CT follows. Maturation of the opal-CT to microquartz and chalcedonic quartz takes place from the nodule centre outwards. It often can be demonstrated that the microquartz has formed by replacement of carbonate and that the chalcedonic quartz and megaquartz are dominantly pore-filling (Figs 9.1 & 9.11). Quartz authigenesis in Silurian limestones of eastern Canada apparently is related to bacterial sulphate reduction and oxidation of organic matter (Noble & Van Stempvoort, 1989). Nodular chert in Ordovician Arbuckle Limestones of Oklahoma apparently formed from dissolution of sponge spicules during shallow burial in marine waters, but the silica recrystallized in meteoric water during exposure of the platform (Gao & Land, 1991). Near-surface direct precipitation of quartz from meteoric water produced metre-sized chert nodules in the Eocene of Egypt (McBride *et al.,* 1999).

The source of biogenic silica in shelf limestones is largely sponge spicules, whereas in deeper-water pelagic limestones, the silica is largely supplied by radiolarians and diatoms. With the Cretaceous chalks of western Europe, a relatively shallow-water pelagic carbonate, silica for the flint nodules was derived mainly from sponges.

Chert nodules also can form by the replacement of evaporites, particularly anhydrite. Length-slow chalcedonic quartz is common in these occurrences (Section 5.5).

9.5 Non-marine siliceous sediments and cherts

Biogenic and inorganic siliceous sediments can form in lakes and ephemeral water bodies and in soils. Diatoms, which can occur in great abundance in lakes, form diatomaceous earths or *diatomites*. Such sediments are accumulating in many sediment-starved, higher-latitude lakes such as Lake Luzern, Switzerland and Lake Baikal, Siberia. During the Pleistocene, diatomites were deposited in many late-glacial and post-glacial lakes in Europe and North America.

Inorganic precipitation of silica can take place where there are great fluctuations in pH. Quartz, with its low solubility in most natural waters, is not affected by pH until values exceed 9, and then with increasing pH, the solubility increases dramatically (Fig. 9.8). In ephemeral lakes of the Coorong district of South Australia, pH values greater than 10 are reached seasonally through photosynthetic activities of phytoplankton. Detrital quartz grains and clay minerals are partially dissolved at these high pH values so that the lake waters become supersaturated with respect to amorphous silica. Evaporation of lake water and a decrease in pH cause silica to be precipitated as a gel of cristobalite, which would give rise to chert on maturation. A Cretaceous example of Coorong-type chert has been described by Chough *et al.* (1996). Another related inorganic process has been described from East African lakes. In very alkaline, sodium carbonate-rich lake waters, silica is leached from volcanic rocks and rock fragments. Exceptionally high concentrations of silica (up to 2500 p.p.m.) are attained, and then lowering of pH by freshwater influxes causes the silica to be precipitated as *magadiite*, a metastable hydrated sodium silicate. This is converted to chert in a relatively short time. Cherts in ancient ephemeral lake successions, such as the Eocene Green River Formation of Wyoming, may well have formed via a sodium silicate precursor, and a Permian example from Italy is described by Krainer & Spötl (1998).

Silica also may be precipitated from hot springs through evaporation and rapid cooling of spring waters to form *sinter*. Silicification of microbes may take place by impregnation of organic tissue (e.g. Renaut *et al.,* 1998). Preservation of microbes in Precambrian cherts also may have occurred in thermal spring environments (e.g. Schopf, 1993), and the famous Rhynie Chert of the Lower Devonian in eastern Scotland, which contains some of the oldest preserved land plants, is also a hot spring deposit (Trewin, 1994).

Silicification of lacustrine (and marine) limestones without any biogenic or detrital silica also may take place through ground-water flow, with silica derived from dissolution of quartz in adjacent formations. Porous zones in the limestones then will be silicified preferentially. See Thiry & Ribet (1999) for an example from the Tertiary of the Paris Basin.

Chert also is precipitated in some soils and *silcrete* is one particular type occurring especially in parts of Australia (e.g. Webb & Golding, 1998) and southern Africa (e.g. Summerfield, 1983). Silcretes mostly appear to form under arid/semi-arid climates, where ground waters are alkaline with a pH above 9, but they can form in humid areas too. Silcretes usually consist of a microquartz cement between sand grains, and microquartz mosaics where they have formed within finer-grained sediments. Megaquartz and fibrous quartz occur within vugs. There may be small canals and tubes from the decay of rootlets. Ancient silcretes have been reported from the Proterozoic of northwest Canada, where they occur upon weathered rhyolite lava flows and volcaniclastic sediments (Ross & Chiarenzelli, 1985).

Further reading

Heaney, P.J., Prewitt, C.T. & Gibbs, G.V. (Eds) (1994) *Silica*. Reviews in Mineralogy 29, Mineralogical Society of America, Washington, DC, 450 pp.

Hein, J.R. (Ed.) (1987) *Siliceous Sedimentary Rock-hosted Ores and Petroleum*. Van Nostrand Reinhold, New York, 304 pp.

Hein, J.R. & Obradovic, J. (Eds) (1989) *Siliceous Deposits of the Tethys and Pacific Regions*. Springer-Verlag, New York, 244 pp.

Hsü, K.J. & Jenkyns, H.C. (Eds) (1974) *Pelagic Sediments: on Land and under the Sea*. Special Publication 1, International Association of Sedimentologists, Blackwell Scientific Publications, Oxford, 447 pp.

Iijima, A., Abed, A.M. & Garrison, R.E. (Eds) (1994) *Siliceous, Phosphatic and Glauconitic Sediments of the Tertiary and Mesozoic*. Proceedings 29th International Geological Congress, Part C. VSP, Utrecht, 450 pp.

Iijima, A., Hein, J.R. & Siever, R. (Eds) (1983) *Siliceous Deposits in the Pacific Region*. Elsevier, Amsterdam, 472 pp.

10 Volcaniclastic sediments

10.1 Introduction

Volcaniclastic sediments are those composed chiefly of grains of volcanic origin, derived from contemporaneous volcanicity. Several different processes operate during magmatic eruptions, giving rise to a range of distinctive volcaniclastic facies. The type of magma — acid, intermediate or basic, determined largely by plate-tectonic setting — is an important factor in the type of volcaniclastic deposit generated. Volcanism also is an important source of sand- and mud-grade material for deposition in many different sedimentary environments.

The study of volcaniclastic rocks does meet with several problems. It is often difficult (and dangerous) to observe modern volcanic processes and mostly the eruptions that can be studied are relatively small-scale compared with the really major, but very infrequent events that deposited the thick volcaniclastic lava piles of the geological record. The techniques used for studying Recent volcaniclastics, such as sieving for grain-size analyses (see Section 2.2.1) are not applicable to their indurated ancient equivalents. Diagenesis is a major factor in altering volcanic glass and minerals, destroying depositional textures and creating matrix. The weathering of volcanic material is also very rapid; in a few years sand-grade ash particles can be reduced to clay. There is also a problem with the preservation of the volcanoes themselves; being topographic highs, they are easily eroded, and also hydrothermally altered. Thus the geological record is often biased towards the more distal parts of the lava–volcanic sediment system.

The term *pyroclast* is applied to any material, regardless of size, ejected from volcanoes. *Tephra* is a collective term for pyroclasts. Pyroclasts are derived chiefly from the magma itself, in which case they usually are composed of volcanic glass, but they may be crystals, if the magma had begun to crystallize before its explosive eruption. Tephra also include lithic fragments, of lava from earlier eruptions and of country rock. Tephra are subdivided on particle size into volcanic dust, ash, lapilli and blocks (Table 10.1). Large fragments of magma ejected in a fluid state are *bombs*. Some have an elongate ellipsoidal shape as a result of rotation during flight. Many pyroclasts are in the form of *pumice*, highly vesicular volcanic glass that may have a porosity of more than 50%. When formed from more basic magma, the term *scoria* is used instead of pumice. Fragmentation of pumice gives rise to glass shards, fragments smaller than the average vesicle size These are a common component of many volcaniclastic sediments and can be recognized by their sickle, lunate or Y-shape, the concave surfaces representing the broken glass-bubble walls (Fig. 10.1). Glass shards are more common in acidic ashes; in more basic volcaniclastics, the glass is present as droplets. In thin-section, glass has a pale to yellow colour and is isotropic. Devitrified glass may show a weak birefringence if replaced by clays or zeolites (Section 10.5). Crystals ejected from volcanoes are usually euhedra or broken euhedra and they may be zoned. Quartz and feldspar crystals are common and in more basic ashes pyroxenes may occur. One particular type of lapilli consists of a spherical to ellipsoidal body composed of a nucleus, which may be a glass shard, crystal or rock fragment, that is enveloped by one or more laminae formed of volcanic ash and dust (Fig. 10.2). These *accretionary lapilli*, as they are called, form during both air-fall and flow processes.

The nature of volcanic explosions and eruptions depends on the volatile content (water and CO_2 especially) and viscosity of the magma. At depth, the volatiles are in solution in the magma but as the latter rises towards the Earth's surface and pressure decreases, the gases exsolve and expand. This causes vesiculation of the magma and produces frothy and foam-like magma, which on ejection and solidification gives rise to the pumice/scoria. Acid magmas contain a higher percentage of volatiles over basic magmas and they are also more viscous, so that acid magmas give rise to more widespread pyroclastic deposits.

Table 10.1 Classification of volcaniclastic grains and sediments, based on grain size

Volcaniclastic grains (tephra)	Volcaniclastic sediments
bombs—ejected fluid	agglomerate
blocks—ejected solid	volcanic breccia
—64 mm—	—
lapilli	lapilli-stone
—2 mm—	—
coarse ash	volcanic sandstone
—0.06 mm—	—
fine ash	volcanic mudstone

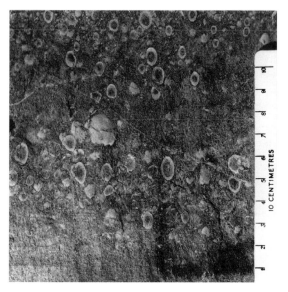

Fig. 10.2 Accretionary lapilli. Borrowdale Volcanic Group, Ordovician. Cumbria, England.

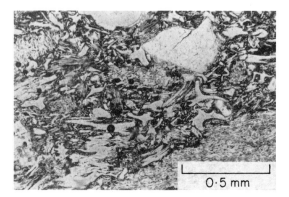

Fig. 10.1 Glass shards with characteristic broken bubble-wall shape, broken crystals and streaked devitrified glass, called fiamme (lower right). Plane-polarized light. Ignimbrite, Tertiary. New Zealand. Courtesy of Hugh Battey.

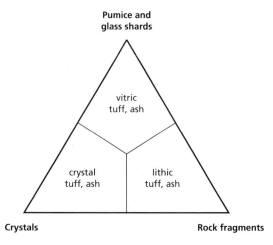

Fig. 10.3 Classification of pyroclastic tuffs on basis of nature of components.

Volcaniclastic sediments composed of ash are *tuffs*; those of lapilli are *lapilli-stones*, those of bombs and/or blocks are *agglomerates* and *volcanic breccias* (Table 10.1). Intermediate varieties are lapilli-tuff and tuff-breccia. On the basis of the proportions of glass, crystals and rock fragments, tuffs can be divided into vitric, crystal and lithic types (Fig. 10.3). Terrigenous clastic or carbonate sediments containing volcanic ash can be described as *tuffaceous*, or referred to as *tuffites*. The intimate mixing of fresh magma with wet unconsolidated sediment produces a *peperite*.

Volcaniclastic sediments are best discussed in terms of their modes of formation. Five principal categories can be distinguished (Table 10.2): autoclastic, pyroclastic-fall, volcaniclastic-flow, hydroclastic and epiclastic deposits. Detailed reviews are provided by

Fisher & Schmincke (1984), Cas & Wright (1987), Lajoie & Stix (1992), Orton (1996) and papers in Fisher & Smith (1991).

10.2 Autoclastic deposits

These are volcanogenic rocks produced by autobrec-

Table 10.2 Main types of volcaniclastic deposit

Autoclastic deposits
   sediment generation during lava flow

Pyroclastic-fall deposits
   formed of tephra ejected from vent

Volcaniclastic-flow deposits (and type of flow)
   (a) ignimbrites (pyroclastic flows)
   (b) surge deposits (pyroclastic surges)
   (c) lahar deposits (volcanic mudflows)

Hydroclastites: fragmented lava through contact with water
   (a) hyaloclastites (non-explosive)
   (b) hyalotuffs (explosive)

Epiclastic deposits
   volcanic material reworked by currents, waves, wind, gravity
      flows, etc.

Fig. 10.4 Flow breccia. Angular blocks of acidic lava in glassy matrix. Temagami Greenstone Belt, Archaean. Ontario, Canada.

ciation of lavas. As a lava flows along, it cools and the upper surface may develop a brittle crust, which fractures and brecciates on top of the moving lava. Jostling between blocks induces a degree of rounding, sorting and comminution. As the lava advances, the brecciated material slides off the front of the flow and is then overridden to give a basal breccia carpet to the flow. Lava itself is mixed in with the upper and lower breccias so that textures can vary from clast-supported to matrix- (lava-) supported. There are variations in the nature of the blocks, generally more vesicular and angular in basaltic lavas and more homogeneous and oblate/rectilinear in intermediate–acidic lavas. In the latter, flow banding and fold structures are common in the lava. In the more distal parts of lava flows, the whole of the deposit may be flow breccia (Fig. 10.4), whereas in more proximal areas, the breccias just occur at the bottom and tops of the more massive lava beds (Fig. 10.5).

10.3 Pyroclastic-fall deposits

These sediments are simply formed through the fallout of volcanic fragments ejected from a vent or fissure as a result of a magmatic explosion. In the majority of cases, explosive volcanoes are subaerial and the material is deposited on land, but if there are subaqueous environments nearby these will also receive the pyroclastic debris. Two types of subaerial fallout are eruption-plume derived fall deposits, which are ejected explosively from a vent, producing a plume of tephra

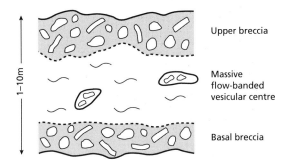

Fig. 10.5 Typical succession through flow-brecciated lava. After Suthren (1985) from other sources.

and gas, and ash-cloud derived fall deposits, resulting in part from ash clouds rising off a moving pyroclastic flow. Pyroclastic falls also can be subaqueous, from underwater volcanic eruptions.

The characteristic features of air-fall deposits are a gradual decrease in both bed thickness and grain size away from the site of eruption, and a good to moderate sorting. Blocks and bombs are deposited relatively close to the vent, whereas ash may be carried many tens of kilometres and dust thousands of kilometres away from the vent. The aerial distribution of the volcanic material is related to the height of the eruption column. This may reach several tens of kilometres, so that the ash is then carried around the Earth by the high-velocity winds in the upper atmosphere. Individual beds of air-fall material typically show normal

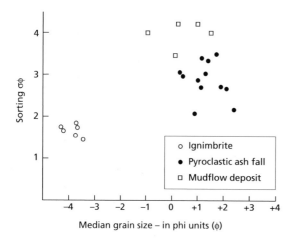

Fig. 10.6 An illustration of the use of grain-size parameters to distinguish between volcaniclastic sediments of different origin. Median grain-size/sorting scatter diagram for ignimbrite, pyroclastic ash fall and mudflow deposits of the Minoan eruption (1470 BC). Santorini, Greece. After Bond & Sparks (1976).

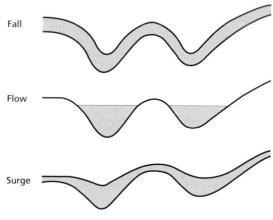

Fig. 10.7 Sketch of typical geometry of pyroclastic-fall, flow and surge deposits: mantle bedding, valley-fill and topographic drape, respectively. After Suthren (1985).

grading of particles, although in some cases, inverse grading of pumice and lithic clasts has occurred. Where deposition takes place in water quite large fragments of low-density pumice may occur towards the top of an air-fall bed, as a result of the pumice floating on the water surface before deposition. Particle-size analysis can be used in the interpretation of pyroclastic-fall deposits and to distinguish between the latter and volcaniclastic-flow deposits (Fig. 10.6; Bond & Sparks, 1976; Pyle, 1989; Barbieri *et al.*, 1989). However, grain-size analysis of ancient volcanogenic deposits is difficult, because of diagenetic alteration (see Section 10.7).

A further feature of air-fall deposits is the development of *mantle bedding*, whereby the tephra layer follows and blankets any original topography (Fig. 10.7), with a similar thickness over topographic highs and lows. Although typical of pyroclastic-flow deposits, welding does occur in air-fall tuffs deposited close to the vent if they derive from high-temperature magma and/or accumulate rapidly.

The impact of bombs may depress or rupture the bedding in air-fall tuffs to produce '*bomb-sags*'. While in the air, pyroclastic material is strongly affected by any prevailing wind so that air-fall deposits frequently are concentrated in a particular direction away from the volcanic vent. Water/steam in the eruption column and rain generated by the eruption can lead to the formation of *accretionary lapilli* (Fig. 10.2) and aggregates, although they often disintegrate on impact. Reworking of air-fall deposits can take place in the subaerial environment by wind and rain, and in the submarine environment by waves and tidal currents. Air-fall pyroclastic deposits can be very important in stratigraphic correlation. Geologically they represent instantaneous deposition and they can cover vast areas. They have been used effectively in this way in the Jurassic and Tertiary of the North Sea (e.g. Morton & Knox, 1990).

10.4 Pyroclastic-flow and -surge deposits

In the subaerial environment, *pyroclastic-flow deposits* are the product of hot gaseous particulate density currents. They generally form through fluidization by magmatic gas and give rise to deposits known as *ignimbrites*. *Pyroclastic-surge deposits* result from highly expanded turbulent gas–water–solid density currents with low particle concentrations, which can move at hurricane speeds. Both of these generally derive from acidic magmas. A further type of subaerial flow is a *lahar* or volcanic mudflow. The transport of volcaniclastic sediment is reviewed in Fisher & Schmincke (1994).

10.4.1 Ignimbrites

These pyroclastic ash-flows are generated by the collapse of eruption columns and they are hot, dense, laminar flows of volcanic debris. Fluidization (an upward movement of gas or water causing the particles to behave as a fluid) is brought about by the expansion of gases exsolved from the magma and of air caught up in the advancing flow. The flows, which are a component of the *nuée ardente* type of eruption, can travel for great distances (up to 100 km), even over flat ground. The deposits of these flows, ignimbrites, are characterized by a homogeneous appearance with little sorting of the finer ash particles, but if coarse lithic clasts are present they may show normal size grading, and large pumice fragments are commonly reversely graded (Fig. 10.8). There generally is a lack of internal stratification, although bedding may develop in more distal regions. Ignimbrites do not mantle the topography, but tend to follow valleys and low ground (Fig. 10.7). Evidence for flow is provided by flattened and stretched pumice and glass fragments, termed *fiamme* (Figs 10.1 & 10.9). Pumice clasts may be concentrated on the top surface of an ignimbrite. Fine ash in a flow is probably derived from comminution of larger tephra

during transport, and some rounding of pumice takes place too. After deposition, the hot, plastic ash particles in the central part of the ignimbrite typically become welded together to form a dense rock. The term *'welded tuff'* was formerly used for ignimbrite. As a result of the more rapid heat loss, the lower and upper parts of an ignimbrite are usually non-welded and so have a higher porosity and are often more friable or weathered there. Ignimbrite beds in a succession may show considerable variation in degree of welding, thickness and grain size. Ignimbrites grade laterally into loose, poorly welded pumiceous tuffs. Fine air-fall tuffs commonly are associated with ignimbrites (Fig. 10.8), and may be the deposits of turbulent dust clouds, which usually are part of *nuée ardente*-type eruptions (see Boudon & Gourgaud (1989) for descriptions of the Mont Pelée *nuée ardente* deposits). Pyroclastic-surge deposits (see below) may form a basal unit to an ignimbrite (Fig. 10.8).

Ignimbrites typically are derived from acid magmas and indeed they may be difficult to distinguish from rhyolite lavas. The ignimbrites can form substantial thicknesses of tuff; individual flows are commonly 10 m or more thick. Ignimbrites also may be intrusive into previously deposited pyroclastic-flow deposits. Well-developed ignimbrite successions occur in Armenia, New Zealand, New Mexico and the Andes (e.g. Silva, 1989).

Related ash-flow tuffs can be deposited subaqueously. This can result from submarine eruptions, or from a subaerially erupted pyroclastic flow travelling

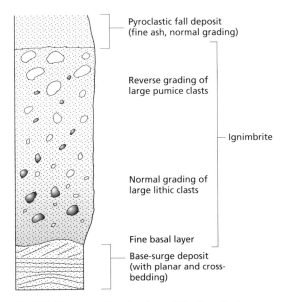

Fig. 10.8 Diagrammatic section through the deposits of an ignimbrite eruption. Pyroclastic surges are not always associated with ignimbrite flows. After Sparks (1976).

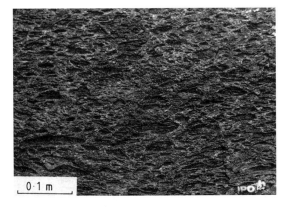

Fig. 10.9 Ignimbrite with aligned stretched glass fragments (fiamme). Borrowdale Volcanic Group, Ordovician. Cumbria, England.

into water. Such deposits are recorded from the Ordovician of North Wales (Kokelaar & Königer, 2000).
Subaqueous-flow units commonly pass up into bedded tuffs, deposited from ash clouds separated off
from the dense pyroclastic flow. Pyroclastic material
deposited in water around shallow subaqueous or
emergent volcanoes may be transported into deeper
water through slumping, debris flows and turbidity
currents to deposit conglomerates (diamictites) and
graded ash beds. Such facies also occur in the Ordovician of North Wales and in the Neogene of California
(Cole & Stanley, 1994). A modern deep-sea volcaniclastic fan system has been described from Réunion,
Indian Ocean (Ollier *et al.*, 1998).

10.4.2 Pyroclastic-surge deposits

Pyroclastic surges are dilute, subaerial, fast-flowing
turbulent mixtures of volcanic particles and gas. Two
types occur: (a) hot, dry surges and (b) cool, wet
surges. The first type usually is associated with pyroclastic flows, usually forming the basal part, so that a
surge deposit may occur beneath an ignimbrite (Fig.
10.8). Many of these *ground surges* are generated by
collapse of an eruption column. The second type is a
base surge and this usually forms during phreatic and
phreatomagmatic eruptions, where magma comes
into contact with water.

Pyroclastic-surge deposits are generally thin and
fine grained, and usually they drape the topography,
but with a thicker accumulation in depressions (Fig.
10.7). Grain size and bed thickness decrease away
from the volcano and erosive bases and channel structures are common. The deposits show evidence of
downflow decrease in turbulence, particle concentration and suspended-load fallout rate and an increase in
traction processes. The distinguishing feature of base-
surge deposits is the presence of stratification, and planar and cross-bedding (Fig. 10.10). Antidune bedding,
a feature of high-velocity flows (see Section 2.3.2), is
not uncommon in base-surge deposits. The low undulating bedforms are preserved and show low-angle
upflow-directed cross-bedding. Maximum bed thickness is generally less than 1 m. Accretionary lapilli may
develop during the surge as a result of accretion of volcanic dust and fine ash onto wet nuclei. Base surges
frequently are associated with maar-type volcanoes,
i.e. those that erupt in lakes. Recent descriptions include Colella & Hiscott (1997) from the Pleistocene of

Fig. 10.10 Base-surge tuffs showing well-developed planar
bedding and both upstream- and downstream-directed cross-
bedding (upper part) formed through antidune development.
Flow from left to right. A volcanic bomb is present, upper left,
with a characteristic sag of the bedding beneath it. Pleistocene.
Eifel, Germany.

Italy and Bull & Cas (2000) from the Devonian of
southeast Australia.

10.4.3 Lahar deposits

Lahars or volcanic mudflows occur on the slopes of
some subaerial volcanoes. Cold lahars are mostly produced by heavy rain falling on unconsolidated ash.
Hot lahars are formed when a pyroclastic flow enters a
lake or river or when air-fall ash is dumped into a
crater lake, which then overflows. Hot lahars developed from pyroclastic surges during the 1980 Mount
St Helens eruption (see Brantley & Waitt, 1988).
Lahar deposits have textures similar to those of mudflows on alluvial fans (Section 2.11.1) and debris flows
in the deep sea (Section 2.11.7): a lack of sorting and
matrix-support fabric (Fig. 10.11). The basal layer
may be reversely graded and large blocks may 'float' in
the ash.

10.5 Hydroclastites: hyaloclastites and hyalotuffs

When extruding lava comes into contact with water,
the rapid chilling and quenching causes fragmentation
of the lava. The surface of the lava flow is chilled and as
the flow moves forward, the surface rind is fragmented
and granulated, allowing more magma to be chilled
and fragmented. This may be a fairly gentle process
or highly explosive. The grains produced by this

Fig. 10.11 Lahar deposit, consisting of large, angular lithic clasts scattered randomly in a matrix of low-density pumice and ash. From the Minoan eruption of Santorini, Greece. Courtesy of Steve Sparks.

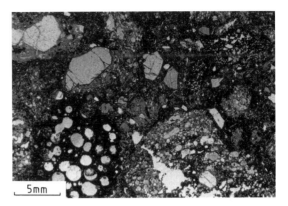

Fig. 10.13 Hyaloclastite consisting of vesicular lava fragments and broken crystals in matrix of fine glass, much of which is being altered to clays. Plane-polarized light. Quaternary. Iceland.

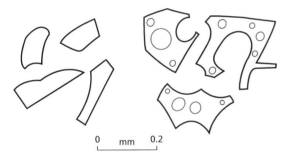

Fig. 10.12 Typical grain shapes for hydroclastic deposits. Left, hyaloclastites; right, hyalotuffs.

magma–water interaction are called *hydroclasts*. Volcaniclastic sediments produced through this process are generally known as hyaloclastites (formerly aquagene tuffs) but a distinction can be made between *hyaloclastites sensu stricto*, formed by non-explosive fragmentation of lava by water, and *hyalotuffs* resulting from explosive magma–water interaction. The main difference between these two types is in the shapes of the lava fragments: less vesicular and more planar surfaces in the case of hyaloclastites, and more vesicular with concave outer surfaces (broken-bubble walls) in the hyalotuffs (Fig. 10.12). The lava grains

vary from a few millimetres to many centimetres in diameter (Fig. 10.13). Such tuffs can be produced where a lava is extruded underwater, where a subaerially erupted lava flows into water, or where lava is erupted subglacially (often the case in Iceland).

In shallow water, apart from chilling, lava fragmentation is also brought about by exsolution of gases from the magma, and this causes vesiculation. A pressure compensation level (PCL) can be recognized, below which the hydrostatic pressure inhibits explosive eruptions, and lava fragmentation takes place purely by the sudden chilling. The depth varies between 500 and 1000 m, depending on the composition and volatile content of the magma. Above the PCL, hydroclastic and pyroclastic deposits are abundant, whereas below, pillow lavas are more common, along with volcanic breccia of broken pillows to small pieces of pillow rind. With submarine basaltic eruptions, there are commonly units of pillow lava passing up into pillow breccia and then hyaloclastites (Fig. 10.14). They reflect the contact of the upper parts of the lava flow with the cold seawater. Many hyaloclastites are without any apparent stratification or any degree of sorting, although in shallow water they can be reworked by waves and currents to give sedimentary structures as with any other clastic sediment. They also may be transported into deep water through slumping and turbidity currents, to give graded beds.

Hyaloclastites and pillow-lava breccias have been

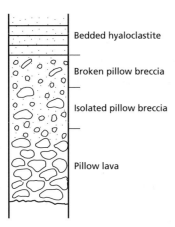

Fig. 10.14 Sketch section through a typical pillow-lava–hydroclastite sequence. Such units are typically several tens of metres in thickness. After Suthren (1985).

recovered from many sites on the ocean floor during deep-sea drilling. They are a characteristic feature of submarine basaltic volcanism and so occur in many sequences of active continental-margin–oceanic settings and in greenstone belts of the Precambrian.

Hyalotuffs are best developed in subaerial or coastal situations where water seeps into a volcanic vent to cause a violent eruption (Surtseyan-type of volcano). The hyalotuffs generally are fine grained, poorly sorted lava fragments lacking bombs or lava droplets.

10.6 Epiclastic volcanogenic deposits

Once deposited by volcanic processes, volcaniclastic sediments can be reworked in the sedimentary environment in the same way as any other sediment. In continental settings, volcanic ash is carried into river systems and lakes by surface runoff, deflated by the wind and incorporated into soils. In the shallow-marine environment, ash will be reworked by waves, tidal and storm currents and mixed with non-volcanogenic material. Redeposition of hyaloclastites in the deep sea by sediment gravity flows has been noted already. Thus all the various depositional sedimentary structures described in Chapter 2 can be found in reworked volcaniclastic deposits. Being relatively soft and friable, volcanic debris is easily broken down into finer grades and rounded by abrasion in moderate- to high-energy environments.

10.7 Diagenesis of volcaniclastic sediments

Volcanic glass is metastable so that, excepting special conditions, it is not preserved in rocks older than mid-Tertiary. Volcanic glass is readily devitrified, altered and replaced during weathering and diagenesis. Volcaniclastic sediments can be difficult to recognize as a result of this alteration. The common alteration products are clay minerals and zeolites, and, in modern submarine basaltic volcaniclastic sediments, palagonite.

The clay minerals that replace volcanic glass are mainly smectites, in particular montmorillonite and saponite in more basic ashes, and kaolinite in feldspathic ashes. Chlorite also may replace basic tephra. Smectite-rich clay beds derived from the alteration of volcanic ash are known as *bentonites*. Some kaolinite-rich mudrocks called *tonsteins* are of volcanic-ash origin. Apart from clay mineralogy, the presence of some glass shards or their pseudomorphs, together with euhedral or zoned phenocrysts, especially of quartz, feldspar or pyroxene, will further confirm a volcanic origin. Trace elements also can be used. Spears *et al.* (1999), for example, demonstrated a rhyodacite volcanic origin for some bentonites and tonsteins from the British Carboniferous on the basis of a distinctive trace-element composition, although there are few glass shards or crystals, other than biotite flakes concentrated at the base of the beds. Diagenetically altered volcaniclastic deposits include the bentonites from the Ordovician and Cretaceous of North America (e.g. Delano *et al.,* 1990) and the 'fuller's earths' from the Mesozoic of Britain.

The common *zeolites* formed by alteration of volcanic glass are analcime, clinoptilolite, phillipsite, laumontite and mordenite. They usually are cryptocrystalline, locally fibrous. Zeolites are commonly developed where ash has fallen into alkaline lakes. Lacustrine sequences of the Eocene Green River Formation, western USA and the Quaternary of East Africa contain such zeolitic horizons. In fact, volcanic ash is not a prerequisite; if the chemistry of a saline alkaline lake is appropriate, then zeolites will be precipitated. Examples include the Triassic Lockatong Formation of eastern USA and the present-day Lake Natron, Tanzania. The zeolite phillipsite is an important constituent of hemipelagic muds on the floor of the Pacific Ocean, having formed from palagonite. Zeolite minerals are discussed by Gottardi & Galli

(1985) and there is a journal, *Zeolites*, devoted to the subject.

Many submarine basaltic tuffs are altered to *sideromelane*, a translucent variety of glass, or to *palagonite*, an amorphous yellow-to-orange mineraloid. Pillow lavas generally possess a rind of palagonite and many hyaloclastite grains have been completely altered to palagonite (hence the older term palagonite tuffs). Processes of alteration taking place during palagonitization are hydration of the glass, oxidation of iron, a loss of some silica and Na and Mg, and an increase in K and Fe. Palagonite itself is not a mineral, but an intergrowth of montmorillonite and phillipsite.

Finally, tuffs may be replaced by silica or calcite. Silica is released on the alteration of glass to clays and zeolites and this can be precipitated as chert by replacement and as a cement. Completely silicified tuffs have been referred to as porcelanite or halleflinta. Calcite is commonly a cement of tuffs, as other sediments, but it also may replace the volcanic grains.

Further reading

Cas, R.A.F. & Busby-Spera, C.J. (Eds) (1991) *Volcaniclastic Sedimentation. Sedimentary Geology* **74**, 362 pp.

Cas, R.A.F. & Wright, J.V. (1987) *Volcanic Successions: Modern and Ancient.* Unwin–Hyman, London, 528 pp.

Fisher, R.V. & Schmincke, H.U. (1984) *Pyroclastic Rocks.* Springer-Verlag, Berlin, 472 pp.

Fisher, R.V. & Schmincke, H.U. (1994) Volcaniclastic sediment transport and deposition. In: *Sediment Transport and Depositional Processes* (Ed. K. Pye). Blackwell Scientific Publications, Oxford, pp. 351–388.

Fisher, R.V. & Smith, G.A. (Eds) (1991) *Sedimentation in Volcanic Settings.* Special Publication 45, Society of Economic Paleontologists and Economic Mineralogists, Tulsa, OK, 257 pp.

Lajoie, J. & Stix, J. (1992) Volcaniclastic rocks. In: *Facies Models: Response to Sea-level Change* (Eds R.G. Walker & N.P. James). Geological Association of Canada, Waterloo, Ontario, pp. 101–118.

Orton, G.J. (1996) Volcanic environments. In: *Sedimentary Environments: Processes, Facies and Stratigraphy* (Ed. H.G. Reading). Blackwell Science, Oxford, pp. 485–567.

Also see recent issues of the journal *Bulletin of Volcanology* and *Journal of Volcanology and Geothermal Research.*

References

Adams, A.E. & MacKenzie, W.S. (1998) *A Colour Atlas of Carbonate Sediments and Rocks under the Microscope.* Manson Publishing, London, 180 pp.

Adams, A.E., Mackenzie, W.S. & Guilford, C. (1984) *Atlas of Sedimentary Rocks under the Microscope.* Longman, Harlow, 104 pp.

Aigner, T. (1984) Dynamic stratigraphy of epicontinental carbonates, Upper Muschelkalk (M. Triassic), South German Basin. *Neues Jahrbuch für Geologische und Palaeontologie, Abhandlungen* **169**, 127–159.

Alexander, C., Davis, R.A. & Henry, V.J. (1998) *Tidalites. Processes and Products.* Special Publication 61, Society of Sedimentary Geology, Tulsa, OK, 172 pp.

Allen, J.R.L. (1982) *Sedimentary Structures: their Character and Physical Basis.* Developments in Sedimentology, 30A & B. Elsevier, Amsterdam, 1258 pp.

Allen, J.R.L. (1986) Pedogenic calcretes in the Old Red Sandstone facies (Late Silurian–Early Carboniferous) of the Anglo-Welsh area, southern Britain. In: *Paleosols: their Recognition and Interpretation* (Ed. V.P. Wright). Blackwell Scientific Publications, Oxford, pp. 58–86.

Allen, P.A. (1997) *Earth Surface Processes.* Blackwell Science, Oxford, 404 pp.

Allen, P.A. & Allen, J.R.L. (2001) *Basin Analysis, Principles and Applications.* Blackwell Science, Oxford.

Allen, P.A. & Homewood, P. (Eds) (1986) *Foreland Basins.* Special Publication 13, International Association of Sedimentologists. Blackwell Scientific Publications, Oxford, 453 pp.

Aller, R.C., Mackin, J.E. & Cox, R.T. (1986) Diagenesis of Fe and S in Amazon inner shelf muds: apparent dominance of Fe reduction and implications for the genesis of ironstones. *Continental Shelf Research* **6**, 263–289.

Alsop, G.I., Blundell, D.J. & Davison, I. (Eds) (1995) *Salt Tectonics.* Special Publication 100, Geological Society of London, Bath, 310 pp.

Amorosi, A. (1995) Glaucony and sequence stratigraphy: a conceptual framework of distribution in siliciclastic sequences. *Journal of Sedimentary Research* **B65**, 419–425.

Anadon, P., Cabrera, L.I. & Kelts, K. (Eds) (1991) *Lacustrine Facies Analysis.* Special Publication, 13, International Association of Sedimentologists. Blackwell Scientific Publications, Oxford, 328 pp.

Anderson, J.B. & Ashley, G.M. (Eds) (1991) *Glacial Marine Sedimentation: Paleoclimatic Sigificance.* Special Publication 261, Geological Society of America.

Anderson, G.M. & McQueen, R.W. (1982) Ore Deposit Models—6. Mississippi Valley-type Pb–Zn Deposits. *Geoscience Canada*, **9**, 108–117.

Aplin, A., Fleet, A.J. & Macquaker, J.H.S. (1999) *Muds and Mudstones: Physical and Fluid-flow Properties.* Special Publication 158, Geological Society of London, Bath, 200 pp.

Armstrong, H.A., Owen, A.W. & Floyd, J.D. (1999) Rare earth geochemistry of Arenig cherts from the Ballantrae Ophiolite and Leadhills Imbricate Zone, southern Scotland: implications for origin and significance to the Caledonian Orogeny. *Journal of the Geological Society* **156**, 549–560.

Arthur, M.A. & Sageman, B.B. (1994) Marine black shales: depositional mechanisms and environments of ancient deposits. *Annual Review of Earth and Planetary Science* **22**, 499–551.

Ashley, G.M. (Chairperson) & others (1990) Classification of large-scale subaqueous bedforms: a new look at an old problem. *Journal of Sedimentary Petrology* **60**, 160–172.

Bahlburg, H. & Floyd, P.A. (Eds) (1999) Advanced techniques in provenance analysis of sedimentary rocks. *Sedimentary Geology* **124**, 1–215.

Baker, J.C., Jell, J.S., Hacker, J.L.F. & Baublys, K.A. (1998) Origin of recent insular phosphate rock on a coral cay, Raine Island, northern Great Barrier Reef, Australia. *Journal of Sedimentary Research* **68**, 1001–1008.

Barbieri, F. *et al.* (1989) Magmatic and phreatomagmatic phases in explosive eruptions of Vesuvius as deduced by grain-size and component analysis of the pyroclastic deposits. *Journal of Volcanology and Geothermal Research* **38**, 287–307.

Barrett, T.J. (1982) Stratigraphy and sedimentology of Jurassic bedded chert overlying ophiolites in the North Apennines, Italy. *Sedimentology* **29**, 353–373.

Barrett, T.J. & Fralick, P.W. (1989) Turbidites and iron-formations, Beardmore–Geraldton, Ontario: application of a combined ramp/fan model to Archean clastic and chemical sedimentation. *Sedimentology* **36**, 221–234.

Barwis, J.H., McPherson, J.G. & Studlick, J.R.J. (1990) *Sandstone Petroleum Reservoirs.* Springer-Verlag, Berlin, 584 pp.

Basu, A. *et al.* (1975) Re-evaluation of the use of undulatory extinction and polycrystallinity in detrital quartz for provenance interpretation. *Journal of Sedimentary Petrology* **45**, 873–882.

Bathurst, R.G.C. (1975) *Carbonate Sediments and Their Diagenesis*. Elsevier, Amsterdam, 658 pp.

Batist, M. & Jacobs, P. (Eds) (1996) *Geology of Siliciclastic Shelf Seas*. Special Publication 117, Geological Society of London, Bath, 345 pp.

Beier, J.A. (1987) Petrographic and geochemical analysis of caliche profiles in a Bahamian Pleistocene dune. *Sedimentology* **34**, 991–998.

Beil, R.J. & Garrison, R.E. (1994) The origin of chert in the Monterey Formation of California (USA). In: *Siliceous, Phosphatic and Glauconite Sediments of the Tertiary and Mesozoic* (Eds A. Iijima, A.M. Abed & R.E. Garrison). 29th International Geological Congress Proceedings Part C. VSP, Utrecht, pp.101–132.

Bellanca, A., Calderone, S. & Neri, R. (1986) Isotope geochemistry, petrology and depositional environments of the diatomite-dominated Tripoli Formation (Lower Messinian), Sicily. *Sedimentology* **33**, 729–743.

Bellanca, A., Masetti, D., Neri, R. & Venezia, F. (1999) Geochemical and sedimentological evidence of productivity cycles recorded in Toarcian Black shales from the Belluno Basin, Southern Alps, Northern Italy. *Journal of Sedimentary Research* **69**, 466–476.

Benison, K.C. & Goldstein, R.H. (2000) Sedimentology of ancient saline pans: an example from the Permian Opeche Shale, Williston Basin, North Dakota, USA. *Journal of Sedimentary Research* **70**, 159–169.

Berg, R.R. (1986) *Reservoir Sandstones*. Prentice Hall, Englewood Cliffs, NJ, 481 pp.

Bergman, K.M. & Snedden, J.W. (Eds) (1999) *Isolated Shallow-marine Sand Bodies: Sequence Stratigraphic Analysis and Sedimentology Interpretation*. Special Publication 64, Society of Sedimentary Geologists, Tulsa, OK, 362 pp.

Berner, R.A. (1981) New geochemical classification of sedimentary environments. *Journal of Sedimentary Petrology* **51**, 359–365.

Besley, B.M. & Kelling, G. (1988) *Sedimentation in a Synorogenic Basin Complex. The Upper Carboniferous of Northwest Europe*. Blackie, Glasgow, 276 pp.

Best, J.L. & Bristow, C.S. (Eds) (1993) *Braided Rivers*. Special Publication 75, Geological Society of London, Bath, 432 pp.

Bhattacharya, J.P. & Walker, R.G. (1992) Deltas. In: *Facies Models: Responses to Sea-level Change* (Eds R.G. Walker & N.P. James). Geological Association of Canada, Waterloo, Ontario, pp. 157–178.

Biddle, K.T. (Ed.) (1991) *Active Margin Basins*. Memoir 52, American Association of Petroleum Geologists, Tulsa, OK, 324 pp.

Biddle, K.T. & Christic-Blick, N. (Eds) (1985) *Strike-slip Deformation, Basin Formation and Sedimentation*. Special Publication 37, Society of Economic Paleontologists and Mineralogists, Tulsa, OK, 386 pp.

Bjørlykke, K. (1988) Sandstone diagenesis in relation to preservation, destruction and creation of porosity. In: *Diagenesis I* (Eds G.V. Chilingar & K.H. Wolf). Elsevier, Amsterdam, pp. 555–587.

Bjørlykke, K. (1989) *Sedimentology and Petroleum Geology*. Springer-Verlag, Berlin, 370 pp.

Bjørlykke, K. & Aagaard, P. (1992) Clay minerals in North Sea Sandstones. In: *Origin, Diagenesis and Petrophysics of Clay Minerals in Sandstones* (Eds D.W. Houseknecht & E.D. Pittman). Special Publication 47, Society for Sedimentary Geologists, Tulsa, OK, pp. 65–80.

Blair, T.C. (1999a) Sedimentary processes and facies of the waterlaid Anvil Spring Canyon alluvial fan, Death Valley, California. *Sedimentology* **46**, 913–940.

Blair, T.C. (1999b) Sedimentology of the debris-flow-dominated Warm Spring Canyon alluvial fan, Death Valley, California. *Sedimentology* **46**, 941–965.

Blair, T.C. & McPherson, J.G. (1999) Grain-size and textural classification of coarse sedimentary particles. *Journal of Sedimentary Research* **69**, 6–19.

Blum, M.D. & Törnqvist, T.E. (2000) Fluvial responses to climate and sea-level change, a review and look forward. *Sedimentology* **47** (Suppl. 1), 2–48.

Bock, B., McLennan, S.M. & Hanson, G.N. (1998) Geochemistry and provenance of the Middle Ordovician Austin Glen Member (Normanskill Formation) and the Taconian Orogeny in New England. *Sedimentology* **45**, 635–655.

Bond, A. & Sparks, R.S.J. (1976) The Minoan eruption of Santorini, Greece. *Journal of the Geological Society of London* **132**, 1–16.

Bosellini, A. (1984) Progradation geometries of carbonate platforms: examples from the Triassic of the Dolomites, northern Italy. *Sedimentology* **31**, 1–24.

Bosence, D.W.J. (1995) Anatomy of a Recent biodetrital mud-mound from Florida Bay, USA. In: *Carbonate Mud Mounds: their Origin and Evolution* (Eds C.L.V. Monty, D.W.J. Bosence, P.H. Bridges & B.R. Pratt). Special Publication 23, International Association of Sedimentologists. Blackwell Science, Oxford, pp. 475–493.

Boudon, G. & Gourgaud, A. (Eds) (1989) *Mont Pelée* (Special Issue). *Journal of Volcanology and Geothermal Research* **38**, 1–214.

Bouma, A.H. (1969) *Methods for the Study of Sedimentary Structures*. Wiley, New York, 458 pp.

Bourque, P.-A. & Boulvain, F. (1993) A model for the origin and petrogenesis of the red stromatactis limestone of Paleozoic carbonate mounds. *Journal of Sedimentary Petrology* **63**, 607–619.

Bown, T.M. & Kraus, M.J. (1987) Integration of channel and floodplain suites, I. Development, sequence and lateral relations of alluvial paleosols. *Journal of Sedimentary Petrology* **57**, 58–601.

Brantley, S.R. & Waitt, R.B. (1988) Interrelations among pyroclastic surge, pyroclastic flow and lahars in Smith

Creek valley during first minutes of 18 May, 1980 eruption of Mount St. Helens, USA. *Bulletin of the Volcanology* **50**, 304–326.

Brenchley, P.J. & Harper, D.A.T. (1998) *Palaeoecology.* Chapman & Hall, London, 402 pp.

Breyer, J.A. (1997) Sequence stratigraphy of Gulf Coast lignite, Wilcox Group (Paleogene), South Texas. *Journal of Sedimentary Research* **67**, 1018–1029.

Bridge, J.S., Collier, R. & Alexander, J. (1998) Large-scale structure of Calmus River deposits (Nebraska, USA) revealed using ground-penetrating radar. *Sedimentology* **45**, 977–986.

Brindley, G.W. & Brown, G. (Eds) (1984) *Crystal Structures of Clay Minerals and their X-Ray Identification.* Mineralogical Society, London, 495 pp.

Bromley, R.G. (1996) *Trace Fossils. Biology, Taphonomy and Applications.* Chapman & Hall, London, 361 pp.

Brookfield, M.E. (1992) Eolian systems. In: *Facies Models: Response to Sea-level Change* (Eds R.G. Walker & N.P. James). Geological Association of Canada, Waterloo, Ontario, pp. 143–156.

Brookfield, M.E. & Martini, I.P. (1999) Facies architecture and sequence stratigraphy in glacially influenced basins: basic problems and water-level/glacier input-point control (with an example from the Quaternary of Ontario, Canada). *Sedimentary Geology* **123**, 183–197.

Brooks, J. & Fleet, A.J. (Eds) (1987) *Marine Petroleum Source Rocks.* Special Publication 26, Geological Society of London, Bath, 444 pp.

Budd, D.A., Saller, A.H. & Harris, P.A. (Eds) (1995) *Unconformities and Porosity in Carbonate Strata.* Memoir 63, American Association of Petroleum Geologists, 313 pp.

Buick, R. & Dunlop, J.S.R. (1990) Evaporitic sediments of early Archaean age from the Warrawoona Group, North Pole, Western Australia. *Sedimentology* **37**, 247–277.

Bull, P.A. (1981) Environmental reconstruction by electron microscopy. *Progress in Physical Geography* **5**, 368–397.

Bull, S.W. & Cas, R.A.F. (2000) Distinguishing base-surge deposits and volcaniclastic fluviatile sediments: an ancient example from the Lower Devonian Snowy River Volcanics, SE Australia. *Sedimentology* **47**, 87–98.

Burchette, T.P., Wright, V.P. & Faulkner, T.J. (1990) Oolitic sand-body depositional models and geometries, Mississippian of southwest Britain: implications for petroleum exploration in carbonate ramp settings. *Sedimentary Geology*, **68**, 87–115.

Burley, S.D., Kantorowicz, J.D. & Waugh, B. (1985) Clastic diagenesis. In: *Sedimentology: Recent Developments and Applied Aspects* (Eds P.J. Brenchley & B.J.P. Williams). Special Publication 18, Geological Society of London, Bath, pp. 189–226.

Burnett, W.C. & Riggs, S.R. (Eds) (1990) *Phosphate Deposits of the World*, Vol. 3, *Neogene to Modern Phosphorites.* Cambridge University Press, Cambridge.

Burns, S.J., McKenzie, J.A. & Vasconcelos, C. (2000) Dolomite formation and biogeochemical cycles in the Phanerozoic. *Sedimentology* **47** (Suppl. 1), 49–61.

Busby, C.J. & Ingersoll, R.V. (Eds) (1995) *Tectonics of Sedimentary Basins.* Blackwell Science, Oxford, 579 pp.

Carter, R.W.G. (1988) *Coastal Environments.* Academic Press, London, 617 pp.

Cas, R.A.F. & Wright, J.V. (1987) *Volcanic Successions. Modern and Ancient* Unwin–Hyman, London, 528 pp.

Casas, E. & Lowenstein, T.K. (1989) Diagenesis of saline pan halite: comparison of petrographic features of modern, Quaternary and Permian halites. *Journal of Sedimentary Petrology* **59**, 724–739.

Chafetz, H.S. (1986) Marine peloids: a product of bacterially induced precipitation of calcite. *Journal of Sedimentary Petrology* **56**, 812–817.

Chamley, H. (1989) *Clay Sedimentology.* Springer-Verlag, Berlin, 623 pp.

Chan, M.A. (1999) Triassic loessite of North-Central Utah: stratigraphy, petrophysical character, and paleoclimate implications. *Journal of Sedimentary Research* **69**, 477–485.

Cheel, R.J. & Leckie, D.A. (1993) Hummocky Cross-stratification. *Sedimentology Review* **1**, Blackwell Scientific Publications, Oxford, pp. 103–122.

Choquette, P.W. & James, N.P. (1987) *Diagenesis in Limestones—3, The Deep Burial Environment.* Geoscience Canada, **14**, 3–35.

Chough, S.K., Kim, S.B. & Chun, S.S. (1996) Sandstone–chert and laminated black shale couplets, Cretaceous Uhangri Formation (SW Korea): depositional events in alkaline lake environments. *Sedimentary Geology* **104**, 227–242.

Clari, P.A. & Martire, L. (1996) Interplay of cementation, mechanical compaction, and chemical compaction in nodular limestones of the Rosso Ammonitico Veronese (Middle–Upper Jurassic, Northeastern Italy). *Journal of Sedimentary Research* **66**, 447–458.

Clark, J.D. & Pickering, K.T. (Eds) (1996) *Submarine Channels. Processes and Architecture.* Vallis Press, London, 150 pp.

Clauzon, G. *et al.* (1996) Alternate interpretation of the Messinian salinity crisis: controversy resolved? *Geology* **24**, 363–366.

Cloud, P. (1983) Banded iron-formation—a gradualist's dilemma. In: *Iron-formation: Facts and Problems* (Eds A.F. Trendall & R.C. Morris). Elsevier, Amsterdam, pp. 401–416.

Cole, R.B. & Stanley, R.G. (1994) Sedimentology of subaqueous volcaniclastic sediment gravity flows in the Neogene Santa Maria Basin, California. *Sedimentology* **41**, 37–54.

Colella, A. & Hiscott, R.N. (1997) Pyroclastic surges of the Pleistocene Monte Guardia sequence (Lipari Island, Italy): depositional processes. *Sedimentology* **44**, 47–66.

Colella, A. & Prior, D. (Eds) (1990) *Coarse-grained Deltas.* Special Publication 10, International Association of Sedimentologists. Blackwell Scientific Publications, Oxford, 368 pp.

Collinson, J.D. (1996) Alluvial sediments. In: *Sedimentary Environments: Processes, Facies and Stratigraphy* (Ed. H.G. Reading) Blackwell Science, Oxford, pp. 37–82.

Collinson, J.D. & Thompson, D.B. (1988) *Sedimentary Structures.* Unwin–Hyman, London, 207 pp.

Colson, J. & Cojan, I. (1996) Groundwater dolocretes in a lake-marginal environment: an alternative model for dolocrete formation in continental settings (Danian of the Provence Basin, France). *Sedimentology* **43**, 175–188.

Cook, P.J. & Shergold, J.H. (Eds) (1986) *Phosphate Deposits of the World. 1, Proterozoic and Cambrian Phosphorites.* Cambridge University Press, Cambridge, 386 pp.

Cooke, R., Warren, A. & Goudie, A. (1993) *Desert Geomorphology.* University College London Press, London, 526 pp.

Cope, J.C.W. & Curtis, C.D. (2000) Palaeobiology meets geochemistry: concretions as tombs. *Journal of the Geological Society* **157**, 163–164.

Cowan, C.A. & James, N.P. (1992) Diastasis cracks: mechanically generated synaeresis-like cracks in Upper Cambrian shallow-water oolite and ribbon carbonates. *Sedimentology* **39**, 1101–1118.

Coward, M.P., Dewey, J.F. & Hancock, P.L. (Eds) (1987) *Continental Extensional Tectonics.* Special Publication 28, Geological Society of London, Bath.

Cox, R. & Lowe, D.R. (1996) Quantification of the effect of secondary matrix on the analysis of sandstone composition and a petrographic–chemical technique for retrieving original framework grain modes of altered sandstones. *Journal of Sedimentary Research* **66**, 548–558.

Crevello, P.D., Wilson, J.L., Sarg, J.F. & Read, J.F. (Eds) (1989) *Controls on Carbonate Platform and Basin Development.* Special Publication 44, Society of Economic Paleontologists and Mineralogists, Tulsa, OK, 405 pp.

Cronan, D.S. (1980) *Underwater Minerals.* Academic Press, London, 362 pp.

Crossey, L.J., Loucks, R. & Totten, M.W. (Eds) (1996) *Siliciclastic Diagenesis and Fluid Flow: Concepts and Applications.* Special Publication 55, Society of Sedimentary Geologists, Tulsa, OK, 222 pp.

Crowell, J.C. (Ed.) (1999) *Pre-Mesozoic Ice Ages: their Bearing on Understanding the Climate System.* Memoir 192, Geological Society of America, Boulder, CO, 112 pp.

Dalrymple, R.W., Boyd, R. & Zaitlin, B.A. (Eds) (1994) *Incised-valley Systems: Origins and Sedimentary Sequences.* Special Publication 51, Society for Sedimentary Geology, Tulsa, OK.

Datta, B., Sarkar, S. & Chaudhuri, A.K. (1999) Swaley cross-stratification in medium to coarse sandstone produced by oscillatory and combined flows: examples from the Proterozoic Kansapathar Formation, Chattisgarh Basin, M. P. , India. *Sedimentary Geology* **129**, 51–70.

Davis, H.R., Byers, C.W. & Pratt, L.M. (1989) Depositional mechanisms and organic matter in Mowry Shale (Cretaceous), Wyoming. *American Association of Petroleum Geologists, Bulletin* **73**, 1103–1116.

Davis, R.A. (Ed.) (1994) *Geology of Holocene Barrier Island Systems.* Springer-Verlag, New York, 500 pp.

De Deckker, P. & Last, W.M. (1989) Modern, non-marine dolomite in evaporitic playas of western Victoria, Australia. *Sedimentary Geology* **64**, 223–238.

Dean, W.E. & Anders, D.E. (1991) Effects of source, depositional environment and diagenesis on characteristics of organic matter in oil shale from Green River Formation, Wyoming, Utah and Colorado. *Bulletin of the United States Geological Survey* **1973**, F1–F16.

Dean, W.E. & Fouch, T.D. (1983) Lacustrine environment. In: *Carbonate Depositional Environments* (Eds P.A. Scholle, D.G. Bebout & C.H. Moore). Memoir 33, American Association of Petroleum Geologists, Tulsa, OK, pp. 97–130.

Delano, L.W. *et al.* (1990) Petrology and geochemistry of Ordovician K-bentonites in New York State: constraints on the nature of a volcanic arc. *Journal of Geology* **98**, 157–170.

Demicco, R.V. & Hardie, L.A. (1995) *Sedimentary Structures and Early Diagenetic Features of Shallow-Marine Carbonate Deposits.* Atlas Series **1**, American Association of Petroleum Geologists, Tulsa, OK, 256 pp.

De Ros, L.F., Sgarbi, G.N.C. & Morad, S. (1994) Multiple authigenesis of K-feldspar in sandstones: evidence from the Cretaceous Areado Formation, São Francisco Basin, Central Brazil. *Journal of Sedimentary Research*, **A64**, 778–787.

Deynoux, M., Kocurek, G. & Proust, J.N. (1989) Late Proterozoic periglacial aeolian deposits on the west African Platform, Taoudeni Basin, western Mali. *Sedimentology* **36**, 531–550.

Dickinson, W.R. (1985) Interpreting provenance relations from detrital modes of sandstones. In: *Provenance of Arenites* (Ed. G.G. Zuffa). Reidel, Dordrecht, pp. 333–361.

Dimberline, A.J., Bell, A. & Woodcock, N.H. (1990) A laminated hemipelagic facies from the Wenlock and Ludlow of the Welsh Basin. *Journal of the Geological Society of London* **147**, 693–701.

Dix, G.R. (1993) Patterns of burial- and tectonically controlled dolomitization in an Upper Devonian fringing-reef complex. Leduc Formation, Peace River Arch Area, Alberta, Canada. *Journal of Sedimentary Petrology* **63**, 628–640.

Dix, G.R. & Mullins, H.T. (1988) Rapid burial diagenesis of deep-water carbonates, Exuma Sound, Bahamas. *Geology* **16**, 680–683.

Donaldson, W.S., Plint, A.G. & Longstaffe, F.J. (1999) Tectonic and eustatic control on deposition and preservation of Upper Cretaceous ooidal ironstones and associated facies: Peace River Arch area, NW Alberta, Canada. *Sedimentology* **46**, 1159–1182.

Donovan, S.K. (Ed.) (1994) *The Palaeobiology of Trace Fossils*. Wiley, Chichester, 308 pp.

Donselaar, M.E. (1989) The Cliff House Sandstone, San Juan Basin, New Mexico: model for the stacking of 'transgressive' barrier complexes. *Journal of Sedimentary Petrology* **59**, 13–27.

Doyle, L.J. & Roberts, H.H. (Eds) (1988) *Carbonate–Clastic Transitions*. Elsevier, Amsterdam, 304 pp.

Dowdeswell, J.A. & Scourse, J.D. (Eds) (1990) *Glaciomarine Environments: Processes and Sediments*. Special Publication 53, Geological Society of London, Bath.

Dravis, J. & Yurewicz, D.A. (1985) Enhanced carbonate petrography using fluorescence microscopy. *Journal of Sedimentary Petrology* **55**, 795–804.

Drivet, E. & Mountjoy, E.W. (1997) Dolomitization of the Leduc Formation (Upper Devonian), Southern Rimbey–Meadowbrook Reef Trend, Alberta. *Journal of Sedimentary Research* **67**, 411–423.

Dunagan, S.P. & Driese, S.G. (1999) Control of terrestrial stabilization on Late Devonian palustrine carbonate deposition, Catskill Magnafacies, New York, USA. *Journal of Sedimentary Research* **69**, 772–783.

Dworkin, S.I. & Land, L.S. (1994) Petrographic and geochemical constraints on the formation and diagenesis of anhydrite cements, Smackover Sandstones, Gulf of Mexico. *Journal of Sedimentary Research* **A64**, 339–348.

Eberli, G.P. & Ginsburg, R.N. (1989) Cenozoic progradation of northwestern Great Bahama Bank: a record of lateral platform growth and sea-level fluctuations. In: *Controls on Carbonate Platform and Basin Development* (Eds P.D. Crevello, J.L. Wilson, J.F. Sarg & J.F. Read). Special Publication 44, Society of Economic Paleontologists and Mineralogists, Tulsa, OK, pp. 339–351.

Einsele, G. (2000) *Sedimentary Basins. Evolution, Facies and Sediment Budget*. Springer-Verlag, Berlin.

Elliott, L.A. & Warren, J.K. (1989) Stratigraphy and depositional environment of Lower San Andres Formation in subsurface and equivalent outcrops. Chaves, Lincoln and Roosevelt Counties, New Mexico. *American Association of Petroleum Geologists, Bulletin* **73**, 1307–1325.

Els, B.G. (1998) The auriferous late Archean sedimentation systems of South Africa: unique palaeo-environmental conditions. *Sedimentary Geology* **120**, 205–224.

El-Tabakh, M., Schreiber, B.C. & Warren, J.K. (1998) Origin of gypsum in the Newark rift basin, eastern North America. *Journal of Sedimentary Research* **68**, 88–99.

Emery, D. & Meyers, K. (1996) *Sequence Stratigraphy*. Blackwell Science, Oxford, 297 pp.

Emery, D. & Robinson, A. (1993) *Inorganic Geochemistry*. Blackwell Science, Oxford, 254 pp.

Enos, P. & Stephens, B.P. (1993) Mid-Cretaceous basin-margin carbonates, east–central Mexico. *Sedimentology* **40**, 539–556.

Esteban, M. & Klappa, C.F. (1983) Subaerial exposure environment. In: *Carbonate Depositional Environments* (Eds P.A. Scholle, D.G. Bebout & C.H. Moore). Memoir 33, American Association of Petroleum Geologists, Tulsa, OK, pp. 1–54.

Eyles, C.H., Eyles, N. & Gostin, V.A. (1998) Facies and allostratigraphy of high-latitude, glacially influenced marine strata of the Early Permian, southern Sydney Basin, Australia. *Sedimentology* **45**, 121–161.

Eyles, N., Eyles, C.H. & Miall, A.D. (1983) Lithofacies types and vertical profile analysis; an alternative approach to the description and environmental interpretation of glacial diamict and diamictite sequences. *Sedimentology* **30**, 393–410.

Fagerstrom, I.A. (1987) *The Evolution of Reef Communities*. Wiley–Interscience, New York, 600 pp.

Fairchild, I.J. (1993) Balmy shores and icy wastes: the paradox of carbonates associated with glacial deposits in Neoproterozoic times. *Sedimentology Review 1*. Blackwell Science, Oxford, pp. 1–16.

Fairchild, I.J., Hendry, G., Quest, M. & Tucker, M.E. (1988) Geochemical analysis of sedimentary rocks. In: *Techniques in Sedimentology* (Ed. M.E. Tucker). Blackwell Science, Oxford, pp. 274–354.

Falk, P.D. & Dorsey, R.J. (1998) Rapid development of gravelly high-density turbidity currents in marine Gilbert-type fan deltas, Loreto Basin, Baja California Sur, Mexico. *Sedimentology* **45**, 331–349.

Fernandez, A., Chauvel, J.J. & Moro, M.C. (1998) Comparative study of the Lower Ordovician ironstones of the Iberian Massif (Zamora, Spain) and of the Armorican Massif (Central Brittany, France). *Journal of Sedimentary Research* **68**, 53–62.

Fielding, C.R. (Ed.) (1993) *Current Research in Fluvial Sedimentology*. *Sedimentary Geology* **85**, 656 pp.

Fielding, C.R. & Webb, J.A. (1996) Facies and cyclicity of the Late Permian Bainmedart Coal Measures in the Northern Prince Charles Mountains, MacRobertson Land, Antarctica. *Sedimentology* **43**, 295–322.

Fisher, Q.J., Raiswell, R. & Marshall, J.D. (1998) Siderite concretions from non-marine shales (Westphalian A) of the Pennines, England: controls on their growth and composition. *Journal of Sedimentary Research* **68**, 1034–1045.

Fisher, R.V. & Schmincke, H.U. (1984) *Pyroclastic Rocks*. Springer-Verlag, Berlin, 472 pp.

Fisher, R.V. & Schmincke, H.U. (1994) Volcaniclastic sediment transport and deposition. In: *Sediment*

Transport and Depositional Processes (Ed. K. Pye). Blackwell Science, Oxford, pp. 351–388.

Fisher, R.V. & Smith, G.A. (Eds) (1991) *Sedimentation in Volcanic Settings*. Special Publication 45, Society of Economic Paleontologists and Mineralogists, Tulsa, OK, 257 pp.

Fleet, A.J, Kelts, K. & Talbot, M.R. (Eds) (1988) *Lacustrine Petroleum Source Rocks*. Special Publication 40, Geological Society of London, Bath, 391 pp.

Flemming, B. & Bartoloma, A. (Eds) (1995) *Tidal Signatures in Modern and Ancient Sediments*. Special Publication 24, International Association of Sedimentologists. Blackwell Science, Oxford, 358 pp.

Flügel, E. (Ed.) (1977) *Fossil Algae. Recent Results and Developments*. Springer-Verlag, Berlin, 375 pp.

Flügel, E. (1982) *Microfacies Analysis of Limestones*. Springer-Verlag, Berlin, 610 pp.

Folk, R.L. & Ward, W. (1957) Brazos River bar: a study in the significance of grain-size parameters. *Journal of Sedimentary Petrology* 27, 3–26.

Folk, R.L., Andrews, P.B. & Lewis, D.W. (1970) Detrital sedimentary rock classification and nomenclature for use in New Zealand. *New Zealand Journal of Geology and Geophysics* 13, 937–968.

Fralick, P. & Barrett, T.J. (1995) Depositional controls on iron formations in Canada. In: *Sedimentary Facies Analysis* (Ed. A.G. Plint). Special Publication 22, International Association of Sedimentologists. Blackwell Science, Oxford, pp. 137–156.

Friedman, G.M. & Johnson, K.G. (1982) *Exercises in Sedimentology*. Wiley, New York, 208 pp.

Fritz, W.J. & Moore, J.N. (1988) *Basics of Physical Stratigraphy and Sedimentology*. Wiley, New York, 371 pp.

Frostick, L.E. & Reid, I. (Eds) (1987) *Desert Sediments: Ancient and Modern*. Special Publication 35, Geological Society of London, Bath.

Frostick, L.E. et al. (Ed.) (1986) *Sedimentation in the African Rifts*. Special Publication 25, Geological Society of London.

Gao, G. & Land, L.S. (1991) Nodular chert from the Arbuckle Group, Slick Hills, SW Oklahoma: a combined field, petrographic and isotopic study. *Sedimentology* 38, 857–870.

Garrels, R.M. (1987) A model for the deposition of the microbanded Precambrian iron formation. *American Journal of Science* 287, 81–106.

Garrels, R.M. & Christ, C.L. (1965) *Solutions, Minerals and Equilibria*. Harper & Row, New York.

Gayer, R. & Petek, J. (Eds) (1997) *European Coal Geology and Technology*. Special Publication 125, Geological Society of London, Bath, 448 pp.

Geldsetzer, H. (Ed.) (1989) *Reef Case Histories*. Memoir 13, Canadian Society of Petroleum Geologists, Calgary.

Gibson, T.G., Bybell, L.M. & Mason, D.B. (2000) Stratigraphic and climatic implications of clay mineral changes around the Paleocene/Eocene boundary of the northeastern US margin. *Sedimentary Geology* 134, 65–92.

Ginsburg, R.N. (1975) *Tidal Deposits. A Casebook of Recent Examples and Fossil Counterparts*. Springer-Verlag, Berlin.

Giresse, P., Wiewiora, A. & Lacka, B. (1998) Processes of Holocene ferromanganese coated grains (oncolites) in the nearshore shelf of Cameroon. *Journal of Sedimentary Research* 68, 20–36.

Given, R.K. & Wilkinson, B.H. (1985) Kinetic control of morphology, composition and mineralogy of abiotic sedimentary carbonates. *Journal of Sedimentary Petrology* 55, 109–119.

Given, R.K. & Wilkinson, B.H. (1987) Dolomite abundance and stratigraphic age: constraints on rates and mechanisms of Phanerozoic dolostone formation. *Journal of Sedimentary Petrology* 57, 1068–1078.

Glasby, G.P. (Ed.) (1977) *Marine Manganese Deposits*. Elsevier, Amsterdam.

Glasmann, J.R. et al. (1989) Diagenesis and hydrocarbon accumulation, Brent Sandstone (Jurassic), Bergen High Area, North Sea. *American Association of Petroleum Geologists, Bulletin* 73, 1341–1360.

Glenn, C.R., Prévôt-Lucas, L. & Lucas, J. (Eds) (2000) *Marine Authigenesis: from Global to Microbial*. Special Publication 66, Society of Sedimentary Geologists, Tulsa, OK.

Glennie, K.W. (Ed.) (1998) *Petroleum Geology of the North Sea*. Blackwell Science, Oxford, 636 pp.

Gluyas, J. & Swarbrick, R.E. (2001) *Petroleum Geoscience*. Blackwell Science, Oxford, 400 pp.

Goldring, R. (1991) *Fossils in the Field*. Longman, Harlow, 218 pp.

Goldsmith, I.R. & King, P. (1987) Hydrodynamic modelling of cementation patterns in modern reefs. In: *Diagenesis of Sedimentary Sequences* (Ed. J.D. Marshall). Special Publication 36, Geological Society of London, Bath, pp. 1–13.

Goldstein, R.H. & Reynolds, T.J. (1994) *Systematics of Fluid Inclusions in Diagenetic Minerals*. Short Course 31, Society of Economic Paleontologists and Mineralogists, Tulsa, OK, 199 pp.

Gonzalez, R. & Eberli, G.P. (1997) Sediment transport and bedforms in a carbonate tidal inlet; Lee Stocking Island, Exumas, Bahamas. *Sedimentology* 44, 1015–1030.

Gortner, C.W. & Laurue, D.K. (1986) Hemipelagic rocks at Bissen Hill, Barbados: sedimentology, geochemistry and depositional environment. *Journal of Sedimentary Petrology* 56, 307–316.

Gottardi, G. & Galli, E. (1985) *Natural Zeolites*. Springer-Verlag, Berlin.

Goudie, A.S. (Ed.) (1999) *Aeolian Environments. Sediments and Landforms*. Wiley, Chichester, 338 pp.

Gradstein, F.M., Sandvik, K.O. & Milton, N.J. (Eds) (1998)

Sequence Stratigraphy: Concepts and Applications. Elsevier, Amsterdam, 439 pp.

Graham, J. (1988) Collection and analysis of field data. In: *Techniques in Sedimentology* (Ed. M.E. Tucker). Blackwell Scientific Publications, Oxford, pp. 5–62.

Gross, G.A. (1983) Tectonic systems and the deposition of iron-formation. *Precambrian Research* **20**, 171–187.

Gupta, S. & Allen, P.A. (1999) Fossil shore platforms and drowned gravel beaches: evidence for high-frequency sea-level fluctuations in the distal alpine foreland basin. *Journal of Sedimentary Research* **69**, 394–413.

Hambrey, M.J. (1994) *Glacial Environments.* University College London Press, London, 296 pp.

Hampson, G., Stollhofen, H. & Flint, S. (1999) A sequence stratigraphic model for the Lower Coal Measures (Upper Carboniferous) of the Ruhr district, north-west Germany. *Sedimentology* **46**, 1199–1231.

Handford, C.R., Loucks, R.G. & Davies, G.R. (Eds) (1982) *Depositional and Diagenetic Spectra of Evaporites—a Core Workshop.* Core Workshop 3, Society Economic Paleontologists and Mineralogists, Tulsa, OK.

Handford, C.R. (1986) Facies and bedding sequences in shelf-storm-deposited carbonates—Fayetteville Shale and Pitkin Limestone (Mississippian), Arkansas. *Journal of Sedimentary Petrology* **56**, 123–137.

Handford, C.R. (1988) Review of carbonate sand-belt deposition of ooid grainstones and application to Mississippian Reservoir, Damme Field, Southwestern Kansas. *American Association of Petroleum Geologists, Bulletin* **72**, 1184–1199.

Hardie, L.A. (Ed.) (1977) *Sedimentation on the Modern Carbonate Tidal Flats of Northwestern Andros Island, Bahamas.* Johns Hopkins University Press, Baltimore, 202 pp.

Hardie, L.A. (1986) Stratigraphic models for carbonate tidal-flat deposition. *Quarterly Journal of the Colorado School of Mines* **81**, 59–74.

Hardie, L.A. (1987) Dolomitization: a critical review of some current views. *Journal of Sedimentary Petrology* **57**, 166–183.

Hardy, R. & Tucker, M.E. (1988) X-ray diffraction. In: *Techniques in Sedimentology* (Ed. M.E. Tucker). Blackwell Scientific Publications, Oxford, pp. 191–228.

Harris, P.M. (Ed.) (1984) *Carbonate Sands—A Core Workshop.* Core Workshop 5, Society of Economic Paleontologists and Mineralogists, Tulsa, OK.

Harris, P.M., Saller, A.H. & Simo, J.A. (Eds) (1999) *Advances in Carbonate Sequence Stratigraphy: Applications to Reservoirs, Outcrops and Models.* Special Publication 63, Society of Sedimentary Geologists, Tulsa, OK, 421 pp.

Hart, B.S. & Plint, A.G. (1995) Gravelly shoreface and beachface deposits. In: *Sedimentary Facies Analysis* (Ed. A.G. Plint). Special Publication 22, International Association of Sedimentologists. Blackwell Science, Oxford, pp. 75–100.

Hartley, A.J. & Prosser, D.J. (Eds) (1995) *Characterization of Deep Marine Clastic Systems.* Special Publication 94, Geological Society of London, Bath, 247 pp.

Harville, D.G. & Fritz, S.J. (1986) Modes of diagenesis responsible for observed succession of potash evaporites in the Salado Formation, Delaware Basin, New Mexico. *Journal of Sedimentary Petrology* **56**, 648–656.

Hattori, I., Umeda, M., Nakagawa, T. & Yamamoto, H. (1996) From chalcedonic chert to quartz chert: diagenesis of chert hosted in a Miocene volcanic succession, central Japan. *Journal of Sedimentary Research* **66**, 163–174.

Haynes, S.J., Boland, R. & Hughes-Pearl, J. (1989) Depositional setting of gypsum deposits, S. W. Ontario: the Domtar Mine. *Economic Geology* **84**, 857–870.

Heaney, P.J., Previtt, C.T. & Gibbs, G.V. (Eds) (1994) *Silica.* Reviews in Mineralogy 29, Mineralogical Society of America, 450 pp.

Heckel, P.H. (1995) Glacial-eustatic base-level—climatic model for late middle to late Pennsylvanian coal-bed formation in the Appalachian Basin. *Journal of Sedimentary Research* **B65**, 348–356.

Hein, J.R. & Obradovic, J. (Eds) (1989) *Siliceous Deposits of the Tethys and Pacific Regions.* Springer-Verlag, New York.

Hendry, J.P., Wilkinson, M., Fallick, A.E. & Haszeldine, R.S. (2000) Ankerite cementation in deeply buried Jurassic sandstone reservoirs of the central North Sea. *Journal of Sedimentary Research* **70**, 227–239.

Heward, A.P. (1989) Early Ordovician alluvial fan deposits of the Marmul oil field, South Oman. *Journal of the Geological Society of London* **146**, 557–565.

Heydari, E. & Moore, C. (1994) Paleoceanographic and paleoclimatic controls on ooid mineralogy of the Smackover Formation, Mississippi Salt Basin: implication for Late Jurassic seawater composition. *Journal of Sedimentary Research* **A64**, 101–114.

Hillgärtner, H. (1998) Discontinuity surfaces on a shallow-marine carbonate platform (Berriasian, Valanginian, France and Switzerland). *Journal of Sedimentary Research* **68**, 1098–1108.

Hird, K. & Tucker, M.E. (1988) Contrasting diagenesis of two Carboniferous oolites from South Wales: a tale of climatic influence. *Sedimentology* **35**, 587–602.

Hoey, T.B. & Bluck, B.J. (1999) Identifying the controls over downstream fining of river gravels. *Journal of Sedimentary Research* **69**, 40–50.

Hoffman, P. (2000) Snowball Earth. *Scientific American*, January, 50–70.

Horbury, A.D. & Robinson, A.G. (Eds) (1993) *Diagenesis and Basin Development.* Studies in Geology 36, American Association of Petroleum Geologists, Tulsa, OK, 274 pp.

Hounslow, M.H. (1987) Magnetic fabric characteristics of bioturbated wave-produced grain orientation in the Bridport Yeovil Sands (Lower Jurassic) of southern England. *Sedimentology* **34**, 117–128.

Hounslow, M.W. (1997) Significance of localized pore pressures to the genesis of septarian concretions. *Sedimentology* 44, 1133–1147.

Houseknecht, D. (1988) Intergranular pressure dissolution in four quartzose sandstones. *Journal of Sedimentary Petrology* 58, 228–246.

Houseknecht, D.W. & Pittman, E.D. (Eds) (1992) *Origin, Diagenesis and Petrophysics of Clay Minerals in Sandstones*. Special Publication 47, Society of Sedimentary Geologists, Tulsa, OK, 282 pp.

Hovorka, S. (1987) Depositional environments of marine-dominated bedded halite, Permian San Andres Formation, Texas. *Sedimentology* 34, 1029–1054.

Hsü, K.J. & Jenkyns, H.C. (Eds) (1974) *Pelagic Sediments: on Land and Under the Sea*. Special Publication 1, International Association of Sedimentologists. Blackwell Scientific Publications, Oxford.

Hsü, K.J. *et al.* (1978) History of the Mediterranean salinity crisis. In: *Initial Reports of the Deep Sea Drilling Project 42, Part 1* (Eds K.J. Hsü , L. Montadert *et al.*) US Government Printing Office, Washington, DC.

Hubert, J.F., Panish, P.T., Chure, D.J. & Prostak, K.S. (1996) Chemistry, microstructure, petrology, and diagenetic model of Jurassic dinosaur bones, Dinosaur National Monument, Utah. *Journal of Sedimentary Research* 66, 531–547.

Huc, A.-Y. (Ed.) (1995) *Paleogeography, Paleoclimate and Source Rocks*. Studies in Geology 40, American Association of Petroleum Geologists, Tulsa, OK, 314 pp.

Hunt, D. & Gawthorpe, R.L. (Eds) (2000) *Sedimentary Responses to Forced Regressions*. Special Publication 172, Geological Society of London, Bath, 383 pp.

Iijima, A., Abed, A.M. & Garrison, R.E. (Eds) (1994) *Siliceous Phosphatic and Glauconite Sediments of the Tertiary and Mesozoic*. 29th International Geological Congress Proceedings Part C. VSP, Utrecht, 450 pp.

Iijima, A., Hein, J.R. & Seiver, R. (Eds) (1983) *Siliceous Deposits of the Pacific Region*. Elsevier, New York.

Illenberger, W. (1991) Pebble shape (and size!). *Journal of Sedimentary Petrology* 61, 756–767.

Ingersoll, R.V. (1983) Petrofacies and provenance of late Mesozoic forearc basins, northern and central California. *American Association of Petroleum Geologists, Bulletin* 67, 1125–1142.

Insalaco, E., Skelton, P.W. & Palmer, T.P. (Eds) (2000) *Carbonate Platform Systems: Components and Interactions*. Special Publication 178, Geological Society of London, Bath, 240 pp.

Isaacs, C.M. (1981) Porosity reduction during diagenesis of the Monterey Formation, Santa Barbara coastal area, California. In: *The Monterey Formation and Related Siliceous Rocks of California* (Eds R.E. Garrison & R.G. Douglas). Pacific Section Publication 15, Society of Economic Paleontologists and Mineralogists, Tulsa, OK, pp. 257–271.

James, H.L. (1992) Precambrian Iron Formation. In: *Diagenesis III* (Eds K.H. Wolf & G.V. Chilingarian). Elsevier, Amsterdam, pp. 543–589.

James, N.P. (1983) Reefs. In: *Carbonate Depositional Environments* (Eds P.A. Scholle, D.G. Bebout & C.H. Moore). Memoir 33, American Association of Petroleum Geologists, Tulsa, OK, pp. 345–462.

James, N.P. & Bone, Y. (1989) Petrogenesis of Cenozoic, temperate water calcarenites, South Australia: a model for meteoric/shallow burial diagenesis of shallow-water calcitic sediments. *Journal of Sedimentary Petrology*, 59, 191–203.

James, N.P. & Choquette, P.W. (1988) *Paleokarsts*. Springer-Verlag, New York.

James, N.P. & Clarke, J.A.D. (Eds) (1997) *Cool-water Carbonates*. Special Publication 56, Society of Sedimentary Geologists, Tulsa, OK, 440 pp.

James, N.P., Narbonne, G.M. & Sherman, A.G. (1998) Molar-tooth carbonates: shallow subtidal facies of the mid-late Proterozoic. *Journal of Sedimentary Research*, 68, 716–722.

Jarvis, I. (1992) Sedimentology, geochemistry and origin of phosphatic chalks: the Upper Cretaceous deposits of NW Europe. *Sedimentology* 39, 55–97.

Jennette, D.C. & Pryor, W.A. (1993) Cyclic alteration of proximal and distal storm facies. Kope and Fairview Formations (Upper Ordovician), Ohio and Kentucky. *Journal of Sedimentary Petrology* 63, 183–202.

Jerram, D.A. (2001) Visual comparators for degree of grain-size sorting in 2-D and 3-D. *Computers in Geosciences* 27.

Jerram, D.A., Mountney, N.P., Howell, J.A., Long, D. & Stollhofen, H. (2000) Death of a sand sea: an active aeolian erg systematically buried by the Etendeka flood basalts of NW Namibia. *Journal of the Geological Society* 157, 513–516.

Johnson, H.D. & Baldwin, C.T. (1996) Shallow clastic seas. In: *Sedimentary Environments: Processes, Facies and Stratigraphy* (Ed. H.G. Reading). Blackwell Science, Oxford, pp. 232–280.

Johnson, H.D. & Levell, B.K. (1995) Sedimentation of a transgressive estuarine sand complex: the Lower Cretaceous Woburn Sands (Lower Greensand), Southern England. In: *Sedimentary Facies Analysis* (Ed. A.G. Plint). Special Publication 22, International Association of Sedimentologists. Blackwell Science, Oxford, pp. 17–46.

Johnson, S.Y. (1989) Significance of loessites in the Maroon Formation (Middle Pennsylvanian to Lower Permian), Eagle Basin, Northwest Colorado. *Journal of Sedimentary Petrology* 59, 782–791.

Johnsson, M.J. & Basu, A. (Eds) (1993) *Processes Controlling the Composition of Clastic Sediments*. Special Paper 284, Geological Society of America, Boulder, CO, 242 pp.

Jones, A., Tucker, M.E. & Hart, J.K. (1999) The description and analysis of Quaternary stratigraphic field sections.

Quaternary Research Association Technical Guide 7, 293.

Kaldi, J. (1986) Sedimentology of sand waves in an oolite shoal complex in the Cadeby Magnesian Limestone Formation (Upper Permian) of eastern England. In: *The English Zechstein and Related Topics* (Eds G.M. Harwood & D.B. Smith). Special Publication 22, Geological Society of London, Bath, pp. 63–74.

Kasper, D.C., Larue, D.K. & Meeks, Y.J. (1987) Fine-grained Paleogene terrigenous turbidites in Barbados. *Journal of Sedimentary Petrology* 57, 440–448.

Katz, B.J. & Pratt, I.M. (Eds) (1993) *Source Rocks in a Sequence Stratigraphic Framework*. Studies in Geology 37, American Association of Petroleum Geologists, Tulsa, OK, 247 pp.

Kelly, J.C. & Webb, J.A. (1999) The genesis of glaucony in the Oligo-Miocene Torquay Group, southeastern Australia: petrographic and geochemical evidence. *Sedimentary Geology* 125, 99–114.

Kemp, A.E.S. (Ed.) (1996) *Palaeoclimatology and Palaeo-oceanography from Laminated Sediments*. Special Publication 116, Geological Society of London, Bath, 272 pp.

Kench, P.S. & McLean, R.F. (1996) Hydraulic characteristics of bioclastic deposits: new possibilities for environmental interpretation using settling velocity fractions. *Sedimentology* 43, 561–570.

Kendall, A.C. (1985) Radiaxial fibrous calcite: reappraisal. In: *Carbonate Cements* (Eds N. Schneidermann & P.M. Harris). Special Publication 36, Society of Economic Paleontologists and Mineralogists, Tulsa, OK, pp. 59–77.

Kendall, A.C. (1988) Aspects of evaporite basin stratigraphy. In: *Evaporites and Hydrocarbons* (Ed. B.C. Schreiber). Columbia University Press, New York, pp. 11–65.

Kendall, A.C. (1989) Brine mixing in the Middle Devonian of Western Canada and its possible significance to regional dolomitization. *Sedimentary Geology* 64, 271–285.

Kendall, A.C. (1992) Evaporites. In: *Facies Models: Response to Sea-level Change* (Eds R.G. Walker & N.P. James). Geological Association of Canada, Waterloo, Ontario, pp. 375–409.

Kendall, A.C. & Harwood, G.M. (1989) Shallow-water gypsum in the Castile Evaporite—significance and implications. In: *Subsurface and Outcrop Examination of the Capitan Shelf Margin, Northern Delaware Basin* (Eds P.M. Harris & G.A. Grover). Core Workshop 13, Society of Economic Paleontologists and Mineralogists, Tulsa, OK, pp. 451–458.

Kendall, A.C. & Harwood, G.M. (1996) Marine evaporites: arid shorelines and basins. In: *Sedimentary Environments: Processes, Facies and Stratigraphy* (Ed. H.G. Reading). Blackwell Science, Oxford, pp. 281–324.

Kendall, C.G. St. C. & Warren, J. (1987) A review of the

origin and setting of tepees and their associated fabrics. *Sedimentology* 34, 1007–1027.

Keogh, S.M. & Butler, R.W.H. (1999) The Mediterranean water body in the late Messinian: interpreting the record from marginal basins on Sicily. *Journal of the Geological Society* 156, 837–846.

Kiessling, W., Flügel, E. & Golonka, J. (1999) Paleoreef Maps: Evaluation of a Comprehensive Database on Phanerozoic Reefs. *Bulletin, American Association of Petroleum Geologists* 83, 1552–1587.

Kirkland, D.W., Denison, R.E. & Dean, W.E. (2000) Parent brine of the Castile Evaporites (Upper Permian), Texas and New Mexico. *Journal of Sedimentary Research* 70, 749–761.

Klein, C. & Beukes, N.J. (1989) Geochemistry and sedimentology of a facies transition: from limestone to iron-formation deposition in the early Proterozoic Transvaal Supergroup, S. Africa. *Economic Geology*, 84, 1733–1773.

Knauth, L.P. (1994) Petrogenesis of chert. In: *Silica*. Reviews in Mineralogy 29, Mineralogical Society of America, Washington, DC, pp. 233–258.

Kneller, B.C. & Buckee, C. (2000) The structure and fluid mechanics of turbidity currents: a review of some recent studies and their geological implications. *Sedimentology* 47 (Suppl. 1), 6–94.

Kocurek, G.A. (1996) Desert aeolian systems. In: *Sedimentary Environments: Processes, Facies and Stratigraphy* (Ed. H.G. Reading). Blackwell Science, Oxford, pp. 125–153.

Kokelaar, B.P. & Howells, M.F. (Eds) (1984) *Marginal Basin Geology: Volcanic and Associated Sedimentary and Tectonic Processes in Modern and Ancient Marginal Basins*. Special Publication 16, Geological Society of London, Bath.

Kokelaar, P. & Königer, S. (2000) Marine emplacement of welded ignimbrite: the Ordovician Pitts Head Tuff, North Wales. *Journal of the Geological Society of London* 157, 517–536.

Krainer, K. & Spötl, C. (1998) Abiogenic silica layers within a fluvio-lacustrine succession, Bolzano Volcanic Complex, northern Italy: a Permian analogue for Magadi-type cherts? *Sedimentology* 45, 489–505.

Kuehl, S.A., Nittrouer, C.A. & De Master, D.J. (1988) Microfabric study of fine-grained sediments: observations from the Amazon subaqueous delta. *Journal of Sedimentary Petrology* 58, 12–23.

Kupecz, J.A., Gluyas, J. & Bloch, S. (Eds) (1997) *Reservoir Quality Prediction in Sandstones and Carbonates*. Memoir 69, American Association of Petroleum Geologists, Tulsa, OK, 311 pp.

Lajoie, J. & Stix, J. (1992) Volcaniclastic rocks. In: *Facies Models: Response to Sea-level Change* (Eds R.G. Walker & N.P. James). Geological Association of Canada, Waterloo, Ontario, pp. 101–118.

Landon, S.M. (Ed.) (1994) *Interior Rift Basins*. Memoir 59,

American Association of Petroleum Geologists, Tulsa, OK, 276 pp.

Lanson, B., Beaufort, D., Berger, G., Baradat, J. & LaCharpagne, J.-C. (1996) Illitization of diagenetic kaolinite-to-dickite conversion series: late stage diagenesis of the lower Permian Rotliegend sandstone reservoir, offshore the Netherlands. *Journal of Sedimentary Research* **66**, 501–518.

Lasemi, Z. & Sandberg, P.A. (1984) Transformation of aragonite-dominated lime muds to microcrystalline limestones. *Geology* **12**, 420–423.

Lasemi, Z., Boardman, M.R. & Sandberg, P.A. (1989) Cement origin of supratidal dolomite, Andros Island, Bahamas. *Journal of Sedimentary Petrology* **59**, 249–257.

Leeder, M.R. (1999) *Sedimentology and Sedimentary Basins*. Blackwell Science, Oxford, 592 pp.

Leggett, J.K. (Ed) (1982) *Trench–Forearc Geology: Sedimentation and Tectonics on Modern and Ancient Active Plate Margins*. Special Publication 10, Geological Society of London, Bath.

Lehmann, C., Osleger, D.A. & Montañez, I. (2000) Sequence stratigraphy of Lower Cretaceous (Barremian–Albian) carbonate platforms of northeastern Mexico: regional and global correlations. *Journal of Sedimentary Research* **70**, 373–391.

Lehrmann, D.J. & Goldhammer, R.K. (1999) Secular variation in parasequence and facies stacking patterns of platform carbonates: a guide to application of stacking-pattern analysis in strata of diverse ages and settings. In: *Carbonate Sequence Stratigraphy: Application to Reservoirs, Outcrops and Models* (Eds P.M. Harris, A. Saller & J.A. Simo) Special Publication 63, Society of Sedimentary Geologists, Tulsa, OK, pp. 187–225.

Leigh, S. & Hartley, A.J. (1992) Mega-debris flow deposits from the Oligo-Miocene Pindos foreland basin, western mainland Greece: implications for transport mechanisms in ancient deep marine basins. *Sedimentology* **39**, 1003–1012.

Lepp, H. (Ed.) (1975) *Geochemistry of Iron*. Dowden, Hutchinson & Ross, Stroudsburg.

Lewis, D.W. (1984) *Practical Sedimentology*. Hutchinson & Ross, Stroudsburg.

Lewis, D.G. & McConchie, D. (1994) *Analytical Sedimentology*. Chapman & Hall, London, 197 pp.

Lindholm, R.C. (1987) *A Practical Approach to Sedimentology*. Unwin–Hyman, London.

Lomas, S.A. (1999) A Lower Cretaceous clastic slope succession, Livingston Island, Antarctica: sand-body characteristics, depositional processes and implications for slope apron depositional models. *Sedimentology* **46**, 477–504.

Loucks, R.G. & Longman, M. (1982) Lower Cretaceous Ferry Lake anhydrite, Fairway Field, East Texas: product of shallow-subtidal deposition. In: *Depositional and Diagenetic Spectra of Evaporites* (Eds C.R. Handford, R.G. Loucks & G.R. Davies). Core Workshop 3, Society

of Economic Paleontologists and Mineralogists, Tulsa, OK, pp. 130–173.

Loucks, R.G. & Sarg, J.F. (Eds) (1993) *Carbonate Sequence Stratigraphy*. Memoir 57, American Association of Petroleum Geologists, Tulsa, OK, 545 pp.

Loucks, R.G., Dodge, M.M. & Galloway, W.E. (1984) Regional controls on diagenesis and reservoir quality in Lower Tertiary sandstones along the Texas Gulf Coast. In: *Clastic Diagenesis* (Eds D.A. MacDonald & R.C. Surdam), pp. 15–46. Memoir 37, American Association of Petroleum Geologists, Tulsa, OK.

Lowe, D.R. & Guy, M. (2000) Slurry-flow deposits in the Britannia Formation (Lower Cretaceous), North Sea: a new perspective on the turbidity current and debris flow problem. *Sedimentology* **47**, 31–70.

Lowenstein, T.K. (1987) Evaporite depositional fabrics in the deeply buried Jurassic Buckner Formation, Alabama. *Journal of Sedimentary Petrology* **57**, 108–116.

Lowenstein, T.K. & Hardie, L.A. (1985) Criteria for the recognition of salt-pan evaporites. *Sedimentology* **32**, 627–644.

Lowenstein, T.K. & Spencer, R.J. (1990) Syndepositional origin of potash evaporites: petrographic and fluid inclusion evidence. *American Journal of Science* **290**, 1–42.

Lucia, F.J. (1995) Rock fabric/petrophysical classification of carbonate pore space for reservoir characterization. *Bulletin, American Association of Petroleum Geologists* **79**, 1275–1300.

Lucia, F.J. (1999) *Carbonate Reservoir Characterization*. Springer-Verlag, Berlin, 226 pp.

Lugli, S., Schrieber, B.C. & Triberti, B. (1999) Giant polygons in the Realmonte Mine (Agrigento, Sicily): evidence for the dessication of a Messinian halite basin. *Journal of Sedimentary Research* **69**, 764–771.

Lukasik, J.J., James, N.P., McGowran, B. & Bone, Y. (2000) An epeiric ramp: low energy, cool-water carbonate facies in a Tertiary inland sea, Murray Basin, South Australia. *Sedimentology* **47**, 851–881.

Lyons, P.C. & Alpern, B. (Eds) (1989a) *Coal*. Elsevier, Amsterdam, 678 pp.

Lyons, P.C. & Alpern, B. (Eds) (1989b) *Peat and Coal*. Elsevier, Amsterdam, 892 pp.

Machel, H.G. (1993) Anhydrite nodules formed during deep burial. *Journal of Sedimentary Petrology* **63**, 659–662.

Machel, H.G. & Mountjoy, E.W. (1986) Chemistry and environments of dolomitization—a reappraisal. *Earth-Science Reviews* **23**, 175–222.

Macintyre, I.G., Reid, R.P. & Steneck, R.S. (1996) Growth history of stromatolites in a Holocene fringing reef, Stocking Island, Bahamas. *Journal of Sedimentary Research* **66**, 231–242.

MacKenzie, W.S. & Adams, A.E. (1994) *A Colour Atlas of Rocks and Minerals in Thin Section*. Manson Publishing, London, 192 pp.

Macquaker, J.H.S. & Gawthorpe, R.L. (1993) Mudstone lithofacies in the Kimmeridge Clay Formation, Wessex Basin, Southern England: implications for the origin and controls of the distribution of mudstones. *Journal of Sedimentary Petrology* 63, 1129–1143.

Macquaker, J.H.S. & Howell, J.K. (1999) Small-scale (<5.0 m) vertical heterogeneity in mudstones: implications for high-resolution stratigraphy in siliciclastic mudstone successions. *Journal of the Geological Society of London* 156, 105–112.

Macqueen, R.W. & Leckie, D.A. (Eds) (1992) *Foreland Basins and Fold Belts*. Memoir 55, American Association of Petroleum Geologists, Tulsa, OK, 460 pp.

Mange, M.A. & Maurer, H.F.W. (1992) *Heavy Minerals in Colour*. Chapman & Hall, London.

Maples, C.G. & West, R.G. (Eds) (1992) *Trace Fossils*. Short Course in Paleontology, 5, Paleontological Society, 238 pp.

Maroulis, J.C. & Nanson, G.C. (1996) Bedload transport of aggregated muddy alluvium from Cooper Creek, central Australia: a flume study. *Sedimentology* 43, 771–790.

Marriott, S. & Alexander, J. (Eds) (1999) *Floodplains: Interdisciplinary Approaches*. Special Publication 163, Geological Society of London, Bath, 284 pp.

Marshall, D.J. (1988) *Cathodoluminescence of Geological Materials*. Unwin–Hyman, London, 146 pp.

Marshall, J.R. (Ed.) (1987) Clastic particles. *Scanning Electron Microscopy and Shape Analysis of Sedimentary and Volcanic Clasts*. Van Nostrand Reinhold, New York, 346 pp.

Marzo, M. & Puigdefabrigas, C. (Eds) (1993) *Alluvial Sedimentation*. Special Publication 17, International Association of Sedimentologists. Blackwell Scientific Publications, Oxford, 586 pp.

Matter, A. & Ramseyer, K. (1985) Cathodoluminescence petrography as a tool for provenance studies of sandstones. In: *Provenance of Arenites* (Ed. G.G. Zuffa). Reidel, Dordrecht, pp. 191–211.

Matter, A. & Tucker, M.E. (Eds) (1978) *Modern and Ancient Lake Sediments*. Special Publication 2, International Association of Sedimentologists. Blackwell Scientific Publications, Oxford, 290 pp.

Maynard, J.B. (1984) Composition of plagioclase feldspar in modern deep-sea sands: relationship to tectonic setting. *Sedimentology* 31, 493–501.

McBride, E.F., Abdel-Wahab, A. & El-Younsy, A.R.M. (1999) Origin of spheroidal chert nodules, Drunka Formation (Lower Eocene), Egypt. *Sedimentology* 46, 733–755.

McBride, E.F., Picard, D.M. & Folk, R.L. (1994) Oriented concretions, Ionian Coast, Italy: evidence of groundwater flow direction. *Journal of Sedimentary Research* **A64**, 535–540.

McHardy, W.J. & Birnie, A.C. (1987) Scanning electron microscopy. In: *A Handbook of Determinative Methods in Clay Mineralogy* (Ed. M.J. Wilson). Blackie, Glasgow, pp. 173–208.

McManus, J. (1988) Grain-size determination and interpretation. In: *Techniques in Sedimentology* (Ed. M.E. Tucker). Blackwell Scientific Publications, Oxford, pp. 63–85.

Melvin, J. & Knight, A.S. (1984) Lithofacies, diagenesis and porosity of the Ivishak Formation, Prudhoe Bay Area, Alaska. In: *Clastic Diagenesis* (Eds D.A. McDonald & R.C. Surdam). Memoir 37, American Association of Petroleum Geologists, Tulsa, OK, pp. 347–365.

Menzies, J. (1996) *Past Glacial Environments*. Butterworth–Heinemann, Oxford, 598 pp.

Miall, A.D. (1995) Description and interpretation of fluvial deposits: a critical perspective. *Sedimentology* 42, 379–389.

Miall, A.D. (1996) *The Geology of Fluvial Deposits*. Springer-Verlag, Berlin, 582 pp.

Miall, A.D. (1997) *The Geology of Stratigraphic Sequences*. Springer-Verlag, Berlin, 433 pp.

Miall, A.D. (2000) *Principles of Sedimentary Basin Analysis*. Springer-Verlag, New York, 616 pp.

Mickelson, D.M. & Attig, J.W. (Eds) (1999) *Glacial Processes Past and Present*. Special Publication 337, Geological Society of America, Boulder, CO, 200 pp.

Midtgaard, H.H. (1996) Inner-shelf to lower-shoreface hummocky sandstone bodies with evidence for geostrophic influenced combined flow, Lower Cretaceous, West Greenland. *Journal of Sedimentary Research* **66**, 343–353.

Miller, J.M.G. (1996) Glacial sediments. In: *Sedimentary Environments: Processes, Facies and Stratigraphy* (Ed. H.G. Reading), Blackwell Science, Oxford, pp. 454–484.

Miller, D.J. & Eriksson, K.A. (2000) Sequence stratigraphy of Upper Mississippian strata in the central Appalachians: a record of glacioeustasy and tectonoeustasy in a foreland basin setting. *Bulletin, American Association of Petroleum Geologists* 84, 210–233.

Miller, J. (1988a) Microscopical techniques: I. Slices, slides, stains and peels. In: *Techniques in Sedimentology* (Ed. M.E. Tucker). Blackwell Scientific Publications, Oxford, pp. 86–107.

Miller, J. (1988b) Cathodoluminescence microscopy. In: *Techniques in Sedimentology* (Ed. M.E. Tucker). Blackwell Scientific Publications, Oxford, pp. 174–190.

Milliken, K.L. (1988) Loss of provenance information through subsurface diagenesis in Plio-Pleistocene sandstones, Northern Gulf of Mexico. *Journal of Sedimentary Petrology* 58, 992–1002.

Milliken, K.L. (1989) Petrography and composition of authigenic feldspars, Oligocene Frio Formation, South Texas. *Journal of Sedimentary Petrology* 59, 361–374.

Milliman, J.D., Freile, D., Steinen, R.P. & Wilber, R.J. (1993) Great Bahama Bank aragonitic muds: mostly inorganically precipitated, mostly exported. *Journal of Sedimentary Research* 63, 589–595.

Molina, J.M., Ruiz-Ortiz, P.A. & Vera, J.A. (1995) Neptunian dykes and associated features, southern Spain. Discussion. *Sedimentology* **42**, 957–969.

Molina, J.M., Ruiz-Ortiz, P.A. & Vera, J.A. (1999) A review of polyphase karstification in extensional tectonic regimes: Jurassic and Cretaceous examples, Betic Cordillera, Southern Spain. *Sedimentary Geology* **129**, 71–84.

Montañez, I.P., Gregg, J.M. & Shelton, K.L. (Eds) (1997) *Basin-wide Diagenetic Patterns: Integrated Petrologic, Geochemical and Hydrologic Considerations*. Special Publication 57, Society of Sedimentary Geologists, Tulsa, OK, 302 pp.

Monty, C.L.V. (Ed.) (1981) *Phanerozoic Stromatolites*. Springer-Verlag, Berlin.

Monty, C.L.V., Bosence, D.W.J., Bridges, P.H. & Pratt, B.R. (1995) *Carbonate Mud Mounds: their Origin and Evolution*. Special Publication 23, International Association of Sedimentologists. Blackwell Science, Oxford, 544 pp.

Moore, C.H. (1989) *Carbonate Diagenesis and Porosity*. Elsevier, Amsterdam, 338 pp.

Moore, S.E., Ferrell, R.E. & Aharon, P. (1992) Diagenetic siderite and other ferroan carbonates in a modern subsiding marsh sequence. *Journal of Sedimentary Petrology* **62**, 357–366.

Morad, S. (1998a) Carbonate cementation in sandstones: distribution patterns and geochemical evolution. In: *Carbonate Cementation in Sandstones* (Ed. S. Morad). Special Publication 26, International Association of Sedimentologists. Blackwell Science, Oxford, pp. 1–26.

Morad, S. (Ed.) (1998b) *Carbonate Cementation in Sandstones*. Special Publication 26, International Association of Sedimentologists. Blackwell Science, Oxford, 511 pp.

Morad, S., Ketzer, J.M. & De Ros, L.F. (2000) Spatial and temporal distribution of diagenetic alterations in siliciclastic rocks: implications for mass transfer in sedimentary basins. *Sedimentology* **47** (Suppl. 1), 95–120.

Moraes, M.A.S. & De Ros, L. (1992) Depositional, infiltrated and authigenic clays in fluvial sandstones of the Jurassic Sergi Formation, Reconcavo Basin, Northeastern Brazil. In: *Origin, Diagenesis and Petrophysics of Clay Minerals in Sandstones* (Eds D.W. Houseknecht & E.D. Pittman). Special Publication 47, Society of Sedimentary Geologists, Tulsa, OK, pp. 197–208.

Morton, A.C. (1987) Influences of provenance and diagenesis on detrital garnet suites in the Palaeocene Forties Sandstone, Central North Sea. *Journal of Sedimentary Petrology* **57**, 1027–1032.

Morton, A.C. & Hallsworth, C.R. (1999) Processes controlling the composition of heavy mineral assemblages in sandstones. *Sedimentary Geology* **124**, 3–29.

Morton, A.C. & Knox, R.W. O'B. (1990) Geochemistry of late Palaeocene and early Eocene tephras from the North Sea Basin. *Journal of the Geological Society of London* **147**, 425–437.

Morton, A.C., Haszeldine, R.S. & Giles, M.R. (1992) *Geology of the Brent Group*. Special Publication 61, Geological Society of London, Bath.

Morton, A.C., Todd, S.P. & Haughton, P.D.W. (Eds) (1991) *Developments in Sedimentary Province Studies*. Special Publication 57, Geological Society of London, Bath.

Mount, J. (1985) Mixed siliciclastic and carbonate sediments: a proposed first-order textural and compositional classification. *Sedimentology* **32**, 435–442.

Mountney, N. & Howell, J. (2000) Aeolian architecture, bedform climbing and preservation space in the Cretaceous Etjo Formation, NW Namibia. *Sedimentology* **47**, 825–849.

Mücke, A. (1994) Post-diagenetic ferruginization of sedimentary rocks, including a comparative study of reddening of red beds. In: *Diagenesis IV* (Eds K.H. Wolf & G.V. Chilingarian). Elsevier, Amsterdam, pp. 361–423.

Mullins, H.T. & Cook, H.E. (1986) Carbonate apron models: alternatives to the submarine fan model for paleoenvironmental analysis and hydrocarbon exploration. *Sedimentary Geology* **48**, 37–79.

Mutti, M. (1994) Association of tepees and palaeokarst in the Ladinian Calcare Rosso (Southern Alps, Italy). *Sedimentology* **41**, 621–641.

Mutti, E. (1999) *An Introduction to the Analysis of Ancient Turbidite Basins from an Outcrop Perspective. Continuing Education Notes* **39**, American Association of Petroleum Geologists, Tulsa, OK, 98 pp.

Mutti, M., Bernoulli, D., Eberli, G.P. & Vecsei, A. (1996) Depositional geometries and facies associations in an upper Cretaceous prograding carbonate platform margin (Orfento Supersequence, Maiella, Italy). *Journal of Sedimentary Research* **66**, 749–765.

Nemec, W. & Steel, R.J. (Eds) (1988) *Fan Deltas. Sedimentology and Tectonic Settings*. Blackie, London.

Nesbitt, H.W. & Young, G.M. (1989) Formation and diagenesis of weathering profiles. *Journal of Geology* **97**, 129–147.

Nichols, G. (1999) *Sedimentology and Stratigraphy*. Blackwell Science, Oxford, 355 pp.

Nichols, G. & Jones, T. (1992) Fusain in Carboniferous shallow-marine sediments, Donegal, Ireland: the sedimentological effects of wildfire. *Sedimentology* **39**, 487–502.

Nicholson, K. (Ed.) (1996) *Manganese Mineralization: Geochemistry and Mineralogy of Terrestrial and Marine Deposits*. Special Publication 119, Geological Society of London, Bath, 352 pp.

Nickling, W.G. (1994) Aeolian sediment transport and deposition. In: *Sediment Transport and Depositional Processes* (Ed. K. Pye). Blackwell Science, Oxford, pp. 293–350.

Noble, J.P.A. & Van Stempvoort, D.R. (1989) Early burial quartz authigenesis in Silurian platform carbonates, New Brunswick, Canada. *Journal of Sedimentary Petrology* 59, 65–76.

Normark, W.R., Piper, D.J.W. & Hiscott, R.N. (1998) Sea-level controls on the textural characteristics and depositional architecture of the Hueneme and associated submarine fan systems, Santa Monica Basin, California. *Sedimentology* 45, 53–70.

North, C.P. & Prosser, D.J. (Eds) (1993) *Characterization of Fluvial and Aeolian Reservoirs*. Special Publication 73, Geological Society of London, Bath, 450 pp.

North, F.K. (1990) *Petroleum Geology*. Unwin–Hyman, London.

Nummedal, D., Pilkey, O.H. & Howard, L.D. (Eds) (1987) *Sea-level Fluctuations and Coastal Evolution*. Special Publication 41, Society of Economic Paleontologists and Mineralogists, Tulsa, OK.

O'Brien, N.R. (1987) The effects of bioturbation on the fabric of shale. *Journal of Sedimentary Petrology* 57, 449–455.

Odin, G.S. (Ed.) (1988) *Green Marine Clays*. Elsevier, Amsterdam, 445 pp.

Ollier, G., Cochonat, P., Lenat, J.F. & Labazuy, P. (1998) Deep-sea volcaniclastic sedimentary systems: an example from La Fournaise volcano, Reunion Island, Indian Ocean. *Sedimentology* 45, 293–330.

Olsen, T.R., Mellere, D. & Olsen, T. (1999) Facies architecture and geometry of landward-stepping shoreface tongues: the Upper Cretaceous Cliff House Sandstone (Mancos Canyon, south-west Colorado). *Sedimentology* 46, 603–625.

Orton, G.J. (1996) Volcanic environments. In: *Sedimentary Environments: Processes, Facies and Stratigraphy* (Ed. H.G. Reading). Blackwell Science, Oxford, pp. 485–567.

Oti, M. & Postma, G. (Eds) (1995) *Geology of Deltas*. Balkema, Rotterdam, 315 pp.

Owen, G. (1996) Experimental soft-sediment deformation: structures formed by the liquefaction of unconsolidated sands and some ancient examples. *Sedimentology* 43, 279–293.

Palmer, T.J., Hudson, J.D. & Wilson, M.A. (1988) Palaeoecological evidence for early aragonite dissolution in ancient calcite seas. *Nature* 335, 809–811.

Paola, C. (2000) Quantitative models of sedimentary basin filling. *Sedimentology* 47 (Suppl. 1), 121–178.

Percival, C.J. (1986) Paleosols containing an albic horizon: examples from the Upper Carboniferous of Northern Britain. In: *Paleosols* (Ed. V.P. Wright). Blackwell Scientific Publications, Oxford, pp. 87–110.

Perry, C.T. (1998) Grain susceptibility to the effects of microboring: implications for the preservation of skeletal carbonates. *Sedimentology* 45, 39–51.

Peryt, T.M., Orti, F. & Rosell, L. (1993) Sulphate platform–basin transition of the Lower Werra Anhydrite (Zechstein, Upper Permian), Western Poland: facies and paleogeography. *Journal of Sedimentary Research* 63, 645–658.

Peryt, T.M., Pierre, C. & Gryniv, S.P. (1998) Origin of polyhalite deposits in the Zechstein (Upper Permian) Zdrada platform (northern Poland). *Sedimentology* 45, 565–576.

Pettijohn, F.J., Potter, P.E. & Siever, R. (1987) *Sand and Sandstone*. Springer-Verlag, New York.

Pickering, K.T., Hiscott, R.N. & Hein, F.J. (1989) *Deep Marine Environments. Clastic Sedimentation and Tectonics*. Unwin–Hyman, London.

Pickering, K.T., Stow, D.A.V., Watson, M.P. & Hiscott, R.N. (1986) Deep-water facies, processes and models; a review and classification scheme for modern and ancient sediments. *Earth-Science Reviews* 23, 75–174.

Pickering, K.T. *et al.* (1999) Glacio-eustatic control on deep-marine clastic forearc sedimentation, Pliocene–mid-Pleistocene (*c.* 1180–600 ka) Kazusa Group, SE Japan. *Journal of the Geological Society* 156, 125–136.

Pierre, C. & Rouchy, J.M. (1988) Carbonate replacements after sulfate evaporites in the Middle Miocene of Egypt. *Journal of Sedimentary Petrology* 58, 446–456.

Pisciotto, K.A. (1981) Distribution, thermal histories, isotopic compositions and reflection characteristics of siliceous rocks recovered by the Deep Sea Drilling Project. In: *The Deep Sea Drilling Project: a Decade of Progress* (Eds J.D. Warme, R.G. Douglas & E.L. Winterer). Special Publication 32, Society of Economic Paleontologists and Mineralogists, Tulsa, OK, pp. 129–148.

Pittman, E.D. (1988) Diagenesis of Terry Sandstone (Upper Cretaceous), Spindle Field, Colorado. *Journal of Sedimentary Petrology* 58, 785–800.

Pittman, E.D. & Lewan, M.D. (Eds) (1995) *Organic Acids in Geological Processes*. Springer-Verlag, Berlin, 500 pp.

Platt, N.H. (1989) Lacustrine carbonates and pedogenesis: sedimentology and origin of palustrine deposits from the Early Cretaceous Rupelo Formation, W. Cameros Basin, N Spain. *Sedimentology* 36, 665–684.

Platt, N.H. & Wright, V.P. (1992) Palustrine carbonates and the Florida Everglades: towards an exposure index for the fresh-water environment. *Journal of Sedimentary Petrology* 62, 1058–1071.

Plink-Björklund, P. & Ronnert, L. (1999) Depositional processes and internal architecture of Late Weichselian ice-margin submarine fan and delta settings, Swedish west coast. *Sedimentology* 46, 215–236.

Pollock, S.G. (1987) Chert formation in an Ordovician volcanic arc. *Journal of Sedimentary Petrology* 57, 75–87.

Posamentier, H.W., Summerhayes, C.P., Haq, B.U. & Allen, G.P. (Eds) (1993) *Sequence Stratigraphy and Facies Associations*. Special Publication 18, International Association of Sedimentologists. Blackwell Scientific Publications, Oxford, 644 pp.

Potter, P.E. & Pettijohn, F.J. (1977) *Paleocurrents and Basin Analysis*. Springer-Verlag, Berlin.

Pratt, B.R. (1998) Molar-tooth structure in Proterozoic carbonate rocks: origin from synsedimentary earthquakes, and implications for the nature and evolution of basins and marine sediment. *Geological Society of America Bulletin* **110**, 1028–1045.

Prave, A.R., Duke, W.L. & Slattery, W. (1996) A depositional model for storm- and tide-influenced prograding siliciclastic shorelines from the Middle Devonian of the central Appalachian foreland basin, USA. *Sedimentology* **43**, 611–629.

Primmer, T.J. *et al.* (1997) Global patterns in sandstone diagenesis: their application to reservoir quality prediction for petroleum exploration. In: *Reservoir Quality Prediction in Sandstones and Carbonates* (Eds J.A. Kupecz, J. Gluyas & S. Bloch). Memoir 69, American Association of Petroleum Geologists, Tulsa, OK, pp. 61–77.

Purdy, E.G. & Waltham, D. (1999) Reservoir implications of modern karst topography. *Bulletin, American Association of Petroleum Geologists* **83**, 1774–1795.

Purser, B.H. & Bosence, D.W.J. (1998) *Sedimentation and Tectonics in Rift Basins: Red Sea—Gulf of Aden*. Chapman & Hall, London, 663 pp.

Purser, B.H., Tucker, M.E. & Zenger, D.H. (1994) *Dolomites: A Volume in Honour of Dolomieu*. Special Publication 21, International Association of Sedimentologists. Blackwell Science, Oxford, 451 pp.

Pye, K. (Ed.) (1994) *Sediment Transport and Depositional Processes*. Blackwell Science, Oxford, 397 pp.

Pye, K. (1995) The nature, origin and accumulation of loess. *Quaternary Science Reviews* **14**, 653–668.

Pye, K. & Krinsley, D.H. (1984) Petrographic examination of sedimentary rocks in the SEM using back-scattered electron detectors. *Journal of Sedimentary Petrology* **54**, 877–888.

Pye, K. & Lancaster, N. (Eds) (1993) *Aeolian Sediments: Ancient and Modern*. Special Publication 16, International Association of Sedimentologists. Blackwell Scientific Publications, Oxford, 167 pp.

Pye, K. & Tsoar, H. (1990) *Aeolian Sand and Sand Dunes*. Unwin–Hyman, London.

Pyle, D.M. (1989) The thickness, volume and grain-size of tephra fall deposits. *Bulletin of Volcanology* **51**, 1–15.

Rachocki, A.H. & Church, M. (Eds) (1990) *Alluvial Fans. A Field Approach*. Wiley, Chichester, 391 pp.

Rahamani, R.A. & Flores, R.M. (Eds) (1985) *Sedimentology of Coal and Coal-bearing Sequences*. Special Publication 7, International Association of Sedimentologists. Blackwell Scientific Publications, Oxford.

Raiswell, R. & Canfield, D.E. (1998) Sources of iron for pyrite formation in marine sediments. *American Journal of Science* **298**, 219–245.

Read, J.F. & Horbury, A.D. (1993) Eustatic and tectonic controls on porosity evolution beneath sequence bounding unconformities and parasequence disconformities on carbonate platforms. In: *Diagenesis and Basin Development* (Eds A.D. Horbury & A.G. Robinson). Studies in Geology 36, American Association of Petroleum Geologists, Tulsa, OK, pp. 155–197.

Reading, H.G. (Ed.) (1996) *Sedimentary Environments. Processes, Facies and Stratigraphy*. Blackwell Science, Oxford.

Reading, H.G. & Collinson, J.D. (1996) Clastic coasts. In: *Sedimentary Environments: Processes, Facies and Stratigraphy*. Blackwell Science, Oxford, pp. 154–231.

Reid, R.P. & Macintyre, I.G. (1998) Carbonate recrystallization in shallow marine environments: a widespread diagenetic process forming micritized grains. *Journal of Sedimentary Research* **68**, 928–946.

Reinhardt, J. & Sigleo, W.R. (Eds) (1988) *Paleosols and Weathering through Geologic Time: Principles and Applications*. Special Paper 216, Geological Society of America, Boulder, CO.

Renaut, R.W., Jones, B. & Tiercelin, J.-J. (1998) Rapid *in situ* calcification of microbes at Loburu hot springs, Lake Bogoria, Kenya Rift Valley. *Sedimentology* **45**, 1083–1103.

Retallack, G.J. (2001) *Soils of the Past*. Blackwell Science, Oxford, 416 pp.

Retallack, G.J. (1997) *A Colour Guide to Paleosols*. Wiley, Chichester, 175 pp.

Ricci-Lucchi, F. (1995) *Sedimentographica: Photographic Atlas of Sedimentary Structures*. Columbia University Press, New York.

Rice, S. (1999) The nature and controls on downstream fining within sedimentary links. *Journal of Sedimentary Research* **69**, 32–39.

Ridgeway, K.D., Trop, J.M. & Jones, D.E. (1999) Petrology and provenance of the Neogene Usibelli Group and Nenana Gravel: implications for the denudation history of the Central Alaska Range. *Journal of Sedimentary Research* **69**, 1262–1275.

Riding, R. (Ed.) (1990) *Calcareous Algae and Stromatolites*. Springer-Verlag, Berlin.

Riding, R. (2000) Microbial carbonates: the geological record of calcified bacterial–algal mats and biofilms. *Sedimentology* **47** (Suppl. 1), 179–214.

Riding, R. & Awramik, S.M. (Eds) (2000) *Microbial Sediments*. Springer-Verlag, Heidelberg.

Rine, J.M. & Ginsburg, R.N. (1985) Depositional facies of a mud shoreface in Surinam, South America—a mud analogue to sandy shallow-marine deposits. *Journal of Sedimentary Petrology* **55**, 633–652.

Robert, P. (1988) *Organic Metamorphism and Geothermal History*. Reidel, Dordrecht.

Robertson, A.H.F. (1990) Sedimentology and tectonic implications of ophiolite-derived clastics overlying the Jurassic Coast Range Ophiolite, northern California. *American Journal of Science* **290**, 109–163.

Robertson, A.H.F. & Hudson, J.D. (1974) Pelagic sediments in the Cretaceous and Tertiary history of the Troodos

Massif, Cyprus. In: *Pelagic Sediments: on Land and Under the Sea* (Eds K.J. Hsü & H.C. Jenkyns). Special Publication 1, International Association of Sedimentologists. Blackwell Scientific Publications, Oxford, pp. 403–406.

Roehl, P.O. & Choquette, P.W. (Eds) (1985). *Carbonate Petroleum Reservoirs*. Springer-Verlag, New York.

Ross, G.M. & Chiarenzelli, J.R. (1985) Paleoclimatic significance of widespread Proterozoic silcretes in the Bear and Churchill Provinces of the northwestern Canadian Shield. *Journal of Sedimentary Petrology* 55, 196–204.

Rossetti, D. (1999) Soft-sediment deformation structures in late Albian to Cenomanian deposits. São Luís Basin, northern Brazil: evidence for palaeoseismicity. *Sedimentology* 46, 1065–1081.

Rouchy, J.M., Pierre, C. & Sommer, F. (1995) Deep-water resedimentation of anhydrite and gypsum deposits in the Middle Miocene (Belayim Formation) of the Red Sea, Egypt. *Sedimentology* 42, 267–282.

Roylance, M.H. (1990) Depositional and diagenetic history of a Pennsylvanian algal-mound complex. Bug and Papose Canyon Fields, Paradox Basin, Utah and Colorado. *Bulletin, American Association of Petroleum Geologists* 74, 1087–1099.

Ruffell, A.H. & Batten, D.J. (1990) The Barremian–Aptian arid phase in western Europe. *Palaeogeography, Palaeoclimatology, Palaeoecology* 80, 197–212.

Ryu, I.-C. & Niem, A.R. (1999) Sandstone diagenesis, reservoir potential, and sequence stratigraphy of the Eocene Tyee Basin, Oregon. *Journal of Sedimentary Research* 69, 384–393.

Saigal, G.C., Morad, S., Bjorlykke, K., Egeberg, P.K. & Aagaard, P. (1988) Diagenetic albitization of detrital K-feldspar in Jurassic, Lower Cretaceous and Tertiary clastic reservoir rocks from offshore Norway, 1. Textures and origin. *Journal of Sedimentary Petrology* 58, 1003–1013.

Saller, A.H., Harris, P.M., Kirkland, B.L. & Mazzullo, S.J. (Eds) (2000) *Geologic Framework of the Capitan Reef*. Special Publication 65, Society of Sedimentary Geologists, Tulsa, OK.

Sandberg, P.A. (1983) An oscillating trend in Phanerozoic non-skeletal carbonate mineralogy. *Nature* 305, 19–22.

Satterley, A.K. (1996) Cyclic carbonate sedimentation in the Upper Triassic Dachstein Limestone, Austria: the role of patterns of sediment supply and tectonics in a platform–reef–basin system. *Journal of Sedimentary Research* B66, 307–323.

Schieber, J. (1986) The possible role of benthic microbial mats during the formation of carbonaceous shales in shallow Mid-Proterozoic basins. *Sedimentology* 33, 521–536.

Schieber, J. (1989) Facies and origin of shales from the mid-Proterozoic Newland Formation, Belt Basin, Montana, USA. *Sedimentology* 36, 203–220.

Schieber, J. (1999) Distribution and deposition of mudstone facies in the Upper Devonian Sonyea Group of New York. *Journal of Sedimentary Research* 69, 909–925.

Schieber, J., Zimmerle, W. & Sethi, P. (Eds) (1998) *Shales and Mudstones*. Schweizerbart'sche Verlagsbuchhandlung, Stuttgart.

Schneidermann, N. & Harris, P.M. (Eds) (1985) *Carbonate Cements*. Special Publication 36, Society of Economic Paleontologists and Mineralogists, Tulsa, OK.

Scholle, P.A. (1978) *A Colour Illustrated Guide to Carbonate Rock Constituents, Textures, Cements and Porosities*. Memoir 27, American Association of Petroleum Geologists, Tulsa, OK.

Scholle, P.A. (1979) *A Colour Illustrated Guide to Constituents, Textures, Cements and Porosities of Sandstones and Associated Rocks*. Memoir 28, American Association of Petroleum Geologists, Tulsa, OK.

Scholle, P.A., Arthur, M.A. & Ekdale, A.A. (1983) Pelagic environments. In: *Carbonate Depositional Environments* (Eds P.A. Scholle, D.G. Bebout & C.H. Moore). Memoir 33, American Association of Petroleum Geologists, Tulsa, OK, pp. 620–691.

Schopf, J.W. (1993) Microfossils of the Early Archean Apex Chert: new evidence for the antiquity of life. *Science* 260, 640–646.

Schreiber, B.C. & El Tabakh, M. (2000) Deposition and early alteration of evaporites. *Sedimentology* 47 (Suppl. 1), 215–238.

Schroeder, J.H. & Purser, B.H. (Eds) (1986). *Reef Diagenesis*. Springer-Verlag, Berlin.

Schubel, K. & Lowenstein, T.K. (1997) Criteria for the recognition of shallow-perennial, saline lake halites based on recent sediments from the Qaidam Basin, Western China. *Journal of Sedimentary Research* 67, 74–87.

Scotchman, I.C. (1991) The geochemistry of concretions from the Kimmeridge Clay Formation of southern and eastern England. *Sedimentology* 38, 79–106.

Scott, A.C. (1987) *Coal and Coal-bearing Strata: Recent Advances*. Special Publication 32, Geological Society of London, Bath.

Selley, R.C. (1988) *Applied Sedimentology*. Academic Press, London.

Shanley, K.W. & McCabe, P.J. (Eds) (1998) *Relative Role of Eustasy, Climate and Tectonism in Continental Rocks*. Special Publication 59, Society of Sedimentary Geologists, Tulsa, OK, 234 pp.

Shen, Y., Schidlowski, M. & Chu, X. (2000) Biogeochemical approach to understanding phosphogenic events of the terminal Proterozoic to Cambrian. *Palaeogeography, Palaeoclimatology, Palaeoecology* 158, 99–108.

Shiki, T., Cita, M.B. & Gorslime, D.S. (Eds) (2000) Seismoturbidites, Seismites and Tsunamites. *Sedimentary Geology*, 135, 1–326.

Shinn, E.A. (1983) Tidal flat environment. In: *Carbonate Depositional Environments* (Eds P.A. Scholle, D.G. Bebout & C.H. Moore). Memoir 33, American

Association of Petroleum Geologists, Tulsa, OK, pp. 173–210.

Shinn, E.A. & Lidz, B.H. (1988) Blackened limestone pebbles: fire at subaerial unconformities. In: *Paleokarst* (Eds N.P. James & P.W. Choquette). Springer-Verlag, New York, pp. 117–131.

Shinn, E.A., Steinen, R.P., Lidz, B.H. & Swart, P.K. (1989) Whitings, a sedimentologic dilemma. *Journal of Sedimentary Petrology* 59, 147–161.

Shukla, V. & Baker, P.A. (Eds) (1988) *Sedimentology and Geochemistry of Dolostones*. Special Publication 43, Society of Economic Paleontologists and Mineralogists, Tulsa, OK.

Silva, S.L. de (1989) Correlation of large ignimbrites: two case studies from the Central Andes of northern Chile. *Journal of Volcanology and Geothermal Research* 37, 93–132.

Simonson, B.M. (1985) Sedimentology of cherts in the early Proterozoic Wishart Formation, Quebec–New foundland, Canada. *Sedimentology* 32, 23–40.

Simonson, B.M. (1987) Early silica cementation and subsequent diagenesis in arenites from four early Proterozoic iron formations of North America. *Journal of Sedimentary Petrology* 57, 494–511. Discussion. *Journal of Sedimentary Petrology* 58, 544–549.

Simpson, J. (1985) Stylolite-controlled layering in a homogeneous limestone: pseudo-bedding produced by burial diagenesis. *Sedimentology* 32, 495–505.

Singer, J.K. *et al.* (1988) An assessment of analytical techniques for the size analysis of fine-grained sediments. *Journal of Sedimentary Petrology* 58, 534–543.

Smith, D.C., Reinson, G.E., Zaitlin, B.A. & Rahmani, R.A. (1991) *Clastic Tidal Sedimentology*. Memoir 16, Canadian Society of Petroleum Geologists, Calgary, 307 pp.

Smith, N. & Rogers, J. (Eds) (1999) *Fluvial Sedimentology VI*. Special Publication 28, International Association of Sedimentologists. Blackwell Science, Oxford, 488 pp.

Smosna, R. (1989) Compaction law for Cretaceous sandstones of Alaska's North Slope. *Journal of Sedimentary Petrology* 59, 572–584.

Solomon, S.T. (1989) The early diagenetic origin of Lower Carboniferous mottled limestones (pseudobreccias). *Sedimentology* 36, 399–418.

Soudry, D. & Southgate, P.N. (1989) Ultrastructure of a Middle Cambrian primary non-pelletal phosphorite and its early transformation into phosphate vadoids, Georgina Basin, Australia. *Journal of Sedimentary Petrology* 59, 53–64.

Southgate, P.N. *et al.* (1989) Depositional environments and diagenesis in Lake Parakeetya: a Cambrian alkaline playa from the Officer Basin, South Australia. *Sedimentology* 36, 1091–1113.

Sparks, R.S.J. (1976) Grain-size variations in ignimbrites and implications for the transport of pyroclastic flows. *Sedimentology* 23, 147–188.

Spears, D.A., Kanaris-Sotiriou, R., Riley, N. & Krause, P. (1999) Namurian bentonites in the Pennine Basin, UK—origin and magmatic affinities. *Sedimentology* 46, 385–401.

Spence, G.H. & Tucker, M.E. (1997) Genesis of limestone megabreccias and their significance in carbonate sequence stratigraphic models—a review. *Sedimentary Geology* 112, 163–193.

Spötl, C. & Pitman, J.K. (1998) Saddle (baroque) dolomite in carbonates and sandstones: a reappraisal of a burial–diagenetic concept. In: *Carbonate Cementation in Sandstones* (Ed. S. Morad), Special Publication 26, International Association of Sedimentologists. Blackwell Science, Oxford, pp. 437–460.

Spötl, C., Houseknecht, D.W. & Longstaffe, F.J. (1994) Authigenic chlorites in sandstones as indicators of high-temperature diagenesis, Arkoma Foreland Basin, USA. *Journal of Sedimentary Research* A64, 553–566.

Stach, E. *et al.* (1982) *Textbook of Coal Petrology*. Gebrüder Borntraeger, Stuttgart.

Stalder, P.J. (1973) Influence of crystalline habit and aggregate structure of authigenic clay minerals on sandstone permeability. *Geologie en Mijnbouw* 52, 217–219.

Stanley, S.M. & Hardie, L.A. (1998) Secular oscillations in the carbonate mineralogy of reef-building and sediment-producing organisms driven by tectonically forced shifts in seawater chemistry. *Palaeogeography, Palaeoclimatology, Palaeoecology* 144, 3–19.

Stoker, M.S., Evans, D. & Cramp, A. (1998) *Geological Processes on Continental Margins: Sedimentation, Mass-wasting and Stability*. Special Publication 129, Geological Society of London, Bath, 362 pp.

Stow, D.A.V. & Faugères, J.C. (Eds) (1998) Contourites, turbidites and process interaction. *Sedimentary Geology*, 115, 1–386.

Stow, D.A.V. & Mayall, M. (2000) Deep-water sedimentary systems: new models for the 21st Century. *Marine and Petroleum Geology* 17, 125–135.

Stow, D.A.V. & Shanmugam, G. (1980) Sequence of structures in fine-grained turbidites; comparison of recent deep-sea and ancient flysch sediments. *Sedimentary Geology* 25, 23–42.

Stow, D.A.V., Reading, H.G. & Collinson, J.D. (1996) Deep seas. In: *Sedimentary Environments: Processes, Facies and Stratigraphy* (Ed. H.G. Reading), Blackwell Science, Oxford, pp. 395–453.

Sturesson, U., Dronov, A. & Saadre, T. (1999) Lower Ordovician iron ooids and associated oolitic clays in Russia and Estonia: a clue to the origin of iron oolites? *Sedimentary Geology* 123, 63–80.

Sugitani, K. *et al.* (1998) Archean cherts derived from chemical, biogenic and clastic sedimentation in a shallow restricted basin: examples from the Gorge Creek Group in the Pilbara Block. *Sedimentology* 45, 1045–1062.

Summerfield, M.A. (1983) Silcrete as a palaeoclimatic

indicator: evidence from southern Africa. *Palaeogeography, Palaeoclimatology, Palaeoecology* **41**, 65–79.

Sun, S.Q. (1994) A reappraisal of dolomite abundance and occurrence in the Phanerozoic. *Journal of Sedimentary Research* **A64**, 396–404.

Sun, S.Q. & Wright, V.P. (1989) Peloidal fabrics in Upper Jurassic reefal limestones, Weald Basin, southern England. *Sedimentary Geology* **65**, 165–181.

Suthren, R.J. (1985) Facies analysis of volcaniclastic sediments: a review. In: *Sedimentology: Recent Advances and Applied Aspects* (Eds P. Brenchley & B.P.J. Williams). Special Publication 18, Geological Society of London, Bath, pp. 123–146.

Sweet, M.L. (1999) Interaction between aeolian, fluvial and playa environments in the Permian Upper Rotliegend Group, UK southern North Sea. *Sedimentology* **46**, 171–187.

Swennen, R., Viaene, W. & Cornelissen, C. (1990) Petrography and geochemistry of the Belle Roche breccia (lower Visean, Belgium): evidence of brecciation by evaporite dissolution. *Sedimentology* **37**, 859–878.

Swift, D.J.P., Oertel, G.F., Tillman, R.W. & Thorne, J.A. (1991) *Shelf Sand and Sandstone Bodies*. Special Publication 14, International Association of Sedimentologists. Blackwell Scientific Publications, Oxford, 295 pp.

Syvitski, J.P.M. (1991) *Principles, Methods and Application of Particle Size Analysis*. Cambridge University Press, Cambridge, 368 pp.

Tada, R. (1991) Origin of rhythmical bedding in middle Miocene siliceous rocks of the Onnagawa Formation, northern Japan. *Journal of Sedimentary Petrology* **61**, 1123–1145.

Talbot, M.R. & Allen, P.A. (1996) Lakes. In: *Sedimentary Environments: Processes, Facies and Stratigraphy* (Ed. H.G. Reading). Blackwell Science, Oxford, pp. 83–124.

Tanner, P.W.G. (1998) Interstratal dewatering origin for polygonal patterns of sand-filled cracks: a case study from late Proterozoic metasediments of Islay, Scotland. *Sedimentology* **45**, 71–89.

Taylor, G.H. *et al.* (1998) *Organic Petrology*. Gebrüder-Borntraeger, Berlin, 704 pp.

Taylor, K.G. & Curtis, C.D. (1995) Stability and facies association of early diagenetic mineral assemblages: an example from a Jurassic ironstone–mudstone succession, UK. *Journal of Sedimentary Research* **A65**, 358–368.

Taylor, K.G. & Macquaker, J.H.S. (2000) Early diagenetic pyrite morphology in a mudstone-dominated succession: the Lower Jurassic Cleveland Ironstone Formation, eastern England. *Sedimentary Geology* **131**, 77–86.

Taylor, K.G. *et al.* (2000) Carbonate cementation in a sequence stratigraphic framework: Upper Cretaceous sandstones, Book Cliffs, Utah-Colorado. *Journal of Sedimentary Research*, **70**, 360–372.

Teal, C.S., Mazzullo, S.J. & Bischoff, W.D. (2000) Dolomitization of shallow marine deposits mediated by sulphate reduction and methanogenesis in normal-salinity seawater, northern Belize. *Journal of Sedimentary Research* **70**, 649–663.

Teichmuller, M. (1987) Recent advances in coalification studies and their application to geology. In: *Coal and Coal-bearing Strata: Recent Advances* (Ed. A.C. Scott). Special Publication 32, Geological Society of London, Bath, pp. 127–169.

Terry, R.D. & Chilingar, G.V. (1955) Summary of 'Concerning some additional aids in studying sedimentary formations' by M. S. Shretsor. *Journal of Sedimentary Petrology* **25**, 229–234.

Testa, G. & Lugli, S. (2000) Gypsum–anhydrite transformations in Messinian evaporites of central Tuscany (Italy). *Sedimentary Geology* **130**, 249–268.

Teyssen, T.A.L. (1984) Sedimentology of the Minette oolitic ironstones of Luxembourg and Lorraine: a Jurassic subtidal sandwave complex. *Sedimentology* **31**, 195–211.

Thiry, M. & Ribert, I. (1999) Groundwater silicification in Paris Basin limestones: fabrics, mechanisms and modeling. *Journal of Sedimentary Research* **69**, 171–183.

Thiry, M. & Simon-Coinçon, R. (Eds) (1999) *Palaeoweathering, Palaeosurfaces and Related Continental Deposits*. Special Publication 27, International Association of Sedimentologists. Blackwell Science, Oxford, 408 pp.

Tillman, R.W. & Weber, K.J. (Eds) (1987) *Reservoir Sedimentology*. Special Publication 40, Society of Economic Paleontologists and Mineralogists, Tulsa, OK.

Tissot, B.P. & Welte, D.H. (1984) *Petroleum Formation and Occurrence*. Springer-Verlag, Berlin.

Toomey, D.F. (Ed.) (1981) *European Fossil Reef Models*. Special Publication 30, Society of Economic Paleontologists and Mineralogists, Tulsa, OK.

Toomey, D.F. & Nitechi, M.H. (Eds) (1985) *Paleoalgology*. Springer-Verlag, Berlin.

Trendall, A.F. (2000) The significance of banded iron formation (BIF) in the Precambrian stratigraphic record. *Geoscientist* **10**, 4–7.

Trendall, A.F. & Morris, R.C. (1983) *Iron-formation: Facts and Problems*. Elsevier, Amsterdam.

Trewin, N. (1988) Use of the scanning electron microscope in sedimentology. In: *Techniques in Sedimentology* (Ed. M.E. Tucker). Blackwell Scientific Publications, Oxford, pp. 229–273.

Trewin, N.H. (1994) Depositional environment and preservation of biota in the Lower Devonian hot-springs of Rhynie, Aberdeenshire, Scotland. *Transactions of the Royal Society of Edinburgh: Earth Sciences* **84**, 433–442.

Trewin, N.H. & Davidson, R.G. (1999) Lake-level changes, sedimentation and faunas in a middle Devonian basin-margin fish bed. *Journal of the Geological Society* **156**, 535–548.

Trichert, J. & Fikri, A. (1997) Organic matter in the genesis of high-island atoll peloidal phosphorites: the lagoonal link. *Journal of Sedimentary Research* 67, 891–897.

Tucker, M.E. (Ed.) (1988) *Techniques in Sedimentology*. Blackwell Scientific Publications, Oxford, 394 pp.

Tucker, M.E. (1991) Sequence stratigraphy of carbonate–evaporite basins: models and applications to the Upper Permian (Zechstein) of northeast England and adjoining North Sea. *Journal of the Geological Society of London* 148, 1019–1036.

Tucker, M.E. (1992) The Precambrian–Cambrian boundary: seawater chemistry, ocean circulation and nutrient supply in metazoan evolution, extinction and biomineralization. *Journal of the Geological Society of London* 149, 655–668.

Tucker, M.E. (1993) Carbonate diagenesis and sequence stratigraphy. *Sedimentology Review, 1,* Blackwell Science, Oxford, pp. 51–72.

Tucker, M.E. (1996) *Sedimentary Rocks in the Field*. Wiley, Chichester, 153 pp.

Tucker, M.E. & Hollingworth, N.T.J. (1986) The Upper Permian Reef Complex (EZI) of North East England: diagenesis in a marine to evaporitic setting. In: *Reef Diagenesis* (Eds J.H. Schroeder & B.H. Purser). Springer-Verlag, Berlin, pp. 270–290.

Tucker, M.E. & Wright, V.P. (1990) *Carbonate Sedimentology*. Blackwell Scientific Publications, Oxford.

Tucker, M.E., Wilson, J.L., Crevello, J.R. & Read, J.F. (Eds) (1990) *Carbonate Platforms: Facies, Evolution and Sequences*. Special Publication 9, International Association of Sedimentologists. Blackwell Scientific Publications, Oxford.

Tucker, R.M. (1981) Giant polygons in the Triassic salt of Cheshire, England: a thermal contraction model for their origin. *Journal of Sedimentary Petrology* 51, 779–786.

Tyson, R.V. (1995) *Sedimentary Organic Matter*. Chapman & Hall, London, 615 pp.

Uddin, A. & Lundberg, N. (1998) Unroofing history of the eastern Himalaya and the Indo-Burman Ranges: heavy-mineral study of Cenozoic sediments from the Bengal Basin, Bangladesh. *Journal of Sedimentary Research* 68, 465–472.

Ulmer-Scholle, D.S. & Scholle, P.A. (1994) Replacement of evaporites within the Permian Park City Formation, Bighorn Basin, Wyoming. *Sedimentology* 41, 1203–1322.

Vanstone, S.D. (1998) Late Dinantian palaeokarst of England and Wales: implications for exposure surface development. *Sedimentology* 45, 19–37.

Van Wagoner, I.C. *et al.* (1988) An overview of the fundamentals of sequence stratigraphy and key definitions. In: *Sea-level Changes—an Integrated Approach* (Eds C.K. Wilgus *et al.*). Special Publication 42, Society of Economic Paleontologists and Mineralogists, Tulsa, OK, pp. 39–45.

Visser, J.N.J. (1997) Deglaciation sequences in the Permo-Carboniferous Karoo and Kalahari basins of southern Africa: a tool in the analysis of cyclic glaciomarine basin fills. *Sedimentology* 44, 507–521.

Walderhaug, O. (1994) Temperatures of quartz cementation in Jurassic sandstones from the Norwegian continental shelf—evidence from fluid inclusions. *Journal of Sedimentary Research* A64(2), 311–323.

Walker & James, N.P. (Eds) (1992) *Facies Models: Response to Sea-level Change*. Geological Association of Canada, Waterloo, Ontario, 409 pp.

Walter, M.R. (Ed.) (1976) *Stromatolites*. Elsevier, Amsterdam.

Warren, J.K. (1999) *Evaporites: their Evolution and Economics*. Blackwell Science, Oxford, 438 pp.

Warren, J.K. & Kendall, C.G. St C. (1985) Comparison of sequences formed in marine sabkha (subaerial) and saline (subaqueous) settings—modern and ancient. *Bulletin, American Association of Petroleum Geologists* 69, 1013–1023.

Watson, A. (1985) Structure, chemistry and origins of gypsum crusts in southern Tunisia and the central Namib desert. *Sedimentology* 32, 855–876.

Weaver, C.E. (1989) *Clays, Muds and Shales*. Elsevier, Amsterdam.

Weaver, P.P.E., Wynn, R.B., Kenyon, N.H. & Evans, J. (2000) Continental margin sedimentation, with special reference to the north-east Atlantic margin. *Sedimentology* 47 (Suppl. 1), 239–256.

Webb, J.A. & Golding, S.D. (1998) Geochemical mass-balance and oxygen isotope constraints on silcrete formation and its palaeoclimatic implications in southern Australia. *Journal of Sedimentary Research* 68, 981–993.

Weimer, P. & Link, M.H. (Eds) (1991) *Seismic Facies and Sedimentary Processes of Submarine Fans and Turbidite Systems*. Springer-Verlag, New York.

Weimer, P. & Posamentier, H.W. (1993) *Siliciclastic Sequence Stratigraphy*. Memoir 58, American Association of Petroleum Geologists, Tulsa, OK, 492 pp.

Weimer, P., Bouma, A.H. & Perkins, R.F. (Eds) (1994) *Submarine Fans and Turbidite Systems*. Gulf Coast Section, Society of Economic Paleontologists and Mineralogists, Austin, TX, 440 pp.

Welton, J.E. (1984) *SEM Petrology Atlas*. Methods in Exploration Series, No. 4, American Association of Petroleum Geologists, Tulsa, OK.

Wendt, J. *et al.* (1997) The world's most spectacular carbonate mud mounds (Middle Devonian, Algerian Sahara). *Journal of Sedimentary Research* 67, 424–436.

Whateley, M.K.G. & Pickering, K.T. (Eds) (1989) *Deltas: Sites and Traps for Fossil Fuels*. Special Publication 41, Geological Society of London, Bath.

Whateley, M.K.G. & Spears, D.A. (Eds) (1995) *European Coal Geology*. Special Publication 82, Geological Society of London, Bath, 500 pp.

Whelan, J.K. & Farrington, J.W. (1992) *Organic Matter. Productivity, Accumulation and Preservation in Recent and Ancient Sediments* Columbia University Press, New York, 533 pp.

Whisonant, R.C. (1987) Paleocurrent and petrographic analysis of imbricate intraclasts in shallow-marine carbonates, Upper Cambrian, Southwestern Virginia. *Journal of Sedimentary Petrology* 57, 983–994.

White, S.H., Shaw, H.F. & Huggett, J.M. (1984) The use of backscattered electron imaging for the petrographic study of sandstones and shales. *Journal of Sedimentary Petrology* 54, 487–494.

Wignall, P.B. (1994) *Black Shales*. Clarendon Press, Oxford, 127 pp.

Wilgus, C.K. *et al.* (Eds) (1988) *Sea-level Changes—an Integrated Approach*. Special Publication 42, Society of Economic Paleontologists and Mineralogists, Tulsa, OK, 407 pp.

Williams, C.A. & Krause, F.F. (1998) Pedogenic–phreatic carbonates on a Middle Devonian (Givetian) terrigenous alluvial–deltaic plain, Gilwood Member (Watt Mountain Formation), northcentral Alberta, Canada. *Sedimentology* 45, 1105–1124.

Willis, B.J. (1997) Architecture of fluvial-dominated valley-fill deposits in the Cretaceous Fall River Formation. *Sedimentology* 44, 735–757.

Wilson, J.L. (1975) *Carbonate Facies in Geologic History*. Springer-Verlag, Berlin.

Winn, R.D. & Armentrout, J. M. (Eds) (1995) *Turbidites and Associated Deep-water Facies*. Core Workshop, Society of Economic Paleontologists and Mineralogists, Tulsa, OK.

Wood, R. (1999) *Reef Evolution*. Oxford University Press, Oxford, 414 pp.

Worden, R. & Morad, S. (2000) *Quartz Cementation in Sandstones*. Special Publication 29, International Association of Sedimentologists. Blackwell Science, Oxford, 342 pp.

Wray, J.L. (1977) *Calcareous Algae*. Elsevier, Amsterdam.

Wright, V.P. (Ed.) (1986) *Paleosols, their Recognition and Interpretation*. Blackwell Scientific Publications, Oxford.

Wright, V.P. & Burchette, T.P. (1996) Shallow-water carbonate environments. In: *Sedimentary Environments: Processes, Facies and Stratigraphy* (Ed. H.G. Reading), Blackwell Science, Oxford, pp. 325–394.

Wright, V.P. & Burchette, T.P. (1998) *Carbonate Ramps*. Special Publication 149, Geological Society of London, Bath, 472 pp.

Wright, V.P. & Tucker, M.E. (1991) *Calcretes*. Reprint Series 2, International Association of Sedimentologists. Blackwell Scientific Publications, Oxford, 352 pp.

Wright, V.P., Platt, N.H. & Wimbledon, W.A. (1988) Biogenic laminar calcretes: evidence of root-mat horizons in palaeosols. *Sedimentology* 35, 603–620.

Yerino, L.N. & Maynard, J.B. (1984) Petrography of modern marine sands from the Peru–Chile Trench and adjacent areas. *Sedimentology* 31, 83–89.

Yoo, C.M., Gregg, J.M. & Shelton, K.L. (2000) Dolomitization and dolomite neomorphism. Trenton and Black River Limestones (Middle Ordovician) Northern Indiana, USA. *Journal of Sedimentary Research* 70, 265–274.

Young, T.P. & Taylor, W.E.G. (Eds) (1989) *Phanerozoic Ironstones*. Special Publication 46, Geological Society of London, Bath.

Zhou, Y. *et al.* (1994) Hydrothermal origin of Late Proterozoic bedded chert at Gusui, Guangdong, China: petrological and geochemical evidence. *Sedimentology* 41, 605–619.

Zuffa, G.G. (Ed.) (1985) *Provenance of Arenites*. Reidel, Dordrecht, 408 pp.

Zuffa, G.G., Cibin, U. & Di Giulio, A. (1995) Arenite petrography in sequence stratigraphy. *Journal of Geology* 103, 451–459.

Index

Page numbers in *italics* refer to figures; those in **bold** refer to tables.

SEDIMENTARY PETROLOGY
An Introduction to the Origin of Sedimentary Rocks

TEL. 0151 231 4022